PRINCIPLES OF MODIFIED-ATMOSPHERE AND SOUS VIDE PRODUCT PACKAGING

PRINCIPLES OF MODIFIED-ATMOSPHERE AND SOUS VIDE PRODUCT PACKAGING

EDITED BY

Jeffrey M. Farber, Ph.D.

Karen L. Dodds, Ph.D.

Bureau of Microbial Hazards
Health Canada

CRC Press
Taylor & Francis Group
Boca Raton London New York

CRC Press is an imprint of the
Taylor & Francis Group, an **informa** business

Principals of Modified-Atmosphere and Sous Vide Product Packaging

First published 1995 by Technomic Publishing Company, Inc.

Published 2019 by CRC Press
Taylor & Francis Group
6000 Broken Sound Parkway NW, Suite 300
Boca Raton, FL 33487-2742

First issued in paperback 2019

ISBN-13: 978-0-367-44888-2 (pbk)
ISBN-13: 978-1-56676-276-2 (hbk)

Main entry under title:
Principals of Modified-Atmosphere and Sous Vide Product Packaging

A Technomic Publishing Company book
Bibliography: p.
Includes index p. 459

Library of Congress Catalog Card No. 95-60888

Contents

CHAPTER 1

Introduction

KAREN L. DODDS—*Health Canada, Ontario, Canada*

CONSUMER demand for a wide variety of fresh, minimally processed, convenience foods has increased rapidly over the last two decades, with no down-turn in sight. Several important factors have influenced this demand, including consumers' concerns and perceptions of health risks associated with the use of preservatives in foods, specifically concerns to reduce the dietary intake of salt due to its implication in coronary heart disease, and concerns to reduce nitrite due to the association of nitrosamines with cancer, the increasing number of families with both parents working, and an increasing disposable income. This book focuses on two technologies being widely used to respond to this demand: modified atmosphere packaging and sous vide technology. Both these technologies extend product shelf life beyond that of truly fresh foods, while usually avoiding the use of chemical preservatives. These technologies have seen incredible growth over the last two decades, both in North America and Europe (see Chapters 2 and 3), and have not yet reached their full potential. Some estimates predict that more than half of all food consumed will be in these categories by the year 2000 (Downes, 1989).

MODIFIED ATMOSPHERE PACKAGING

A general definition of modified atmosphere packaging (MAP) simply means packaging a food so that the atmosphere is different from the normal composition of air (78.08% N_2, 20.96% O_2, 0.03% CO_2, variable amounts of water and traces of inert gases). Under this general definition, several packaging techniques are included, such as vacuum packaging, controlled atmosphere packaging and modified atmosphere or gas packaging. Vacuum packaging is perhaps the most common method of modifying the internal package atmosphere. In vacuum packaging, the product is placed in a package of low O_2 permeability, the air is evacuated and the package sealed without deliberate replacement with another gas mixture. Typically, an atmosphere with elevated levels of CO_2 develops as the food product and contaminating microorganisms consume residual O_2 and produce CO_2. Controlled atmosphere packaging includes systems to maintain a certain atmosphere around the product. At present, this application tends to be limited to storage in the warehouse or in bulk packs. Modified atmospheres may be established by gas flushing at the time of packaging, or by allowing air-packed products, mainly fruits and vegetables, to generate a modified atmosphere as a result of natural respiration. As well, additives incorporated into the package, such as O_2 scavengers or emitters or CO_2 absorbers, are capable of modifying the atmosphere after packaging (see Chapter 14).

The use of modified atmospheres to extend the shelf life of foods has a considerable history. Initially, commercial applications were largely confined to controlled atmosphere storage and transport of bulk commodities. Much of the original work was done in Australia and New Zealand in the early 1930s, when beef and lamb carcasses were first shipped to the U.K. under CO_2 storage (Gill, 1986). Controlled atmosphere storage warehouses were designed and implemented in the 1940s and 1950s to extend the shelf life of fresh apples (Zagory and Kader, 1988). The use of MAP technology received a significant boost when the retail chain of Marks and Spencer introduced a wide range of fresh MAP meat products in the U.K. in 1981.

Controlled atmosphere packaging for fresh fruits and vegetables was named one of the 10 most significant innovations in food science between 1939 and 1989 by the Institute of Food Technologists (Anon., 1989). The use of reduced levels of O_2, selective mixtures of gases or increased CO_2 to limit respiration and ethylene production, thereby delaying ripening and decaying and increasing the shelf lives of refrigerated produce, has had a significant impact on the fresh produce industry. Development of packaging systems and films that are selectively permeable to specific gases was a key element in the development of this application and its extension to a wide variety of products.

Modified atmospheres extend the shelf life of foods by inhibiting chemical, enzymatic and microbial spoilage. This allows preservation of the fresh state of the food without the temperature or chemical treatments used by other shelf life extension techniques such as canning, freezing, dehydration and other processes.

The gases used in most MAP applications include CO_2, O_2 and N_2, but different products require different combinations of these gases (Table 1.1). The role of CO_2 is primarily to inhibit the growth and metabolism of microorganisms. CO_2 selectively inhibits the growth of gram-negative bacteria (Enfors and Molin, 1981), which typically grow relatively rapidly and produce the off-flavours and odours associated with spoilage of many foods, such as the spoilage of meat associated with growth of *Pseudomonas* spp. Bacteria such as the lactic acid bacteria are relatively unaffected by CO_2, and they continue to grow but tend to decrease the rate of spoilage. The ability of a CO_2 atmosphere to extend the shelf life of beef was demonstrated as early as 1882, when beef was kept for 18 d in hot summer weather with daytime temperatures up to 32°C (Kolbe, 1882; cited by Dixon and Kell, 1989). By 1938, 26% of the beef from Australia and 60% of that from New Zealand was shipped under CO_2 (Dixon and Kell, 1989). For fresh produce, the level of CO_2 must be carefully determined, as exposure to levels above a tolerance limit can cause physiological damage to the produce (Zagory and Kader, 1988). CO_2 is both water and lipid soluble and thus dissolves in the product to some extent. The addition of O_2 depends upon the product. O_2 is primarily used for fresh red meats to maintain the desirable colour. Again, for fresh produce, the level of O_2 must be carefully determined, as levels which are too low may increase anaerobic respiration and lead to development of off-flavours due to the production of ethanol and acetaldehyde. O_2 is not used for fish high in fat

Table 1.1. Recommended gas mixtures for different food products.

Food Products	N_2 (%)	CO_2 (%)	O_2 (%)	Air (%)
Oily fish	60–40	40–60	<1	
White fish		60	40	0–20
Crustaceans		80–100		
Red meat		15–30	70–85	
Poultry	70–80	20–30	1–2	
Bakery and pasta	20–50	50–80		
Cheese	30–100	0–70		
Vegetables		0–15		
Fruit		0–20		
Coffee	100			
Potato chips	100			

content, such as mackerel and herring, to prevent rancidity problems, and it is not used for cured meats since it has a detrimental effect on product colour. N_2 is an inert tasteless gas with low solubility in water which is often used as a filler.

Almost any type of food product is currently available packed under a modified atmosphere, including red meats and poultry, both raw and cooked (see Chapters 6 and 7), fish (see Chapter 8), a wide range of vegetable and fruit products (see Chapter 9), fresh pasta, breads, cakes, pizza and other baked goods (see Chapter 10), combination products and whole meals, snack foods, and dried foods. The length of refrigerated shelf life which can be obtained is impressive (Table 1.2).

A major concern regarding the widespread use of MAP is its safety. The ability of MAP to inhibit spoilage organisms is well documented, but many pathogenic organisms are less affected (Genigeorgis, 1985; Hintlian and Hotchkiss, 1987; Farber, 1991). Therefore, there is concern that a MAP product may become potentially hazardous before it is overtly spoiled. Indeed, the delay in spoilage gives many contaminating pathogens an extended period of growth. This question is especially important for the psychrotrophic pathogenic organisms such as *Listeria monocytogenes*, nonproteolytic *Clostridium botulinum*, *Aeromonas hydrophila* and *Yersinia enterocolitica* which can all grow at refrigeration temperatures. Further, it is well documented that it is difficult to maintain adequate refrigeration throughout the food chain, particularly at the retail and consumer level. Retail display cases are often at

Table 1.2. Typical shelf life of MAP products.

Product	Temp. (°C)	Atmosphere	Shelf Life (d)	Ref.
Beef	4	20% CO_2, 80% O_2	>15	1
Pork	4	100% CO_2	>24	2
Chicken	1.5	50–80% CO_2	≥23	3
Trout	1.7	80% CO_2, 20% N_2	~20	4
Dungeness crab	1.7	80% CO_2, 20% air	≥25	5
Submarine sandwich	4	30% CO_2, 70% N_2	30	6
Cheeseburger	4	20% CO_2, 80% N_2	30	6
Broccoli	4	10% CO_2, 11% O_2 79% N_2	21	7

References: 1) Shay and Egan, 1987; 2) Enfors et al., 1979; 3) Gardner et al. 1977; 4) Barnett et al., 1987; 5) Parkin and Brown, 1983; 6) Farber, 1991; 7) Berrang et al., 1990.

temperatures of 7 – 10°C, while the temperature in home refrigerators may even exceed 10°C (Conner et al., 1989). Therefore, pathogens which can grow at temperatures of mild abuse, such as 5 – 12°C, are also a concern. These include *Salmonella*, proteolytic *C. botulinum, Clostridium perfringens, Bacillus cereus* and *Staphylococcus aureus*. Holding product at abuse temperatures also means the growth rate of psychrotrophic pathogens will be increased. Various studies have shown that toxigenesis by *C. botulinum* can precede spoilage of MAP products, especially under conditions of temperature abuse (Lambert et al., 1991).

There are several benefits to the use of MAP beyond shelf life extension. MAP products can be prepared centrally and distributed, ready packed, to retailers. The product can often be produced more economically due to large scale production and better utilization of labour and equipment. MAP can increase product utilization, improve inventory control, provide better distribution of product to target markets, and reduce space costs. The sales appeal of the product may be enhanced due to attractive colour and presentation. The use of sealed packs improves hygiene throughout the distribution and retail chain.

SOUS VIDE TECHNOLOGY

It is not surprising that the technology now known as sous vide was developed in France, a country known for gastronomy. While the literal translation of the French term into English simply means "under vacuum," sous vide has implications beyond that. The commonality in sous vide products is that foods, either raw or precooked, are packaged under vacuum in hermetically sealed bags, then cooked or heated, and finally cooled and refrigerated (see Chapters 3 and 10). The majority of them share the following characteristics: 1) they are precooked, packaged products which require little or no additional heat treatment prior to consumption; 2) they have an extended refrigerated shelf life; 3) they are not usually protected by conventional preservation systems, such as a low pH or low water activity; and 4) they are marketed in sealed packages or containers. The sous vide process was developed, many will say "reintroduced," by a chef, Georges Pralus, to improve the flavour, appearance and aroma of prepared foods. This system was initially used for many hospitals, catering units and restaurants, but it is now being extended towards production for retail sale.

Cooking foods in their own juices in a hermetically sealed bag improves product quality for several reasons. Cooking a product in water or steam allows more efficient transfer of thermal energy than in air, and also allows better control of cooking temperature. Therefore, the heating requirement is often reduced and better product consistency and nutritional value are maintained. Because water does not need to be added and steam is not allowed to escape, the original flavour of the product is better preserved. As well, if the product is seasoned with herbs or spices, the organoleptic qualities are improved. While the cooking was originally intended to improve the organoleptic qualities of the product, it can serve as a pasteurization step, inactivating the vegetative microbial population in the product. Finally, because the product is kept in the hermetically sealed bag, there is a reduced risk of post-processing contamination. These last two factors generally result in an extension of refrigerated shelf life.

Typically, the sous vide process includes six steps (Figure 1.1). The first involves preparation of ingredients and/or components. There is general agreement that only top quality fresh raw ingredients should be used. The preparation may include blending or cooking of sauces; grilling or searing of meat, poultry or seafood; blanching of pasta, vegetables, etc. The components are then assembled and packed in

Step 1: Initial preparation of top-quality ingredients (mixing, cooking, blanching, etc.)

↓

Step 2: Packaging (Ingredients weighed, assembled and placed in package)

↓

Step 3: Air extraction and hermetic sealing

↓

Step 4: Cooking/pasteurization

↓

Step 5: Rapid chilling

↓

Step 6: Refrigeration of end product

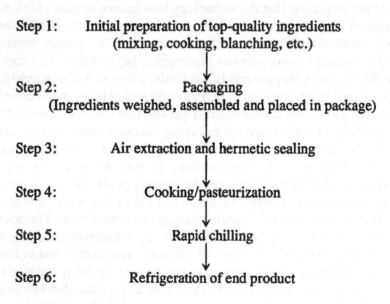

FIGURE 1.1 Typical steps in a sous vide process (from Smith et al., 1990).

preformed or thermoformed pouches or trays, before the packages are sealed under vacuum. The vacuum-sealed food is cooked under precise conditions in steam chambers, retorts, water-submersion tanks or cascading water-fall baths. Unlike conventional canned foods, the thermal process applied to sous vide products is minimal, resulting in a product which is not shelf stable and which requires refrigeration or freezing. The product is then rapidly chilled to stop the cooking process, and finally stored under controlled refrigeration or frozen conditions.

A recent study showed the importance of storage temperature on the shelf life of sous vide products (Simpson et al., 1994a). A formulated spaghetti and meat sauce product was subjected to heat processing at 65 or 75°C equivalent to a 5D or 13D reduction of *Enterococcus faecium*. When stored at 5°C, products had a shelf life of ≥35 d, irrespective of processing treatment. However, for products stored at 15°C, products were visibly swollen after 14 or 21 d, depending on the processing treatment.

As with MAP products, there are safety concerns associated with sous vide foods (Conner et al., 1989; Notermans et al., 1990; Mossel and Struijk, 1991; Rhodehamel et al., 1992; Lund and Notermans, 1993). The heat treatment applied to these products varies, but most are heated at temperatures ranging from 75 to 85°C for several minutes or longer. If the pasteurization process is carried out adequately, all vegetative bacteria present are killed. However, spores, including spores of *C. botulinum*, can survive. As well, *L. monocytogenes* is more heat resistant than many vegetative organisms and there are concerns that it may survive some of the milder treatments, or that post-processing contamination will occur. Many of these products have refrigerated shelf lives extending to six weeks, or even more.

Notermans et al. (1990) examined the potential for botulinum toxigenesis in 11 commercially available sous vide products, including ravioli, moussaka, tortellini, canneloni with meat, canneloni with vegetables, lasagna, lasagna bolognese, tagliatelle with pork, tagliatelle with chicken, tagliatelle with mushrooms and chili con carne. After inoculation with nonproteolytic *C. botulinum* spores, products were stored at 8°C. After four weeks botulinum toxin was detected in two products.

More recently, challenge studies with two formulated sous vide products, spaghetti and meat sauce and a cream rice with salmon product, were completed. The spaghetti was inoculated with proteolytic *C. botulinum* spores, heat processed at 75°C for 36 min, previously determined to achieve a 13D reduction in *Enterococcus faecium*, and then stored at 15°C (Simpson et al., 1995). For samples with a pH of 5.25 or 5.5, toxigenesis preceded spoilage. A mixture of type E strains was used as the inoculum for the rice and salmon product (Simpson et al., 1994b). This product was packaged and heat processed in the same manner as the spaghetti product, and stored at 5, 10 or 15°C. The product proved very susceptible to toxin production. Toxin was detected after 42 d incubation at 10°C at pH 5, the lower pH limit for growth of nonproteolytic *C. botulinum*. Even at 5°C, toxin was detected in samples with a pH of 6.3 after 21 d, and after 28 d for product with a pH of 5.6 or 5.9. All toxic samples stored at 10 or 15°C were visibly spoiled, whereas toxic samples stored at 5°C were not. As the storage temperature increased, the time interval between obvious spoilage and toxin detection was decreased.

These safety concerns have led to the development of Regulations in some countries and Guidelines or Codes of Practice in others (see Chapter 14, Lund and Notermans, 1993). These documents typically relate the degree of heat treatment to the potential shelf life of the product; the greater the heat treatment, the longer the permissable shelf life.

FUTURE

The future for both MAP and sous vide products is very promising; certainly there is no evidence that consumer demand for these types of products is decreasing. While currently one of the major barriers is the concern regarding the safety of the end product (see Chapter 4), food professionals, both in industry and regulatory agencies, are developing viable strategies for dealing with this issue.

The use of the Hazard Analysis Critical Control Point or HACCP approach is probably the single most important strategy for controlling the safety of these products (see Chapter 15). The HACCP approach involves seven principles (Codex, 1993; Sperber, 1991). 1) Hazard analysis and risk assessment: the potential hazards associated with food production at all stages are identified and the likelihood of occurrence is assessed. For example, nonproteolytic *C. botulinum* spores are frequently associated with marine products (Dodds, 1993). 2) Identification of critical control points (CCPs) in the process: CCPs are any points, procedures or operational steps that can be controlled during production to eliminate a hazard or minimize its likelihood of occurrence. In MAP products, the composition of the atmosphere would likely be a CCP. In sous vide products, the heat treatment applied after packaging would be a CCP. 3) Establishing critical limits for each CCP: target levels and tolerances to ensure each CCP is under control must be determined. To ensure the heat treatment of a sous vide product is under control will require several limits, including the initial product temperature, product weight, cooking medium temperature and cook time. 4) Establishing a monitoring schedule for CCPs: the system should ensure control of each CCP by scheduled testing, measurement or observation. Many CCPs, such as the processing temperature, should be monitored continuously. Others, such as the fill weight, should have a specified frequency of monitoring. 5) Establishing corrective action to be taken when a deviation occurs at a CCP: when monitoring indicates a CCP is out of control, corrective actions should be clearly defined beforehand. This step is a clear advantage of the HACCP approach, as proactive measures are taken to avoid production of unacceptable product. For example, if monitoring establishes that the processing temperature is decreasing, a longer cook time can be used until the process temperature increases to specification. 6) Establishing a record-keeping system to document the HACCP approach. 7) Establishing verification procedures: these may be carried out by the manufacturer or by a regulatory agency to determine that the operations within the plant conform to the designed HACCP plan. A main advantage of the HACCP approach for these products is that it is product specific, which means it will cover the wide variety of products included in the MAP and sous vide categories.

Another approach being used by many processors is the use of additional barriers to increase the microbiological safety of MAP and sous vide products (see Chapter 11). It is well recognized that refrigeration is the main barrier controlling both the quality and safety of these products. Baker and Genigeorgis (1990), examining the effects of variables such as storage temperature, type of inoculated spore, and atmosphere, determined that storage temperature alone was responsible for 75% of the variation. The weakness of this barrier is also well recognized. Therefore, other barriers such as the incorporation of lactic acid bacteria, or their metabolites, or the use of different chemical preservatives, are being considered. This builds on the "hurdle approach" first suggested by Leistner and Rodel (1976).

Maintenance of the cold chain is essential to the success of this industry. In Europe, this problem is not as difficult as transportation distances are not great. However, in North America, effective relationships must be developed between manufacturers, distributors and retailers to ensure these products are delivered as quickly and as cold as possible.

The continued improvement and use of time-temperature monitors (see Chapter 17) to provide visual indicators of the temperature history of a product may provide a valuable service to this sector of the food industry.

Finally, one approach rapidly gaining in sophistication, is the use of microbial modelling. This approach involves the development of mathematical models to predict the behaviour, including growth, survival or inactivation, of different microorganisms in specific environments. Indeed, it is the rapid growth and wide product types involved in MAP and sous vide technology, that has played a major role in the growth of microbial modelling. The traditional approach of empirically assessing the safety of a food product is almost impossible to apply to the wide variety of products now being developed. The use of predictive modelling allows product development to proceed on a much sounder basis, and allows the required safety factors to be incorporated from the outset.

Another exciting area in the future of these products are the advances in packaging technology (see Chapters 13 and 14). New, sophisticated films are continually being developed to address specific problems. As described earlier, MAP of fresh produce is problematic because respiration continues and therefore it is difficult to maintain proper levels of O_2 and CO_2. Films are being developed which are highly permeable and microperforated, which may be suitable for highly respiring produce such as mushrooms and broccoli (Choi, 1991). A very exciting development has been the production of temperature compensating films (Anon., 1992). These films have a temperature point at which their permeability characteristics change. Thus, a film's permeability could be set to match changes in produce respiration at a certain temperature. As well, many processors are now incorporating gas absorbers and emitters to more carefully tailor the atmosphere within their final product. For example, solid CO_2 tablets can be dispensed into a package before closing, which will sublime after the package is sealed to prevent package collapse due to the solubility of CO_2 in the food (Milani, 1991). O_2 absorbers, usually iron-based, can be packaged in gas permeable materials in the form of small sachets and used in vacuum or modified atmosphere packages (Abe and Kondoh, 1989). They are useful for eliminating residual amounts of O_2, and absorbing traces of O_2 which penetrate the packaging film after sealing.

REFERENCES

Abe, Y. and Kondoh, Y. 1989. Oxygen absorbers, In *Controlled/Modified Atmosphere/Vacuum Packaging of Foods*, A. L. Brody, ed., Food & Nutrition Press, Inc., Conn., pp. 149–159.

Anon. 1989. "Top 10 food science innovations 1939–1989," *Food Technol.*, 43(9):308.

Anon. 1992. "Temperature compensating films for produce," *Prepared Foods* (Sept.):95.

Baker, D. A. and Genigeorgis, C. 1990. "Predicting the safe storage of fresh fish under modified atmospheres with respect to *Clostridium botulinum* toxigenesis by modeling length of the lag phase of growth," *J. Food Prot.*, 53:131–140.

Barnett, H. J., Conrad, J. W. and Nelson, R.W. 1987. "Use of laminated high and low density polyethylene flexible packaging to store trout (*Salmo gairdneri*) in a modified atmosphere," *J. Food Prot.*, 50:645–651.

Berrang, M. E., Brackett, R. E. and Beuchat, L. R. 1990. "Microbial, color and textural qualities of fresh asparagus, broccoli, and cauliflower stored under controlled atmosphere," *J. Food Prot.*, 53:391–395.

Choi, S. O. 1991. "Orego ultra-high gas permeability films for fresh produce packaging," *CAP'91*, Schotland Business Research Inc., Princeton, N.J.

Codex Alimentarius Commission. 1993. Alinorm 93/13. "Draft principles and application of the hazard analysis critical control point (HACCP) system."

Conner, D. E., Scott, V. N., Bernard, D. T. and Kautter, D. A. 1989. "Potential *Clostridium botulinum* hazards associated with extended shelf-life refrigerated foods: A review," *J. Food Safety*, 10:131–153.

Dixon, N. M., and Kell, D. B. 1989. "The inhibition by CO_2 of the growth and metabolism of micro-organisms," *J. Appl. Bacteriol.*, 67:109–136.

Dodds, K. L. 1993. *Clostridium botulinum* in foods, In *Clostridium botulinum: Ecology and Control in Foods*, A. H. W. Hauschild and K. L. Dodds, eds., Marcel Dekker, Inc., N.Y., pp. 53–68.

Downes, T. W. 1989. "Food packaging in the IFT Era: Five decades of unprecedented growth and change," *Food Technol.* 43(9):228–240.

Enfors, S. O. and Molin, G. 1981. The effect of different gases on the activity of microorganisms, In *Psychrotrophic Microorganisms in Spoilage and Pathogenicity*, T. A. Roberts, G. Hobbs, J. H. B. Christian, N. Skovgaard, eds., London, Academic Press, pp. 335–343.

Enfors, S. O., Molin, G. and Ternstrom, A. 1979. "Effect of packaging under carbon dioxide, nitrogen, or air on the microbial flora of pork stored at 4°C," *J. Appl. Bacteriol.*, 47:197–208.

Farber, J. M. 1991. "Microbiological aspects of modified-atmosphere packaging technology–A review," *J. Food Prot.* 54:58–70.

Gardner, G. A., Denton, J. H., and Hartley, S. E. 1977. "Effects of carbon dioxide environments on the shelf-life of broiler carcasses," *Poultry Sci.*, 56:1715–1716.

Genigeorgis, C. A. 1985. "Microbial and safety implications of the use of modified atmospheres to extend the storage life of fresh meat and fish," *Int. J. Food Microbiol.*, 1:237–251.

Gill, C. O. 1986. The control of microbial spoilage in fresh meats, In *Advances in Meat*

Research. Volume 2, A. M. Pearson, and T. R. Dutson, eds., AVI Publishing Co., Inc., Conn., pp. 49–88.

Hintlian, C. B., and Hotchkiss, J. H. 1987. "Comparative growth of spoilage and pathogenic organisms on modified atmosphere-packaged cooked beef," *J. Food Prot.*, 50:213-223.

Lambert, A. D., Smith, J. P. and Dodds, K. L. 1991. "Effect of headspace CO_2 concentration on toxin production by *Clostridium botulium* in MAP, irradiated fresh pork," *J. Food Prot.*, 54:588–592.

Lambert, A. D., Smith, J. P. and Dodds, K. L. 1992. "Physical, chemical and sensory changes in irradiated fresh pork packaged in modified atmosphere," *J. Food Sci.*, 57:1294–1299.

Leistner, L. and W. Rodel. 1976. The stability of intermediate moisture foods with respect to microorganisms, In *Intermediate Moisture Foods*, R. Davis, G. G. Birch, and Parker, K. J., eds. Applied Science Publishers, London pp. 120–134..

Lund, B. M. and Notermans, S. H. W. 1993. Potential hazards associated with REPFEDS, In *Clostridium botulinum: Ecology and Control in Foods*, A. H. W. Hauschild and K. L. Dodds, eds., Marcel Dekker, Inc., NY, pp. 279–303.

Milani, M. 1991. "Two phase MAP: a new controlled shrinking process," *CAP'91*, Schotland Business Research Inc., Princeton, N.J.

Mossel, D. A. A. and Struijk, C. B. 1991. "Public health implication of refrigerated pasteurized ('sous-vide') foods," *Int. J. Food Microbiol.*, 13:187–206.

Notermans, S., Dufrenne, J. and Lund, B. M. 1990. "Botulism risk of refrigerated, processed foods of extended durability," *J. Food Protect.*, 53:1020-1024.

Parkin, K. L. and Brown, W. D. 1983. "Modified atmosphere storage of Dungeness crab (*Cancer magister*)," *J. Food Sci.*, 48:370–374.

Rhodehamel, E. J., Reddy, N. R. and Pierson, M. D. 1992. "Botulism: the causative agent and its control in foods," *Food Control*, 3:125–143.

Shay, B. J. and Egan, A. F. 1987. "The packaging of chilled red meats," *Food Technol. Aust.*, 39:283–285.

Simpson, M. V., Smith, J. P., Simpson, B. K., Ramaswamy, H. and Dodds, K. L. 1994a. "Storage studies on a sous-vide spaghetti and meat sauce product," *Food Microbiol.*, 11:5–14.

Simpson, M. V., Smith, J. P., Dodds, K. L., Ramaswamy, H. S., and Blanchfield, B. 1994b. "Challenge studies of a sous-vide rice and salmon product with *Clostridium botulinum*," Submitted for publication.

Simpson, M. V., Smith, J. P., Dodds, K. L., Ramaswamy, H. S., Simpson, B. K. and Blanchfield, B. 1995. "Challenge studies with *Clostridium botulinum* in a sous-vide spaghetti and meat sauce," *J. Food Prot.* 58:229–234.

Smith, J. P., Toupin, C., Gagnon, B., Voyer, R., Fiset, P. P. and Simpson, M. V. 1990. "A hazard analysis critical control point approach (HACCP) to ensure the microbiological safety of sous vide processed meat/pasta product," *Food Microbiol.*, 7:177–198.

Sperber, W. H. 1991. "The modern HACCP system," *Food Technol.*, 45(6):116–118.

Zagory, D. and Kader, A. A. 1988. "Modified atmosphere packaging of fresh produce," *Food Technol.*, 42(9):70–77.

CHAPTER 2

A Perspective on MAP Products in North America and Western Europe

AARON L. BRODY—*Rubbright·Brody, Inc., U.S.A.*

MODIFIED atmosphere packaged (MAP) foods comprise a substantial and growing proportion of American and European food supplies, but mostly in distribution packaging. Vacuum, modified atmosphere and controlled atmosphere packaging are all part of the same technology, and so the total market for MAP food includes fresh and minimally processed foods packaged under vacuum or altered gaseous environment.

More than 75% of the beef in the United States and Canada is broken into primal and subprimal cuts, vacuum packaged and shipped from meat packers to retailers and hotel/restaurant/institutional (HRI) outlets. Analogous ratios hold in Europe. Up to one-quarter of ground beef in the United States is distributed under reduced O_2 packaging. About one-third of fresh poultry in North America is master packed under modified atmosphere (MA) for refrigerated distribution to retail grocery and HRI outlets. In the United States, precooked poultry is packaged for retail sale under modified atmosphere or vacuum and marinated poultry is packaged under vacuum. Ground poultry is packaged for retail sale under elevated O_2, yet another form of modified atmosphere. Most cured or processed meat and cheese products in retail outlets are packaged either under vacuum or inert atmosphere (i.e., no O_2). Most cured and cooked meats for delicatessen or supermarket service counter use, i.e., retail dispensing, are vacuum packaged for refrigerated distribution. More than 4% of all fresh vegetables in the United States and 75% of California strawberries are distributed under modified atmosphere conditions and up to US$40 million worth of precut vegetables for retail are packaged under altered atmosphere. In Europe, many meat packers and thousands of retail stores display MAP retail cuts of red meat prepared in centralized factories. In the United States, only about 200 of all of the 31,000 large retail groceries use vacuum packaged fresh beef and pork, but several companies are in developmental programs in 1993.

Hundreds of European bakeries employ controlled atmosphere packaging (CAP) to extend the distribution cycle for soft bakery goods such as breads and cakes. In the United States and Canada, bakeries and sandwich makers are commercial with this technology. Sausage, hamburger and other tortilla-roll and biscuit-based sandwiches are packaged under MA for refrigerated distribution in the United States. Considerable quantities of pizza kits and pizza crusts are packaged with the bread portion under MA. The sauce portion, when present, is hot filled to produce a reduced O_2 interior.

Fresh pasta and now pasta sauces are distributed under MAP in the United States, Canada, and Europe. Large quantities of precooked sous vide foods are used in restaurant chains in France, and several American and Canadian firms are in the business. Over 250 United States kitchens/plants use cook/chill technology for bulk pumpable foods to deliver low O_2 environment products.

Thus, altered gaseous environmental packaging ranges from fresh meat to precooked meals and embraces most refrigerated/perishable foods in refrigerated, and to some extent, ambient temperature distribution.

A BRIEF CHRONOLOGY OF ALTERED GASEOUS ENVIRONMENTAL FOOD PACKAGING

Reducing food temperature reduces microbiological, enzymatic and biochemical activity rates characteristic of catabolic deterioration. Both tissue respiration and microbial respiration and growth processes are reduced when O_2 is decreased and respiratory gases such as CO_2 increased.

Catabolic aerobic respiration is the basis for the degradation of plant materials. Plant product respiration is decreased by increasing CO_2 and water, independent of or coupled with reduction of O_2 and removal of ethylene.

Many years ago, mutton, beef and lamb shipped from Australia to England was often cooled by solid CO_2, leading to shelf lives longer than meat held under ice. These observations provided a basis for increasing CO_2 in transport for fresh meats.

In the 1930s, fresh apples and pears stored in enclosed warehouses were benefited by the natural respiratory activities of the fruit which reduced the O_2 and increased the CO_2 sufficiently to significantly slow degradation. The resulting fruit could be used up to a half year after the harvest—a doubling of shelf life. Natural respiration controlled atmosphere storage for apples and pears became widely used during the 1940s and 1950s in both New York and the Pacific Northwest.

Starting in 1957, Whirlpool Corporation's food scientists developed methods to control the atmospheres surrounding respiring foods. The patents were applied to bulk and industrial distribution of fresh fruit and meat products under the name, Tectrol. Total control of warehouses employs burning hydrocarbon gas to reduce the O_2 and scrubbers to remove excess CO_2. By 1970, hundreds of fruit warehouses throughout the United States and Europe had Tectrol systems to extend the shelf life of the contents, a situation which has since expanded.

Because of Whirlpool's strategic objectives, the patents were sold and the business reformed under the name, Transfresh. Most developments since the transfer have come from Transfresh.

The Tectrol concept has since been expanded to transport containers to control the internal gas content, to deliver bulk packages of fresh vegetables to HRI and retail outlets and pallet loads of strawberries to distributors. In 1990, the process was refined into complete control of container loads and, prior to 1992, for retail packages of cut fresh vegetables—with national distribution in 1992. Prior to the 1960s, vacuum shrink packaging, used to protect turkeys in the frozen distribution, was applied to fresh meat by Cryovac.

Cryovac vacuum skin packaging of fresh meats is based on a virtual absence of O_2, an environment in which meat spoilage is retarded. Simultaneously, however, because of the absence of O_2, the original purple myoglobin color of fresh red meat is retained. During the 1960s, Cryovac and Iowa Beef Packers joined in what was then a new concept of slaughtering cattle at a central location. Rather than shipping hanging carcasses, which was common practice at the time, Iowa Beef Packers broke the carcasses into primal cuts to be vacuum skin packaged in high gas barrier multilayer plastic bags. The vacuum retarded microbiological growth, and the bag retarded the passage of water vapor, so that weight loss due to water evaporation was almost eliminated in refrigerated distribution. The pouches were packed in corrugated fibreboard shipping cases, accounting for the name "boxed beef," for shipment to HRI and then retail supermarket outlets for cutting to consumer size cuts.

Vacuum packaging of nitrite cured meats such as frankfurters, hams, salami, etc., was originally developed by in-line thermoforming webs of flexible O_2 barrier polyester-based packaging material, placing meat within the formed web and heat sealing with a second gas barrier material under vacuum. The system later was changed into nylon/polyvinylidene chloride/polyolefin forming web and polyester/polyvinylidene chloride/ polyolefin flat closure web. This system was later applied to cured cheese. These widely used commercializations of vacuum packaging are usually not discussed in the context of MA packaging.

During the 1960s, researchers of the Flour Milling and Baking Research Association in the United Kingdom, studied CO_2 as a mechanism to retard mold growth of soft bakery products. Their research remained dormant until, in the 1970s, the German government passed a regulation requiring bakers to declare food additives on the label. Rather than signal the presence of preservatives to consumers, commercial bread bakers began sealing their bread in packages containing CO_2 to deliver the requisite shelf life. During the 1980s, MA packaging principles were applied to sandwiches in the United States. Such sandwiches containing nitrite-cured meat fillings, e.g., sausage, retained their quality under MAs for many weeks at refrigerated temperatures. Instructions on the package are to heat the product before eating – a process that overcomes the staling effects in the bakery product. This concept has been extended to hamburgers and tortilla sandwiches.

A similar process, i.e., with instructions to heat before eating, was developed for MA packaged French breads in the 1980s in England. It has since been introduced all over Europe, in Australia and then in the United States.

In Canada, after much intensive research and development, MA packaging of crumpets was commercialized.

In the 1960s, investigations demonstrated that fresh meat could be preserved under refrigeration, with its desirable red colour, if a high CO_2/high O_2 environment was present. It was previously believed that high O_2 accelerates microbiological growth. The research demonstrated that elevated CO_2 coupled with high O_2 was sufficiently effective in retarding microbiological growth, so that undesirable effects of high O_2 could be overcome. These findings were commercialized in West Germany on thermoform/vacuum/gas flush/seal packaging equipment using high O_2 barrier materials. Several European packaging firms including Multivac, Kramer & Graebe and Dixie-Union in Germany, Akerlund & Rausing in Lund, Sweden, Ono in France and Otto Nielson in Denmark developed thermoform/vacuum/gas flush/seal systems for centralized red meat packaging of retail cuts of fresh meat in Europe.

In the early 1980s, French chef, George Pralus, studied the relationship of comprehensive thermal input on food quality, and thus developed the "sous vide" or "vacuum" process for restaurant use. In this total system, the sum of the heat of cooking, cooking after packaging to pasteurize, and reheating is the same as if a fine chef had cooked the original fresh ingredients. One of the thermal inputs, however, is intended to reduce the microbiological count of carefully selected and cleanly-processed ingredients. Vacuum packaging reduces oxidative deterioration, compacts the package and permits rapid heat transfer in and out of the food. Strict distribution temperature control at less than 36° F (2.2°C) and very short distribution cycles of less than 21 d virtually eliminate the risks of adverse microbiological growth. A related technology, cook/chill generating internal vacuum within flexible pouches, has been growing to deliver bulk and now retail size precooked ready-to-eat sauces and entrées for HRI and now retail outlets.

Pasteurization or hot fill will inactivate aerobic spoilage microorganisms but not the heat resistant spores of pathogenic anaerobes. This leaves distribution temperature control as the only barrier to microbiological growth and toxin production in sous vide vacuum packages. This is often regarded as a hazard relative to such products.

During the 1980s, a New York entrepreneur integrated sanitation of ingredients and process, with an MA in O_2 barrier packages, to enable him to distribute wet (30%) moisture pasta under chilled conditions. These practices permitted extension of safe quality distribution of the pasta up to 40 d. Further, the technology stimulated others to apply the principles for entrées, side dishes and meat/fish salads. More recently (in the 1990s), to complement the pasta, sauces are being hot filled into plastic packages to create low O_2 conditions, in order to prolong the shelf life of these sauces under refrigerated conditions. To further complement the pasta, the "Italian" pizza entered refrigerated distribution under MA packaging in the United States in the 1990s. (Actually refrigerated altered gas environment packaged pizzas had been marketed in the United States and Europe as far back as the 1970s.) During the 1950s, hot filled soups packaged in plastic tubs for refrigerated distribution made their debut. These reduced O_2 packaged soups are now made by several American companies.

In Sweden during the early 1980s, Alfa Laval's scientists combined good sanitation, vacuum packaging, water-attenuation plus microwave pasteurization, rapid chilling and controlled distribution at temperatures below 36°F (2.2°C) to successfully deliver precooked pastas and omelets to retail stores.

Microwave plus hot air pasteurization was developed in Italy and the Netherlands in the 1980s, and is now being used in combination with MAP and refrigerated distribution to prolong quality retention time of prepared pastas and meals.

NORTH AMERICA

Fresh Meat

Of the nearly 40 billion pounds of red meat produced in the United States annually, approximately 63% is beef and 36% is pork. About two-thirds of the pork and 12% of beef becomes nitrite cured meats such as ham, bacon and sausage. Twenty-one billion pounds of beef are converted to cuts and packaged at retail level in the United States. Almost all of these beef cuts are fabricated and packaged for retail sale in supermarket backrooms, employing thermoformed foamed polystyrene trays with PVC flexible film overwraps. Another 4 billion pounds of ground beef are ground and sold in retail stores in the United States, also largely in thermoformed, foamed polystyrene trays with PVC film wraps. About 2 billion pounds of ground beef are coarsely ground in federal government inspected factories, packaged in flexible keeper chub casing, and distributed in low O_2 conditions under refrigeration to retail stores where final grinding occurs. Several hundred million pounds are finely ground in federally regulated plants, packaged in chubs under low O_2 conditions directly for retail sale. Under reduced O_2, the internal meat colour is purple, but converts to cherry red after opening. For retail sale, all packaging is opaque to hide the purple myoglobin colour which is regarded as undesirable by consumers.

Centralized packaging of retail beef cuts began in the United States in the 1980s, with relatively little activity until 1987. In that year, packer Excel and the Kroger supermarket chain, using vacuum shrink packaging in high O_2 barrier materials, fabricated central factory packaged red meat cuts, which were distributed under refrigeration to as many as 1,000 of Kroger's 1,700 retail supermarket outlets. The absence of desirable bright red colour on the low O_2 package beef was among the consumer-driven influences which caused a decrease to fewer than 200 stores by 1992. The vacuum also led to excess purge or liquid in the package, which was unsightly.

The gas barrier packaging material costs were three times greater than conventional foamed polystyrene trays plus PVC film overwrap. The negative economic effects of in-store versus central labour, in-store versus central waste, and an almost devastating red meat out-of-stock situation in retail stores, were more than offset by the capital investment in packaging equipment and the higher variable costs of packaging materials. However, the total systems economic benefits of more than $0.02 per pound were not enough to foster expansion of the program.

Numerous subsequent industry studies demonstrating the technical, marketing, and economic benefits of centralized packaging and branding of red meat have not been sufficient to drive any sustained effort by fresh meat interests in the United States or Canada since the Excel/Kroger experiment.

The Excel program paralleled by ventures from other packers into vacuum-packaged fresh pork, most of which have been discontinued, and by several packers in Canada with high O_2 MA packaging of beef cuts, which was also terminated. Except for H.E.Butt (H.E.B.) and Indiana Packers in the United States, centralized packaging of fresh red meat under MA/vacuum was negligible in the United States in 1992.

One of the more significant driving forces for the successful centralized packaging of fresh meat in Europe has been the retailer. In the United States, the stimulus is meat packer driven and controlled. Meat industry standards require extended shelf life of only about 6 d in Europe. For a packer to distribute retail meat cuts throughout the United States from a central location to retail outlets, a minimum of 15 d is required; 9 d to reach the retail display case, 4 d at retail and 2 d before use. Altered O_2 packaging technology is at its marginal limits to achieve 15 d.

During the 1980s, H.E.B, a small Texas supermarket chain, installed a Cryovac vacuum skin packaging (VSP) machine in its Austin, Texas, meat fabrication plant to supply its 180 stores with 25 different retail red meat cuts ranging from filet mignon to boneless roasts. The Cryovac vacuum skin packaging unit can produce up to 20 consumer-ready meat packages per minute. Prior to its Central Packaging venture, the H.E.B. operation supplied vacuum-packaged primal cut beef to its stores for further backroom processing and packaging.

H.E.B. management report their reasons for implementing the VSP system: 1) better utilization of in-store labor; 2) shortages of skilled meat cutters; and, 3) reduction of out-of-stock situations.

In initiating this system, H.E.B. had advantages over other retailers: 1) H.E.B's in-store meat cutters were non-union and unlikely to oppose central processing and packaging; 2) H.E.B had an existing central beef fabrication facility and distribution system, among the few remaining; and, 3) being a small chain, all retail outlets were within 200 miles of the processing plant.

In 1989, H.E.B. began packaging in trays with a peelable O_2 barrier that could be removed just prior to shelf display to allow meat to be exposed to air, to bloom and to thus regain its bright red colour. Actual experience indicated that after the barrier layer was peeled away, O_2 diffusion through the newly exposed film was not uniform and so not all surfaces were exposed equally, resulting in a blotched appearance. Today H.E.B. uses only high O_2 barrier film to retain internal vacuum.

Each package carries a "freeze-by" date which is 25 d from packaging. H.E.B. management cites product colour as the major disadvantage of the program. Repeat sales are high but initial consumer trials remain an obstacle. Less than 10% of H.E.B.'s beef sales are in vacuum skin packaging and the program is still considered in the "trial stage."

In 1990, Indiana Packers began a program with Garwood Packaging using a new version of the Australian Flavorlock system. The system, developed in Australia, was a three web thermoform/vacuum/gas flush/seal system creating two chambers in the final package. The meat was placed in the lower O_2 barrier chamber and covered with a thin highly gas permeable stretch film which held the meat in place regardless of the attitude. A third gas barrier film was heat sealed to the other flange of the base tray, thus creating an upper chamber containing a CO_2/N_2 mixture to extend refrigerated shelf life. Systems were installed for offal in Australia and for fish in the United Kingdom.

After moving to the United States, the system was changed to a preformed tray deposit/vacuum/gas flush which has been installed at Indiana Packers for fresh pork cuts. The base tray is opaque polyester and the intermediate film a liner of low density polyethylene. In the new system, a thermoformed rigid transparent polyester overclosure is sealed to the base. As with the original H.E.B. package, the idea is for the retail store operator to remove the top closure and expose the gas permeable film, which then permits air in to reoxygenate the pork for appearance purposes. In 1992, the system was being applied in some Kroger stores in Indiana. Garwood Packaging claims more than 16 d refrigerated shelf life for pork cuts packaged using their system.

In 1992, the concept of high O_2 packaged red meat was once again transferred from Europe to the United States with ground poultry and retail beef and pork cuts in bulk quantities in North America for warehouse club stores. After several years of relative dormancy, centralized packaging of retail beef and pork is being reexamined and is in about a dozen developmental programs. The marketing drives are apparent: beef consumption has been decreasing significantly, with the industry's advertising efforts being unable to halt or slow the decline. Beef's less-than-healthy perception by consumers has contributed to this shrinkage of what once represented over a quarter of supermarket sales and a higher proportion of their profits.

Further, with the success of alternative retail grocery outlets such as club and convenience stores, fresh meat sales are beginning to shift away from traditional supermarkets.

Thus, the new developments in fresh meat centralized packaging are acknowledgements by retailers of the needs to change their meat marketing strategies.

Technical and marketing tests are under way in the United States and Canada with a number of different concepts:

- High CO_2/high O_2 packaging of beef cuts and ground beef in high O_2 barrier foam polystyrene/ethylene vinyl alcohol trays heat sealed with ethylene vinyl alcohol containing film.
- High CO_2/high O_2 packaging of beef cuts and ground beef in foamed polystyrene trays overwrapped in high gas barrier shrink film.
- Master packing of PVC overwrapped foam polystyrene trays of beef cuts and ground beef in high gas barrier film pouches under high CO_2/low O_2. Opening the master bag exposes the gas permeable primary packages to air which blooms the meat (being tested by club stores such as Sam's).
- Vacuum skin packaging similar to that by Excel/Kroger and H.E.B. being tested for beef by Safeway in Western Canada.

Poultry

Until the 1980s in the United States, almost all poultry was delivered whole dressed in ice to retail stores. Beginning in the 1980s, United States per capita poultry consumption has exceeded that of beef or pork, partly because of crust freezing replacing ice pack. One driving force has been that poultry offerings have changed from whole to further processed such as prepackaged cut-up birds. Almost all poultry at the retail level is packaged in foamed polystyrene trays overwrapped with plasticized PVC film. More than half of the poultry in the United States is master or bulk packed at central processing plants after prepackaging in retail sizes with about a third under altered internal gas, either O_2 removal or vacuum, followed by CO_2 back flush. All poultry, regardless of the gas environment, is distributed under tightly temperature controlled conditions, close to 32°F (0°C). Cut-up poultry that is not gas packaged is almost always chilled by exposure to temperatures below −30°F (−1.1°C) to crust freeze the surface followed by tempering to a temperature at or near the freezing point of the flesh, i.e., about 26° to 28°F (−3° to −2°C). With distribution at these temperatures, plus gas packaging, shelf lives of 15–18 d have been achieved.

Packaging materials are generally gas permeable polyolefin films that permit O_2 from the air to enter and thus obviate any potential adverse effects of anaerobic pathogenic microbiological growth.

In the past, at the retail level, whole birds were removed from the chopped ice by the butcher, cut and repackaged in the backroom in foamed polystyrene trays overwrapped with plasticized PVC, followed by a store label. As a result of the introduction of chill pack and/or MAP in master packs, the concept of "shelf ready" poultry emerged. About one-third of cut-up poultry is processed and prepackaged in conventional foamed polystyrene trays overwrapped with printed PVC film, multipacked in polyethylene master bags which are evacuated and backflushed with CO_2. Master packs are distributed under refrigeration to retail stores where the meat department employees remove the individual packages from the master bags and place them on the shelf. A significant amount of cut-up poultry is master packaged without back gas flushing, i.e., employing crust freezing temperatures to achieve extension of shelf life.

Marinated poultry is vacuum packaged in high gas barrier retail packages for refrigerated distribution. Precooked poultry is factory packaged under MA. A high barrier polystyrene film/ethylene vinyl alcohol/polystyrene foam lamination has been developed for thermoform or deposit/vacuum/gas flush/seal packaging of precooked poultry. Upwards of 200 million packages annually of MAP precooked poultry are distributed at the retail level in the United States.

Since 1990, ground poultry has been packaged under elevated O_2 plus elevated CO_2 to both extend shelf life and retain desirable colour. Most such systems employ foamed polystyrene trays which are overwrapped in horizontal form/fill/seal machines such as Ilapak with high gas barrier shrink films. The packages then have the appearance of in-store packaged meat, but have actually been prepared in centralized processing plants.

In the United States, fresh poultry in added-value form with printed packaging represents a major demonstration of the power of centralized packaging and branding, on what was previously only a commodity. Fresh poultry is also one of the more significant applications of deliberate modified gaseous environments within packages specifically to effect prolonged quality retention. Probably half of all poultry products reaching the consumer have been under CO_2 at some time during distribution.

The poultry system also reflects the basic concept that MA is only a pivot, that is, the careful selection of raw materials, sanitary processing and food temperature control are all together indispensable to technical success with altered gaseous environment. MAP functions only when coupled with the other processing steps.

Fresh Vegetables and Fruits

MAP technology was slow in penetrating the $55 billion fresh produce market in the United States, but it now represents up to US$4 billion in sales, projected by industry sources to double to US$8 billion by 2000.

In California and Florida, MA wrapping of pallets of flats of retail prepackaged strawberries is used for more than 80% of the total crop. The pallet wrap is usually a high ethylene vinyl acetate copolymer/low density polyethylene film box that is sealed around the strawberries. Similar systems are being used for shipments of berries from Latin America.

Transfresh and others have developed packaging systems below the level of bulk for fresh fruits and vegetables. The packaging material most often used is high ethylene vinyl acetate copolymer/low density polyethylene film, characterized by a high gas but low water vapor permeability. This film structure has been used for MA bag packaging of broccoli and cauliflower florets for HRI distribution using high surface-to-volume ratio bags, to permit additional O_2 to enter to obviate respiratory anaerobiosis. In bag form, these systems are commercial for packaging of various types of prepared lettuce: cleaned and cored; trimmed and precut lettuce. Among the other fresh prepared vegetables being packaged under reduced O_2 generated either actively or passively are sliced mushrooms, red and green peppers, diced and sliced onions, cabbage, raddicchio, endive, spinach—almost every vegetable of commercial consequence, except tomatoes.

Consumer-size uncoated oriented polypropylene film pouches of cut green salad with passive MA through natural respiration are in widespread commercial distribution. Temperature control in cutting, packaging and distribution is indispensable. Other vegetables packaged in retail sizes include cabbage, cole slaw, red peppers, and carrot and celery sticks. Since 1988, mineral-filled films which attempt to control gas transmission through the film have been applied to packages of broccoli and cauliflower florets, asparagus, cut lettuce and strawberries for both HRI and retail packs.

The largest use of MA/vacuum produce packaging by far is for precut lettuce. About 15% of the total California iceberg lettuce crop is cleaned, cut and prepackaged for HRI salad-use and the packaging is air, vacuum, quasi-vacuum or reduced O_2 MA. Air packaging in closed packages leads to a passive internal MA environment in chilled distribution.

Significant quantities of lettuce are distributed from California by loading in truck bodies or containers lined with polyethylene film and flushed with N_2 gas to generate an internal MA. In this manner, lettuce may be shipped from the West Coast growing areas throughout the United States and Canada.

As previously indicated, beginning in the early 1980s, major quantities of lettuce were precut into salad-ready form and packaged in polyethylene bags and pails for shipment under refrigeration to HRI installations. The mid-1980s introduction by the McDonald's fast food operations of salad programs in their outlets led to the delivery of altered environment packages of lettuce packaged in bulk sizes to their retail shops. At the packer level, the lettuce is cut under sanitary cold conditions, packed in high EVA/polyethylene film bags; a vacuum is drawn, and the bag is sealed. The bags are distributed under refrigeration in a sufficiently short time such that O_2 extinction conditions are not necessarily established.

Vacuum packaging of fresh broccoli and cauliflower florets retains the quality of the respiring contents sufficiently to allow for refrigerated distribution to HRI outlets and thus a larger market.

Almost 750 million pounds of lettuce are MA packaged in about 150–200 million packages annually in the United States, still mostly for HRI, but increasingly in retail sizes. In addition to the cut lettuce, over 150 million packages of other freshly prepared vegetables are also MA packaged for distribution to HRI and other food outlets. Nearly 200 produce packers in the United States and Canada are involved in this business, representing an amazing growth since none were involved in 1985. In 1992, prepackaged fresh produce represented less than 4%, a small but rapidly growing fraction, of the total fresh produce in the United States and Canada.

Quantities of shrink skin packaged individual fruits and vegetables have been prepared in the United States with some exported. Fifty million pounds of citrus fruit, largely lemons, and much smaller quantities of peppers and cucumber, etc., are individually shrink skin packaged.

Not truly CA/MA packaging, but related, is the "Natural Pak" process for the "Tom AH Toes" brand of fresh tomatoes. Unripe tomatoes are stored at 64°–72°F (17°–22°C) under 3–5% O_2 and elevated CO_2 until they are ripe. In MA environments, the unripe tomatoes may be held for prolonged times without suffering the usual chill damage and are then repackaged in normal air for retail distribution at ambient temperature situations.

There is increasing commercial application of knowledge of MAP benefits on fresh fruits and vegetables in bulk storage and shipment and now in consumer packaged added-value produce.

Soft Bakery Products

High-value soft bakery products such as breads, rolls, cakes and muffins, are subject to quality loss from moisture loss; staling, i.e., crystallization of starch leading to texture hardening; and mold growth. Moisture loss is drying independent of staling. Microbial growth in soft bakery foods is largely surface mold growth.

Moisture loss can be virtually prevented by packaging in low water vapor permeability materials such as low density polyethylene film. Use of such materials retains water vapor within the package, but can create surface moisture conditions optimum for mold growth.

The most common reduced O_2 packaging machine for soft bakery goods is horizontal form/fill/seal using single web PVDC-coated OPP film to create a horizontal pouch. This system employs continuous gas flushing while forming and sealing. The system is used largely in Europe for relatively simple breads such as German rye meal, as well as for frankfurter and hamburger rolls.

In twin-web thermoform/vacuum/gas flush/seal equipment, after formation of the bottom cavity, the bakery food is inserted. The filled cavity is evacuated and back flushed with CO_2 or CO_2 plus N_2. Thermoform vacuum/gas flush/seal systems require more expensive high gas barrier packaging materials and equipment than horizontal form/fill/seal, but deliver better seals. German Multivac, Tiromat and Dixie-Vac thermoform/vacuum/gas flush/seal machinery is used for MA packaging of specialty bakery goods such as pumpernickel, crumpets and par-baked breads and rolls.

Pan-baked or brown-and-serve breads are made possible by MA packaging. The baker partially bakes the loaves and places them into a thermoformed gas barrier plastic (usually nylon) cavity, on a thermoform/vacuum/gas flush/seal machine. Still hot and expelling the CO_2 from baking, the packages are sealed. The products in CO_2 saturated packages have a three- to six-month ambient temperature shelf life. The consumer is instructed to place the product in a conventional conduction/convection oven where a brown crust is developed and the crumb remoisturized. The result is comparable to freshly-baked crusty French baguettes or rolls. Only one or two American bakers are using this system which is used in niche marketing in other countries. This system is commercially used in Europe and Australia.

In Canada, crumpets, somewhat analogous to English muffins in the United States, are commercially MA packaged on thermoform/vacuum/gas flush/seal equipment with CO_2 flushing for national distribution. At least two bakers are involved.

Sandwiches are being packaged under MA using similar systems, but the sandwiches are distributed under refrigeration to preserve the fillings. Fewer than a half dozen small Canadian firms are involved. Several small American companies are MA packaging sandwiches including hamburgers for refrigerated distribution on a regional basis. These products may appear in processed meat or special refrigerated displays. In some instances, the products may be distributed frozen and thawed for display at retail level. The most widely commercialized applications are sausage and biscuits, tortilla and egg, and precooked hamburgers. About 90 d actual refrigerated shelf life is achieved by MA packaging on thermoform/vacuum/gas flush/seal using a mix of N_2/CO_2, with a package label expiration date of 45 d. Since fillings usually contain nitrite cured meats, problems of microbiological safety are minimized. The instruction to consumers to heat prior to eating drives moisture from the meat into the biscuit or roll to refresh it. Many such products are nationally distributed in the United States, accounting for several hundred million packages annually.

About a dozen American bakers are using MA packaging for specialty soft bakery products and specialties such as "high fiber/high protein breads," filled croissants, parbaked French baguettes, etc. Although the MA packaging concept makes sense, MA packaged soft bakery goods are uncommon in United States distribution.

Since 1990, lines of yeast-raised baked shells primarily to function as pizza crusts have been distributed under MA in high barrier flexible packages in the United States. As much as $20 million worth of such soft bakery goods for further commercial processing are in ambient temperature distribution.

MA packaging of soft bakery products in Europe is much more widely used with more than 200 bakeries applying the concept. Further, the good shelf life of soft bakery products in Europe is often prolonged by complementing MAP with steam or microwave pasteurization to destroy surface molds. Bulk quantities up to pallet loads are heated in this manner to destroy mold spores present within packages.

The use of MAs with or without thermal pasteurization, to prolong quality retention of soft bakery products is technologically and economically very sound.

Pasta

Thirty percent moisture unfilled or filled pastas are marketed in dairy, delicatessen or produce departments and from pasta cases. Without MA, "fresh" pastas would have a shelf life of just a few days. MA prolongs refrigerated shelf life up to 90 d. Small companies such as Gourmet Pasta, Los Angeles; Famiglia, Jersey City; Romance, Wisconsin; Trios, Boston; plus two nationals are producing nearly 300 million packages of fresh pastas annually. The two majors, Contadina owned by Nestlé, and DiGiorno owned by Kraft General Foods, are in national distribution, with Contadina using initial MA packaging, and DiGiorno using N_2 plus an internal O_2 scavenger to maintain a zero O_2 interior. The products are cooked and, under sanitary conditions, packaged on thermoform/fill/vacuum/gas flush/seal systems using semi-rigid PVC/PVDC base and flexible PET/PVDC heat seal closure. The low water activity of the pasta itself inhibits growth of pathogenic microorganisms, but this hurdle is complemented in meat and cheese-filled pastas by some ingredients which can function as antimicrobial agents. Both excellent sanitation and tight distribution control contribute to microbiological safety.

Since 1990, most pasta makers have supplemented their pasta lines with sauces hot-filled into semi-rigid plastic tubs or flexible barrier pouches. Hot-filling plus cooling generates a partial vacuum within packages which, like pasta, are distributed under refrigerated conditions. This market has grown to about $500 million retail value in 1992, with further growth anticipated during the remainder of the 1990s.

Pizza

In the United States, pizza is a $20+ billion market, of which about 10% is frozen. Freshly made hot pizza, taken home or delivered home from local pizzerias or retail chains such as Pizza Hut, Little Caesar's and Domino's, competes with frozen pizza. Supermarkets, HRI delicatessens and convenience stores stock freshly prepared chilled pizza to accommodate take-out food customers. Approximately 5% of pizza is refrigerated prepackaged usually in the retail outlet, but this percentage is growing as pizza consumption increases.

Increasingly, pizza bakers are employing MAP, with another small group applying MA to pizza kits (which has also stimulated applications in other ethnic foods such as gyros). Beginning in 1991, Contadina introduced MA packaged parbaked pizza crusts in kits with sauce, cheese and even cured meat topping for refrigerated distribution on a national scale.

Salads

Prepared mayonnaise-based meat, seafood, and vegetable salads were an $800 million market in 1992 with good growth projections. Most such salads are produced by local and regional companies and packaged in air in bulk for delicatessens and HRI retail outlets. Salads for supermarket delicatessens are often prepared in-store, by central commissaries or by one of more than 75 local and regional, and single national suppliers which package bulk. About 5% of total U.S. prepared salads is prepackaged in unit portion size for retail sale. To date, there has been almost no actual United States' implementation of MAP for prepared protein or mayonnaise-based salads.

Other Products

Among the other foods being MA or vacuum packaged by only as few as one to three packers in each of the United States/Canada are: 1) hard boiled eggs in bulk pouches for use in HRI salad applications; 2) Chinese food including chow mein and egg rolls; 3) Mexican tortillas; 4) parfried potatoes; 5) entrées for convenience stores, largely sauced pastas–this product category surges forward and then ebbs with considerable interest accompanying introduction, with sales often failing because of quality, or concern about the microbiological safety of the products in distribution channels. Among the significant operators which have tested the concept of MA packaged entrées are McDonald's and WaWa (convenience stores), both using Keystone Foods as their source; and 6) sous vide for HRI outlets–as of 1992, only two or three U.S. and perhaps one Canadian company are producing such products despite the very heavy trade and technical journal publicity. As with MA packaged entrées, the quantities fluctuate because of entries and exits by companies. The highly publicized microbiological concerns have caused most of the participants to freeze the product for distribution to HRI outlets, so that the major benefit to such retailers is portion control. To date, the only significant retail grocery sales have been in Eastern Canada.

EUROPE

MAP has been more successful in Europe for retail sales while in the United States MAP is mostly used to deliver foods to retailers and HRI outlets. The most significant driving force in Europe has been that, until 1992, packaging of refrigerated foods has been influenced by retailers with objectives of selling more high margin fresh or refrigerated products and reduce in-store losses and waste due to limited shelf life. Further, geographic distances generally are shorter in Europe than in North America, thus permitting tighter control over distribution and its attendant temperature. European retailers often dictate the development of total systems to maintain quality and foster delivery of the desired quality and shelf life results. European retailers were successful in using their influential positions to dictate product and even process quality specifications and distribution requirements to their packaged food suppliers. Large retail grocery chains such as Sainsbury, Tesco, Safeway, Asda and Marks & Spencer in the United Kingdom; Euromarche, Carrefour and Lenor in France; Irma in Denmark; Spar in the Netherlands; and Tengelmann in Germany have dominated the refrigerated foods market and their derivative packaged foods.

During the 1980s, the commercialization of MA packaging for refrigerated food products was fastest in the United Kingdom, which accounts for about half of the European market for MA packaged foods. France is probably second with about 25% of the European MA packaged food market.

The success of food packaging in both the British and French retail food markets is often attributed to the presence of tightly structured retailing organizations. Although the proportions of retail food sales through chain retail groceries in Belgium, Denmark and the Netherlands are comparable to those in most other European nations, the relatively low populations in these countries might account for their relatively modest market shares of the total European MA packaged food markets.

Fresh Meats

In Europe the move to centralized packaging of fresh red meat cuts was stimulated largely by retail chains. Europe's first successful centralized fresh red meat prepackaging occurred in the 1970s in Denmark's Irma chain, which owns and operates its own fresh meat processing and packaging plant so as to supply its retail outlets throughout the country. Irma applied sanitation plus high O_2 MA packaging to extend the refrigerated shelf life of retail red meat cuts. All retail outlets are within 250 miles of the packing plant which employs its own vehicles for daily delivery to its retail outlets thus maintaining total distribution control for 5 d cycles. The chain is also able to market brand meats using this system.

Following Irma's lead and even using its products initially, the English chain, Marks & Spencer, introduced MA packaged fresh red meat cuts to its retail outlets in the 1970s under its St. Michael's brand. Marks & Spencer has been using high O_2/high CO_2 MA packaging for fresh meats in its outlets throughout the United Kingdom. This system is now the standard for British fresh red meat prepackaging which represents over a third of all retail fresh meat in that country. Marks & Spencer takes an active role in processing and packaging specifications and quality control of prepackaged meat. The products, including ground beef, are shipped directly from the packer to Mark & Spencer's retail outlets daily and carry a "sell-by date" of 6 d from packaging. No Marks & Spencer retail grocery outlet is more than 300 miles from the meat packer supplying it.

Three major United Kingdom grocery retailers, Tesco, Sainsbury and Safeway, have followed Marks & Spencer's lead in case-ready MA packaged fresh meat, although about half of their fresh meat at retail is still conventionally cut and packaged in the store backrooms.

No major British meat packer made a commitment to MA or any other centralized meat system except as a contract packer for a retailer. Distribution of MA packaged meat in Europe seldom exceeds 250 miles. European refrigerated meat distribution systems are generally efficient, cost effective and well controlled. European package labels indicate about 6 d of shelf life from packaging to expiration date. Fewer than 15% of European meat purchasers subsequently freeze their meats, whereas in the United States up to three-quarters of all fresh meat purchases are frozen by consumers.

Penetration of the fresh red meat market in Europe by MA packaged meat is estimated at about 35–40% for the United Kingdom; 30–40% for Denmark; less than 10% for France and only 2% for Germany. One reason for the low figure in Germany is the higher proportion of nitrite cured pork consumption in that country.

Fresh Produce

United Kingdom

Since the late 1980s, the MA packaging market in fresh vegetables has demonstrated significant growth. Nearly 300 million MA packages of fresh fruit and vegetables were sold in Europe in 1990. The U.K. market for CA/MA/vacuum-packaged fresh produce is still small when compared to the French market, which initially enjoyed dramatic growth in the precut salad vegetable market in the late 1980s but experienced a decline in 1990. Acceptance has been based on the desire for convenience and consumer perception that these products are fresh and natural. The recent decline in France has been attributed to the rapid industry expansion, which led to compromises with quality and subsequent shakeouts in the business.

The prepared fresh salad vegetable market was initiated in the United Kingdom by Marks & Spencer based on the success observed in France with precut lettuce, raddicchio, etc. The British market for prepared salad packs is about 30+ million packages annually, packaged by about 20 producers. About 25% of the packages are passive MA. Cut vegetables are packaged in oriented polypropylene film pouches and respire to consume O_2 and generate CO_2, thus producing an "equilibrium" MA to reduce the respiration rate and prolong the refrigerated shelf life. Other British companies involved in the prepared produce market include Kent Salads, Bourne Salads, Waterfall, Kane Foods, Hunter-Saphir, D. W. Muncey, Shieldness Produce, Dalgety, etc. Except for Kane Foods and Waterfall which produce salad lines under the "Crispa" and "Mr. Fresh" labels, respectively, the majority of these companies exclusively supply the major supermarket chains with private label products. Tinsley has supplied more than 75% of the mixed salad lines for Marks & Spencer.

The U.K. is following France in the application of the technology and market mix from precut lettuce and mixed green leafy salads into other vegetables such as potatoes, beets, shredded cabbage and carrots, etc. Heavier vegetables are now often packaged in semi-rigid twin web tray packages, thermoformed from a base web of PVC/PE and closed with a gas permeable flexible film.

France

France is the largest market in Europe for pre-prepared salad and other vegetables. In France the "ready-to-eat" fresh vegetable market, known as "la quatrieme gamme," increased by a factor of more than 40 during the 1980s. Nearly fifty producers are now involved. The "quatrieme gamme" sector of the fresh produce market accounted for about 3% of the total French fresh vegetable market in 1987. Of the products, prepared and chopped lettuce is the biggest seller, accounting for 80% of volume. The French market is dominated by three brands; "5eme Saison," "Salade Minute," and "La Florette," which account for half the volume and over 60% of the value.

Essential to the operation of this market is a distribution system which can ensure refrigerated 24-hour delivery of fresh produce anywhere in France.

The majority of prepared salad products are marketed in pillow pouches using oriented polypropylene (OPP) film packaged on vertical form/fill/seal machines such as Ilapak. These packages permit passive equilibrium MAs around the cut vegetables.

In MA packaging of "heavier" vegetables, Soleco (Societe Legumiere du Cotentin) markets the "La Florette" label for thirty-five different products, ranging from 125 g packages of parsley and onions to 500 g packages of sliced potatoes. The potato line includes pan French fries and sliced gratin dauphinois.

Prepared Foods

Among the significant MA packaged prepared food products in the United Kingdom are 30% moisture pasta under reduced O_2 plus elevated CO_2, mostly meat/cheese filled varieties; seafood, both fresh and precooked; entrées; and soft bakery foods. France is the world leader in sous vide packaged foods which are prepared, vacuum packaged, thermally pasteurized and distributed refrigerated and not frozen as in the United States. Most sous vide food products in France are for HRI distribution.

French legislation permits label expiration times of up to 21 d for sous vide products heated to an internal core temperature of 149°F (65°C) and with a pasteurization value of over 100, always with distribution temperatures below 37°F (3°C). French sous vide processors are under government regulation, with each organization mandated to control its own production and safety.

During its earlier years, sous vide was used primarily in French restaurants which packaged and cooked individual portions under vacuum in their own kitchens and stored them under refrigeration for later use. Up to one-third of French restaurants have either sous vide packaging or cooking equipment.

In 1988, sous vide represented more than 15% of refrigerated prepared foods sold by French hypermarkets and supermarkets. Larger stores sell over 75% of all consumer prepackaged sous vide products.

Cafeteria-style chain restaurants are major users of sous vide meals. Flunch, a major chain restaurant, which was one of the first institutional users, generates one-third of their sales from sous vide products. As the benefits of this technology became more evident, their parent company, Auchan, created Le Petit Cuisinier as a primary supplier of sous vide to Flunch. Le Petit Cuisinier distributes nearly 80% of its production to Flunch. Another major sous vide producer is SABIM, producing for Casino retail grocery chain. Fleury Michon, the current retail grocery market leader, entered solely in the consumer segment.

In Germany, the initial emphasis of MA packaging of prepared foods was on three products: fish, prepared salads and pizzas.

In both Belgium and the Netherlands, MA packaging has been developed largely for fresh vegetables, soft bakery goods such as hamburger buns and brown-and-serve French bread and pasta.

In Italy, refrigerated freshly prepared pasta has insufficient shelf life for supermarket distribution, and so MAP fresh pasta with refrigerated shelf life of up to 40 d is growing in Italy. Recently introduced have been steam pasteurized vacuum-packaged pasta with even longer refrigerated shelf life. One firm has implemented microwave energy sterilized pasta which could be distributed at ambient temperatures, but is distributed refrigerated to meet the consumer perception of fresh.

In the Netherlands, two companies are packaging pasta-ready meals under MA and are also using microwave energy to pasteurize the packaged product prior to refrigerated distribution, largely to HRI outlets such as airlines.

In Belgium, one firm has installed a pressurized microwave sterilization unit for MA packaged pasta-ready meals and is distributing them at ambient temperatures.

Microwave pasteurization is also employed in Europe to extend the shelf life of soft bakery goods, some of which are gas flush packaged prior to the heat process.

CONCLUSION

In the United States, up to ten billion MA packages of food representing up to 100 billion pounds of food are produced and distributed annually. The growth rate from a virtually zero base in the early 1980s is approaching 10% annually. Among the most significant MA packaged items in North America are fresh pasta, sandwiches, fresh poultry (for distribution), precooked poultry and cut vegetables (for both distribution and retail). In Europe, the most significant products are fresh red meats, cut fresh salad vegetables, fresh pasta, entrées and soft bakery goods.

By all measures, both volume and growth are present, driven by consumer demand for fresh foods and by awareness of the technology, which is really less than a decade into commercialization.

APPENDIX

T. W. Kutter, Inc. (Kraemer & Graebe)
Division of Tetra Alfa
91 Wales Avenue
Avon, MA 02322

Multivac
11021 Northwest Pomona Avenue
Kansas City, MO 64153

Reiser Packaging (Dixie Union)
725 Dedham Street
Canton, MA 02021

REFERENCES

Anonymous 1988. "Modified atmosphere packing: The quiet revolution begins," West Chester, Pennsylvania, USA, Packaging Strategies.

Anonymous 1990. "Modified atmosphere packaging," Ottawa, Canada, Food Development Division, Agriculture Canada.

Brody, A. L. 1984. *Proceedings of CAP 84, First International Conference on Controlled/Modified Atmosphere/Vacuum Packaging*, Chicago, Illinois, Schotland Business Research, Princeton, New Jersey.

Brody, A. L. 1986. *Proceeding of CAP 86, Second International Conference on Controlled/Modified Atmosphere/Vacuum Packaging*, Teaneck, New Jersey, USA, Schotland Business Research, Princeton, New Jersey.

Brody, A. L. 1987. *Proceeding of CAP 87, Third International Conference on Controlled/Modified Atmosphere/Vacuum Packaging*, Itasca, Illinois, USA, Schotland Business Research, Princeton, New Jersey.

Brody, A. L. 1988. *Proceeding of CAP 88, Fourth International Conference on Controlled/Modified Atmosphere/Vacuum Packaging*, Glen Cove, New York, USA, Schotland Business Research, Princeton, New Jersey.

Brody, A. L. 1989. Controlled/Modified Atmosphere/Vacuum Packaging of Foods, Trumbull, Connecticut, USA, Food & Nutrition Press.

Brody, A. L. 1990. *Proceeding of CAP 90, Fifth International Conference on Controlled/Modified Atmosphere/Vacuum Packaging*, San Jose, California, USA, Schotland Business Research, Princeton, New Jersey.

Brody, A. L. 1991. *Proceeding of CAP 91, Sixth International Conference on Controlled/Modified Atmosphere/Vacuum Packaging*, San Jose, California, USA, Schotland Business Research, Princeton, New Jersey.

Brody, A. L. and Shepherd, L. 1987. *Controlled/Modified Atmosphere Packaging: An Emergent Food Marketing Revolution*, Princeton, New Jersey, USA, Schotland Business Research.

Bruce, J. 1991. "Is European fresh product packaging transferrable to North America – the meat market analogy," *Proceedings of '91, Food & Beverage Packaging Expo & Conference*, New Orleans, Louisiana, USA, Schotland Business Research.

Buchner, N. and Day, B. F. "Packaging in modified atmosphere: Vacuum, gas flushing, packaging with gas mixtures controlled atmosphere," *Proceedings Behr's SEMINARE*, Munich, Germany.

Day, Brian. 1990. "Perspective of modified atmosphere packaging of fresh produce in western Europe," *Proceedings of CAP '90, Fifth International Conference on*

Controlled/Modified Atmosphere/Vacuum Packaging, San Jose, California, USA, Schotland Business Research.

Day, B. 1990. *Proceedings, International Conference on Modified Atmosphere Packaging*, Gloucestershire, U. K., Campden Food & Drink Research Association.

Day, B. 1992. "An update on controlled atmosphere/modified atmosphere/vacuum packaging developments in Europe," *Proceedings, Food Plas 1992*, Plastics Institute of America.

Day, B. 1992. *Guidelines for the Manufacture and Handling of Modified Atmosphere Packed Food.* Campden Food & Drink Research Association, U.K.

Hrdina-Dubsky, D. L. 1989. "Sous vide finds its niche," *Food Engineering International*, 14(7).

Parry, R. T. 1993. *Principles and Applications of Modified Atmosphere Packaging of Foods.* London, U. K., Blackie.

Vaughn, T. 1991. "Today in centralized meat packaging USA–If anything," *Proceedings of CAP 91, Sixth International Conference on Controlled/Modified Atmosphere/Vacuum Packaging*, San Diego, California, USA, Schotland Business Research.

CHAPTER 3

Current Status of Sous Vide in Europe

TOON MARTENS–*ALMA Universiteits, Belgium*

HISTORY

THE idea of cooking foods in sealed containers, or at least pasteurizing foods within hermetically sealed bags, pouches or other containers is not new. In earlier days, people also knew the benefits of protecting food from air. Usually the product was covered with lard, oil or paraffin. Wrapping food in a pig's bladder ("en papillote") before cooking for aroma concentration is a precursor of the "sous vide" system as it is used nowadays. What is new about the "sous vide" method is the minimal heat treatment adapted to the type of product and the use of plastic packaging materials with the right mechanical barrier and thermal resistance properties.

Plastic films became available in the 1960s and have been used in the meat industry for a long time. Another application of plastic films to pack food, was the retortable pouch (NASA, U.S.A.). The first application of "vacuum" packaging in large scale catering dates from the 1960's when the Nacka system (Sweden) and the AGS system (U.S.A.) were developed. In the Nacka system food is prepared and cooked by various traditional cooking methods, but the temperature in the food products is monitored to ensure a core temperature of at least 80°C. Cooked products are then transferred in their hot state to plastic bags, air is extracted and the bag is sealed to make it air-tight. Sealed bags are then placed in boiling water (100°C) for 3 min. The hot bags are then immediately transferred to a conveyer which takes them through a cooling tunnel with running water, the temperature of which is 10°C at first, then reduced to 4°C. The packaged food is dried and stored in a refrigerated room at a temperature of 4°C or less. When needed, normally within 2–21 d, the food is removed from chilled storage and reheated to at least 80°C (food core temperature) prior to service (Light and Walker, 1990).

The Nackà system was not taken up on a world-wide scale but it did attract attention, notably in Sweden, Germany and Switzerland and was used, especially by professional companies producing foods for institutional caterers.

The technology of heating foodstuffs in plastic films has been known and applied in France since the late 1960s. In 1968, the GATINEAU procedure was developed for the in-pack sterilization of potatoes. In 1972, the first in-pack cooked ham appeared.

The experience from the Nacka-system was imported to France in 1974 and was named the DELPHIN procedure. This system was developed in the SEPIAL laboratory and was applied in the catering sector.

In 1974, Georges Pralus started to apply the sous vide technique in gastronomy. He experimented with plastic films (CRYOVAC) which were used in the meat industry to minimize the cooking loss of "pâté de foie gras." Since 1980, Georges Pralus has been very active in promoting the vacuum cooking technique. More than 5,600 people from 36 countries have taken his course. In 1985 he published his book *Cuisson sous vide, une histoire d'amour* (Pralus, 1985) which sold about 10,000 copies. Thanks to the enthusiasm of Georges Pralus and the support of the companies GRACE CRYOVAC, ROSIERES and MULTIVAC, the "sous vide" technique was introduced to the chefs of the world.

Sous vide packaging and in-pack pasteurization for prepared meals was started in 1974–76 by the companies FLEURY MICHON and GUYOMAC'H. In the same period three companies developed a system for cooking meat in-pack at low temperature. But, by regulation in France, the maximum shelf life of the prepared meals was 6 d and this blocked all industrial development. In 1984 a licence was obtained for a longer shelf life product. After this, the number of "sous vide" products increased very quickly in France and Belgium, but not in the more northern European countries: U.K., Germany, Denmark. In the latter countries the safety concern is much higher, but the quality concern is much lower.

However, in addition to legislative authorities being very sceptical about sous vide, many researchers were paralysed by the "botulinum-syndrome." Safety was defined as commercial sterility (12-D concept). A lot of research was done on optimisation of heating conditions for sterilization, not only for cans but also for products packed in flexible films (Martens, 1976, 1980). But the result was a sterile product with a bad taste. Some researchers like Gousseault have worked for a long time on the microbiological aspects of sous vide products and have proved that this type of product is safe if produced and kept at the right temperature.

Sous vide has a long history in Europe, but the future looks brighter. The demands of the consumer for fresh and tasteful convenience products have made both small and large scale catering, as well as the agro-industry, aware of the increased interest of consumers in these types of products.

PRESENT MARKET AND POTENTIAL GROWTH

Global Trends

Product Design Criteria

As a consequence of demographic evolutions, higher spending power, new kitchen equipment (microwave), family composition, ethnic diversity, etc., product design criteria for foodstuffs have changed. "Sous vide" products fit very well into this new evolution of consumer demands.

The evolution in consumer trends in the EEC for food products can be summarized (Keuning, 1990): 1) food products with high quality and low energy content; 2) products with low fat content and rich in complex carbohydrates; 3) convenience foods in innovative packaging; 4) "natural" food processing, emphasis on "natural ingredients"; 5) more products and packaging materials that can be used in a microwave; and, 6) also a trend towards higher quality and natural products was detected in the catering sector.

In the U.S.A. similar trends have emerged (Fox, 1989).

The design criteria for food products in the 1990s are: high quality, high convenience, free of additives and preservatives, organically grown, low energy content, fresh, low cost and packed in an environmental friendly material that is tamper-evident and tamper-resistant.

The acceptance of sous vide in high class restaurants proves that there is little doubt about the improvement of sensorial quality. The better retention of the food's aroma allows the use of lower amounts of taste enhancers. Chemical preservatives are not typically used in sous vide. However the discussion about cost, safety and environmental aspects is open.

The general trends in the U.S.A. and Europe are similar, but there are considerable differences between Anglo-Saxon countries and the Latin countries. In France, Belgium, etc., much more attention is spent on taste and aroma, while in the U.S.A. and U.K. product safety is much more important. This explains why there is such a difference in the adoption of the "sous vide" technique in different countries.

Although the interest for exotic dishes grows, taste will remain regional and therefore opportunities exist for small and medium sized companies in this latter area.

Market of Fresh Prepared Products

Southern and Cakebread (1991) estimated that the market share of chilled meals was 12.3% (versus 82.1% for frozen meals and 5.6% for sterilized meals) in 1988 and had grown to 17.8% in 1991. A detailed analysis of the market share of prepared meals by conservation method in Belgium (Martens, 1992) gives a different view. The Nielsen-numbers are compared with the consumer panel data of the Agricultural-Economical Institute in Table 3.1.

The differences between the numbers can be explained by the conclusion of the C5 distribution chain (specialized shops, markets, etc). In Belgium, a growing number of butchers are selling prepared meals, e.g., lasagna, to compensate for the decreasing sales of meat.

This consumer panel data (Martens, 1992) illustrates very nicely the evolution of the market of fresh prepared meals. In the period 1987–1990 the consumption of freshly prepared meals grew from 1,592 kg/person to 2,651 kg/person (+65%) and the value increased from 261 BEF (Belgian Francs) to 533 BEF/person (+204%). The percentage of families that bought prepared meals at least one time per year increased from 36.0 to 44.4%. The U.K. chilled food sector grew 95% in the period 1985–1990. Few numbers are available on the market share of sous vide produced prepared meals or meal components.

Table 3.1. Market share of different conservation methods in Belgium (Martens, 1992)

Conservation Method	Nielsen	Consumer Panel Data
fresh	22	67
frozen	48	20
sterilized	30	12

In 1989, 2.5% of the prepared meals in France were vacuum cooked. This represented some 12,000 tons, 8,000 tons for catering and 4,000 tons for supermarkets (Grigaut, 1990). In 1991, 12,500 tons were produced for catering and 8,000 for supermarkets (Fiess, 1991). These numbers do not include sous vide products from the catering and restaurant sector. According to Pralus, 10 to 15% of the cooks in French restaurants are using "sous vide."

Zanussi Food Service Equipment estimates that the market share of pre-cooked food in restaurants will grow from the present 30% to 41.6% within 5 years. In large restaurants the share of pre-cooked food will grow from 55 to 60%, but the biggest increase in the use of pre-cooked food will be traditional gastronomy and specialized restaurants which is now 15%. In Spain the sous vide trend is estimated at 5% in institutional catering and alimentary industry, and 3% in catering companies. In the U.K. sous vide is a relative newcomer to the catering scene. Cook and chill is quite popular with the local authorities for school meals, old people homes and hospitals. The Marriot is using the sous vide system in sixteen four- and five-star hotels in the U.K., Commonwealth Hotels International uses sous vide in fourteen of its Holiday Inns, and it features it in 15 to 20% of its restaurant foodservice. The Meal Ticket (a small catering firm) uses sous vide in many of the contracts for executive and boardroom catering. All these numbers show that there is a big increase in Europe in the use of sous vide in hotels, restaurants, catering and agro-industries, but there are no systematic and global numbers available.

Recently, Rouxl Brothers closed down their sous vide factories and some others will follow. The main reason is the overpricing of "sous vide" products. Due to the economic recession, price is a critical success factor and to obtain low prices efficient production is needed. ALMA University Restaurants (Leuven, Belgium) have proven that sous vide in combination with automation of production can improve the quality, safety and cost of foodservice meals.

Identification of Consumers of Sous Vide Products

An analysis of the food consumption panel of the Agricultural-Economical Institute in Belgium gives a more detailed profile of the consumer of prepared meals (Table 3.2). The scale is constructed so that the mean consumption is 100.

From this data the profile of the consumer of prepared meals can be drawn: a single person or dual career family, young and living in a city.

Table 3.2. Profile of consumers of prepared meals (Martens, 1992).

Number of persons per family	1	2	3	4	>4
Consumption score	124	93	102	119	78

Age	<36	36–50	51–65	>65
Consumption score	118	110	99	48

Profession	Working Woman	Employee	Independent	Housewife
Consumption score	137	117	150	87

Area	Rural	Urbanized	Regional Cities	Big Cities
Consumption score	62	89	118	120

From this data the profile of the consumer of prepared meals can be drawn: a single person or dual career family, young and living in a city.

Identification of Producers

In a recent EC-survey (Schellekens and Martens, 1992) more than 100 producers of sous vide products were identified in the European Common Market (Table 3.3).

Sous vide producers are small or medium sized companies. The total dollar value of these products was in the range of 350,000–950,000 US$. Table 3.4 summarizes the product and market information given by sous vide producers.

TECHNOLOGY

Introduction

The type of equipment is very dependent on the size of the batches. For large scale batches of the same products sterilization equipment is used. When a wide variety of products have to be produced, preference is given to steaming cabinets. In restaurants air-o-steam ovens are used.

Table 3.3. Producers of "sous vide" products in EC.

Country	# Producers
France	67
Belgium	20
Netherlands	8
Ireland	5
United Kingdom	2
Germany	3
Spain	3
TOTAL	108

Table 3.4. Product and market information (Schellekens and Martens, 1992).

			% of Total Number of Producers
Type of products:	meal components		87.5%
	complete meals		50%
Portions:	individual		75%
	multiportions		87.5%
Type of consumers:	supermarket		50%
	specialized shops		25%
	catering	schools	62.5%
		companies	62.5%
		catering organizations	75%
Geographical distribution of consumers:		same region	25%
		same country	50%
		international	50%
Forecast of:	growth of sales of "sous vide" products in the next 3 years		24%
	increase in number of products		17%
	expansion of distribution	in the region	5%
		in the country	31%
		in the EC	40%

Sous vide equipment can be classified according to the scale of operation and heating method.

Industrial Equipment

Heating with an Air/Vapour Combination

Packaged products are heated by injecting vapour in the enclosed space. The steam is normally produced in a central installation. A fan assures the continuous movement of the mixture during the whole heating cycle. These systems heat products from 60 to 100°C.

There are a lot of different systems with air/vapour on the market. Manufacturers of this kind of equipment include: THIRODE, CAPIC, BODSON, which are all French companies, and FESSMANN, ATMOS, VEMAG (German companies). This equipment is derived from the meat industry (ham cooking, smoking). Advantages of these systems are their versatile application (unpacked products as well as packaged products can be cooked) and low investment cost. The main disadvantage of these systems is that the temperature distribution and surface heat transfer coefficient are not well controlled and the thermal efficiency is low. Depending on the type of the product, ballooning can occur if any air is left in the bag.

Another type of cooking equipment which uses a mixture of air and vapour is the system manufactured by the company LAGARDE. This computer-controlled system which was originally developed for sterilization processes, can work at overpressure. This has the advantage of preventing swelling of packaged products. Since cooking and refrigeration occur in the same unit, thermal efficiency is high. Vapour is injected into the enclosed space; the vapour mixes with the air that is present and this mixture is collected by a fan and guided along the walls of the installation. Then, the vapour/air mixture passes over the baskets that contain the product. The temperature in this closed system can be regulated with an accuracy of 0.5°C according to the equipment specifications, but this accuracy is very dependent on the level of air in the steam/air mixture. Process time may also be controlled by using a thermometer within the product.

When the heating cycle is completed, refrigeration starts automatically. Refrigeration can start with a pre-refrigeration by sprinkling with water to quickly cool the vessel. Cooling to 10°C in the core of the product in less than 2 h can be accomplished by sprinkling with ice water (0 to 2°C), produced by a central ice water producing unit, or by cryogenic cooling where CO_2 is injected in the room, lowering the temperature to −20°C; after which the temperature is equilibrated to 4°C.

This type of equipment can be programmed to achieve a low temperature when the heating/refrigeration is finished, permitting a cycle to be done at night and product unloaded the next morning without interrupting the chill chain. This type of equipment is used in many central kitchens, as well as many food manufacturers, e.g., Plaisir à la Carte, Nouvelle Gastronomie Française located in France.

Advantages of this system are the overpressure, the versatility of the process, the efficient energy utilization and, last but not least, being fully automated, this system permits highly reproducible cycles.

Heating/Cooling with Water

The THERMIX-system was developed by ARMOR INOX (France). It is an integrated system consisting of: 1) a hot water tank with a heating appliance (the heating source can be gas or electricity); 2) one or more tanks for immersing the product; and 3) an ice water tank with an aggregate to produce ice water. The product is moved in and out of the tanks with a tackle.

When the product is loaded in the tank, hot water descends and mixes with cold water to obtain the programmed temperature. When the tank is full, the heating time is controlled by a probe measuring the core temperature. The pasteurization and cook values of the product may be calculated. When the cooking has finished, the tank is emptied by pumping and part of the lukewarm water is recycled in the hot water tank. Tap water is then introduced (15−20°C) for initial cooling. When this is finished, the tank is emptied again and is filled with ice water (0−2°C). This water will in part be recycled.

This type of equipment has several advantages. First of all, it is an integrated, complete system. The client will not be preoccupied with energy or ice water production. Furthermore, several tanks can be used, allowing one to process different products simultaneously. This way, the system is well adapted for the production of small quantities of variable products. The process is controlled by a microcomputer with onscreen visualization and a printout of the cycle parameters. Control of the process is easy. The main disadvantage of this system is the consequence when a pouch leaks; the heating and cooling water as well as other pouches become contaminated.

The French companies AGIS, FREALIM and NOUVELLE CARTE use the cook tank system.

Another type of cooking/cooling tank is the AFREM-system (France). In this system, two tanks are used; one for heating and the other for refrigeration. With a tackle system the products are transported automatically from one tank to another.

The CAPKOLD-system (U.S.A.) is sometimes promoted as a sous vide system. But, in reality, it is a cook and chill method. The products are traditionally cooked, vacuum packed and then cooled in a tumbler.

Streaming Water Heating

Heating is performed with streaming water in the STERIFLOW apparatus from the company BARRIQUAND STERIFLOW (France). The apparatus is derived from the pressure-sterilization industry.

The products are loaded on baskets and rolled into the vessel. The low quantity of water stocked in the base of the vessel is pumped through the heat exchanger where it is brought up to temperature and is then streamed in a closed circuit and at very high flow rates around the containers. For four baskets, 400 litres of water are recycled every 9 sec at a flow rate of 160 m^3/h, ensuring a rapid flow.

The opening of the heating valve is controlled in relation to the pre-programmed temperature value (accuracy: $\pm 0.5\,°C$). Condensates are evacuated via a steam trap and can be returned to the boiler. The overpressure is regulated independently from the temperature by the injection or exhaust of compressed air according to the pre-programmed pressure values.

The cooling water is entirely separate from the pasteurizing water and can neither soil nor recontaminate the containers. This means that any kind of water, cold water or ice water can be used for cooling. The circulation of cold water in the heat exchanger is controlled automatically in relation to the programmed cooling curve. The pressure is controlled exactly as during the heating phase. At the end of the cooling phase, the circulating pump stops, the quantity of water collects at the bottom of the vessel and the vessel is put under atmospheric pressure.

The major advantage of this system is the separation of the two circuits and the low energy consumption due to the limited amount of water in the vessel. For equipment that works at overpressure, like LAGARDE and STERIFLOW, costs are elevated because of the need for vapour production and ice water production. Only elevated production capacity can justify these high investment costs.

Continuous Microwave Ovens

A lot of effort is being directed towards developing microwave ovens and combined ovens to ensure even temperature distribution and accurate temperature control. Energy absorption and transmission in foods are determined by mutual interactions between the microwave power source and the food product. Some energy may be reflected from product surfaces without being absorbed, causing standing node and antinode wave patterns. These cause high and low incident power levels in the microwave cavity that result in uneven energy distribution at product surfaces, and hot and cold spots within the product. This problem can be minimized by using wave stirrers which distribute microwaves more uniformly at product surfaces or by rotating the product in the microwave oven. More uniform heating can also be obtained in ovens with multiple feed systems (IFT, 1989).

To cope with the fact that microwaved products often show an uneven temperature distribution, combination type ovens (microwave and vapour) have been developed. This way a more evenly heated product can be obtained. In the LUW (Landbouwuniversiteit Wageningen, Netherlands) the use of a continuous system is being studied. The products are guided through a tunnel on a conveyer belt. Microwave energy is responsible for the rapid temperature increase, while conventional heat is used to hold the products at the specific cooking temperatures. A sensor measures the unabsorbed radiation and automatically adjusts the field intensity. This can be particularly useful for a diverse product range. This process saves money and energy (Lengkeek, 1990).

At Delta Daily Food (Nieuw Vennep, The Netherlands) ready meals are heated in a tunnel oven equipped with four special compartment magnetrons, each containing twelve tubes to ensure a precise temperature setting. The meals are pasteurized at 72°C for a few minutes. After the magnetron section, the meals pass through an oven of 75°C for 4 to 5 min, as a sort of second pasteurization (Van Dijk, 1990).

The combination of microwaves and air-under-pressure is actually being used by a small Belgian company (P & T Foods, Hulshout, Belgium) for the sterilization of pasta products. The system contains about fifty microwave generators working at a frequency of 2450 MHz. The energy is focused on fourteen microwave guides at the interior of the apparatus. The whole process is under the control of a PLC (programmable logic controller). An infrared camera detects the temperature of each tray and transmits it to the PLC that controls the microwave guides. The microwave generators are under control of the PLC. To achieve even heating, the trays have an upward domed base in the centre of each compartment (two separate compartments: one for pasta and one for the sauce). Based on the same principles, it should be possible to develop systems for the continuous and automated production of vacuum-cooked products.

The sensory properties of foods cooked using microwave energy can be different from those that have been cooked conventionally. It is the difference between exterior heating (conventional oven) and heating throughout (microwave oven) that could cause a difference between the results. There is much scope for further sensory analysis to compare and quantify these differences.

BONNET (France) has developed a small scale multi-energy tunnel, a tunnel using different heating sources including microwave, steam, and infrared, that can be used in central kitchens.

Equipment in the Catering Sector

As the vacuum cooking method relies on the strict control of temperature to produce highly flavoured tender products (i.e., often using low temperatures for long times), temperature devices should be accurate. Cooking can be carried out in a hot water bath so that the vacuumed bags or pouches are immersed in (agitated) water of a controlled temperature. Some cooks use their traditional kettles, but due to the poor temperature control, this method is not recommended.

Another popular method is the use of the steam combination oven. These devices are simple oven cabinets which can either use convection dry heat alone (forced hot air) or low pressure steam injection in addition to convection heating at temperatures below 100°C. The equipment is supplied by a wide range of manufacturers: JUNO, ZANUSSI (Italy), HOBART (Germany), ELOMA (Germany), LEVENTI (Netherlands), RATIONAL (Germany). Temperature distribution in air-o-steam systems is better than in forced air ovens, but even in these ovens temperature distribution is not homogeneous (Table 3.5). However, most systems follow the same basic principles and are designed to hold standard gastronorm trays or packaged products on shelves. The capacity ranges from five to forty trays.

Table 3.5. Range of times (minutes) required to heat standarized "sous vide" packs from 20 to 75° C core temperature (Sheard and Rodger, 1993).

Oven*		Size	Fastest	Slowest	Difference
A	Lower oven	3-grid			
	Shelves 2, 4, 6		17.0	40.5	23.5
	Shelves 3, 5, 7		16.0	38.0	22.0
A	Upper oven	3-grid			
	Shelves 2, 4, 6		21.0	51.5	30.5
	Shelves 3, 5, 7		22.0	51.0	29.0
B		3-grid			
	Shelves 2, 4, 6		22.5	40.5	18.0
	Shelves 3, 5, 7		31.0	53.0	22.0
C		6-grid	9.0	26.5	17.5
D		6-grid	21.5	45.5	24.0
E		6-grid	15.0	35.0	20.5
F		10-grid	21.5	63.5	42.0
G		10-grid	20.0	42.0	22.0
H		10-grid	15.5	47.0	31.5
I		10-grid	31.5	85.5	54.0
J		10-grid	20.0	52.0	32.0

*Letters stand for a brand of oven.

After cooking, direct chilling in an ice water bath is both cheap and efficient. Agitated systems are preferred since these maximize heat transfer with rapid chilling as a consequence. The method has been shown to be up to four times faster than air-blast chilling. Two different systems exist. The first uses a specially designed paddle device which agitates the food packed in bags (i.e, the bags are massaged and kneaded during the chill cycle) to maximize heat transfer. The second method provides for the rapid movement of water around the system and the use of a heat exchanger to retain the low temperature of the coolant water. The controls on ice water tank chillers are much the same as those described for air-blast chillers and many are fully programmable.

An air-blast (electro-mechanical) chilling machine is usually constructed as a cabinet and food to be chilled is placed inside (often on trays – gastronorms – stacked on prespaced rungs in a trolley). Pre-cooled air, sucked into the machine and cooled using a compressor, is pumped into the cabinet at high velocity and around the inside of the unit (sometimes being deflected by baffles) so that it blows more evenly over the hot food to ensure rapid cooling. Various machine sizes ranging from 10 to 200 kg are available. European blast chillers suppliers are FOSTER (U.K.), WILLIAMS (U.K.), ACFRI (France), FRIGINOX (France), IRINOX (France), FRIULINOX (France), ZANUSSI (Italy) and ASSKÜHL (Germany).

Many air-blast chillers are supplied with a range of controls and recording devices. These include temperature controls which control and monitor air cabinet temperature, thermometer read-out for food probe thermometers, automatic defrost systems with defrost signal lights, pre-programmable chill cycle control, time control systems and alarm systems to signal when the device is faulty or not operating. These controls are vital if proper process control is to be implemented and, thereby, quality assured. Air-blast chillers are much simpler to operate than water baths and leaking pouches do not present any problem.

After chilling, the products are stored in cold rooms or chill cabinets. Both utilize enclosed areas which are cooled by compressor units and are built to specification to retain cold air at a temperature between 0 and 3 °C. However, different researchers have shown that the temperature control of storage rooms in many cases cannot be held within these specifications.

Reheating may take place in the bag with the product still under vacuum, i.e., in a hot water tank or a combination steam oven, or after the contents of the bag have been emptied into a separate container for heating on top of a cooker or inside an oven. Alternatively, after puncturing the bag, the product could be reheated in a microwave oven. The film should be pierced to allow steam to escape. Otherwise pressure can accumulate in the pack and eventually cause an explosion.

Penetration and heating of foods by microwave energy can vary tremendously depending on food type, shape and thickness, and is not limited by low conductivity. In contrast to microwave ovens, conventional heating methods transfer thermal energy from product surfaces toward their centre 10–20 times more slowly. Plastics are transparent to microwaves; they do not heat directly in a microwave field.

In any event, the heat treatment during reheating should not be a prolongation of the cooking cycle, since product quality will decrease. Products should be heated at a temperature lower than the original cooking temperature.

Conclusion

The surface heat transfer coefficient plays an important role in most heat transfer operations in "sous vide" systems. In order to have a homogeneous temperature distribution and higher energy efficiency in the heating or cooling cabinet and inside the product, equipment and control systems have to be redesigned.

To improve the homogeneity of temperature distribution without losing the advantages of speed, microwaves have to be combined with conventional heat sources and intelligent control systems have to be built that take into account product properties, product geometry, load, etc. Intelligent control systems will also reduce the risk of application of the wrong time-temperature conditions that will be different for every product.

Little is known about the thermal efficiency of different types of equipment. The "sous vide" system is very energy intensive. For cost and environmental reasons, the thermal efficiency of different solutions should be investigated. Our total energy costs rose by 25% due to the "sous vide" system as compared to the classic cook-hot hold method.

Packaging Equipment

Industrial Packaging Machines

Thermoforming consists of heating the sheet (or thick film) and then forming it into the desired shape by forcing it into a cool mold by one or more of several methods including air pressure, vacuum and plug assist. The choice between the different methods depends on the firmness of the sheet and the desired firmness of the package. Well known suppliers are KRÄMER & GREBE (Germany) and MULTIVAC (Switzerland).

Requirements for bottom sheets are higher than for the upper sheets since the bottom sheet is stretched to give the desired form. When the material is too heavily stretched, the sheet becomes too thin and can eventually rupture. Techniques to avoid these problems make use of a plug combined with a vacuum to obtain the right form and thickness.

After forming, the trays are filled with product, vacuumed, sealed and cut. This form of packaging is widely used and different molds can be used. It is important to mention that the level of the vacuum is dependent on the temperature (Table 3.6) and the viscosity of the product.

For products where liquid is added (e.g., "coq au vin"), only a small vacuum can be obtained. The air that is left causes ballooning. In Table 3.7, the relationship between vacuum level and temperature is illustrated (De Baerdemaeker and Nicolaï, 1993).

Table 3.6. Dependence of vacuum level on temperature.

Temperature (°C)	Pressure (mbar)	Pressure (Torr)	Vacuum (%)
100	1013	760	0
90	701	526	31
80	474	355	53.3
70	312	234	69.2
60	199	150	80.3
50	123	93	87.8
40	74	56	92.6
30	42	32	95.8
20	23	17.5	97.4
15	17	13	98.03
10	12	9	98.75
5	9	6.5	99.15
0	6	5	99.35

Table 3.7. Partial pressures of air (in 10^5 Pa) and water vapour for different heating temperatures after packing at 3°C (De Baerdemaeker and Nicolaï, 1993).

Operational pressure at closing of package	.1000	.2506	.4000
partial pressure air (mbar)	.0924	.2240	.3924
partial pressure water vapour (mbar)	.0076	.0076	.0076
Heating to 80°C			
partial pressure air (mbar)	.1182	.310	.502
partial pressure water vapour (mbar)	.4740	.474	.474
Total pressure	.5922	.784	.976
Heating to 90°C			
partial pressure air (mbar)	.122	.320	.516
partial pressure water vapour (mbar)	.700	.700	.700
Total pressure	.822	1.020	1.216

Chamber Vacuum Machines

This type of equipment is mainly used in the small scale catering sector. Flexible pouches (mainly CRYOVAC) are used. After filling the pouches, they are placed in the chamber, a vacuum is applied and the pouches sealed. The vacuum applied varies from 99 to 99.9%. Depending on the type of machine, the amount of vacuum applied can be controlled, (single or double sealing bars are available) and atmospheric pressure can be slowly reintroduced.

Allewijn and Norder (1992) have recently given an extensive overview of vacuum and packaging machines. Some of the manufacturers involved include MULTIVAC (Switzerland), TURBOVAC (Netherlands), ROSCHMATIC and HENKOVAC (Netherlands).

Packaging Material

Requirements

The use of a vacuum greatly inhibits the progress of oxidation reactions which generally lead to the deterioration of foodstuffs during storage, either before or after cooking, and inhibits the growth of aerobic microorganisms. Furthermore, post-process recontamination is avoided.

The use of a vacuum serves to pull down the plastic film tightly onto the food surface. If an air or gas space was left in the bag or pouch this would act as a layer of insulation. After vacuum packing the plastic film merely acts as a skin on the surface of the food and does not unduly inhibit heat transfer. Heat transfer through the packaging material is governed by its thermophysical properties.

Due to the applied vacuum, the food can be subjected to a pressure effect. This may be an explanation for the more uniform water distribution in vacuum packed asparagus and consequently more even heating.

The packaging material must fulfil several tasks. For this reason the material must conform to several specifications. In Table 3.8 abbreviations for various plastics are given.

Resistance to High Temperatures

The plastic to be used is highly dependent on the temperatures attained during the heating/reheating of the product. For example, LDPE should not be used at temperatures above 85°C. When temperatures above

Table 3.8. Abbreviation of various plastics.

PE	polyethylene
PA	polyamide
PP	polypropylene
PS	polystyrene
PC	polycarbonate
LDPE	low density polyethylene
LLDPE	linear low density polyethylene
MDPE	medium density polyethylene
HDPE	high density polyethylene
PET	polyethylene terephthalate
CPET	crystallizable polyesters
PPO	polyphenyloxide
PVC	polyvinyl chloride
PVDC	polyvinylidene chloride
EVOH	ethylene vinyl alcohol
VC	vinyl chloride
VDC	vinylidene chloride
EVA	ethylene-vinyl acetate copolymer
EPS	expanded polystyrene

85°C are reached, MDPE, PP, PA or PET can be used. HDPE can be used for temperatures up to 100°C. Above this temperature, the HDPE-film becomes opaque. But even if the chemical composition and layer thickness is the same, resistance to temperature can be very different. The orientation of a PE-film has an important effect. Of course, the price/quality of a film is a very critical factor.

A study of Castle et al. (1990) revealed that reheating ready-prepared foods packaged in plastic pouches, trays or dishes in the microwave oven according to the manufacturers' instructions, resulted in temperatures in the range of 61 – 121°C. This can result in a build-up of pressure within the package which may cause it to explode.

For products pasteurized at temperatures of 70 – 80°C, stored between 0 and 3°C, and with a shelf life of less than 21 d, PP, HDPE and CPET can be used. When the shelf life exceeds 21 d, plastics based on PP and a resin barrier assure the impermeability. Resins of PS melt at temperatures around 82°C. At temperatures above 82°C, this material loses its rigidity and produces a typical styrene odour. When PPO is mixed with the PS, the rigidity is maintained and the product keeps its permeability to microwaves. For each 1% of PPO, the temperature of resistance increases 1.5°C.

Recently, microwave-"active" packaging materials, commonly called susceptors, have been introduced for use in microwave ovens for the browning of foods such as pizzas, potato chips, popcorn and pastries. These materials have a construction consisting of plastic (typically a PET food contact layer), aluminum and adhesive on a paperboard substrate. The susceptor functions by partially absorbing and partially transmitting microwave energy. The ratio of absorbed/transmitted energy is controlled by varying the thickness or composition of the metal layer. The absorbed energy is re-radiated at longer wavelengths in the thermal range, thus simulating the radiation encountered by foods in a conventional radiant oven. High local temperatures (in excess of 200°C) can be encountered. This high temperature can, on occasion, lead to some disruption of the PET (crazing), charring of the adhesive and paperboard layers with the release of pyrrolysis volatiles, as well as cause a build-up of pressure within the package.

Impermeability to Gases

Another requirement is the impermeability to gases (O_2, CO_2 and vapour). Plastics mentioned in this respect are PVDC, EVOH and PA. PVDC retains barrier properties in moist conditions. Moisture-sensitive EVOH barriers are typically damaged by immersion in water-baths. EVOH suppliers now claim that improved grades of EVOH and cloaking of the material in a coextrusion, allow EVOH to approximate the performance of PVDC.

Recently, "flexible glass" has been developed to combine the high barrier properties of glass with the flexibility of plastics. Silicon-dioxide or silicon-monoxide is vacuum deposited on the plastic, giving a layer of about 1 μm "glass" on top of the plastic. Also, ceramic compounds can be vacuum deposited on a plastic. The effect of these extra layers on the thermophysical properties of the packaging are not described. The permeability of the plastics is described as cm^3 (or gram)/m^2/24 h at 25°C at a 75% relative humidity. For prepared meals the permeability has to be lower than 50 cm^3/m^2/24 h (Rozier et al., 1990).

Restricted Migration of Plastic Constituents

Migration of low molecular weight compounds from the packaging into the product has to be controlled. These compounds can be rest-monomers, additives or processing-aids. On the other hand, aromas from the product, especially oil, water, alcohol and essential oils, can migrate into the packaging material.

The EC standards of food contact materials include a limit on the migration of total constituents of food contact plastics (overall migration) of 10 mg/dm^2 or 60 mg/kg (90/128/EEC), as well as specific migration limits for substances from a permitted list of monomers and other starting substances. A list of plastics additives is also in preparation. It does not apply to "materials and articles composed of two or more layers, one or more of which does not consist exclusively of plastics, even if the one intended to come into direct contact with foodstuffs does consist exclusively of plastics."

Test conditions for the determination of overall migration are specified in Directive 82/711/EEC, together with specifications for food simulants, the choice of the appropriate simulant being based on a given classification of foods (85/572/EEC). Test conditions are stipulated in Directive 82/711/EEC for most situations, depending on the intended temperature of use and contact time. However, for a period of contact of less than 2 h at a temperature exceeding 121°C, the test conditions are allowed to be "in accordance with national laws."

For a number of different plastics and for different chemical species, no quantitative effect on migration could be observed as a result of microwave energy over and above that expected from the heating effect alone. Therefore, it is proposed that testing be carried out using conventional heating. This avoids the particular difficulties associated with establishing reproducible conditions in a microwave oven. Microwave-active materials are a special case where the rapid generation of high local temperatures can only be achieved in a microwave oven.

Sufficient Mechanical Strength

The mechanical strength of the film, and in particular of the seam, should be high enough to withstand all manipulations (heating, cooling, transporting, etc.) without rupture. In the case of leaks, contamination of the product with microorganisms is possible. For this to occur, two conditions must be met. Firstly, there must be a leakage pathway. Secondly, microorganisms must be present in high numbers in the surrounding environment. Another factor that greatly increases the chance of contamination is the presence of water, as it acts as a transport medium for the micro organisms. Furthermore, during heating of vacuum-packed products, fluid can enter through the leaks. This is seen as vapour in the package.

Sealing irons that are too hot can be responsible for melting, contracting, and crumpling of the film, giving rise to leaks. Dirty sealing irons can also be responsible for the loss of seal integrity in hermetically sealed packages.

There are a few methods to check bags and trays for leaks. The oldest method is the "bubble test." The package is immersed in water and the escape of bubbles is checked. Another method is to put the package under a bell-glass. Subsequently, air is withdrawn until bubbles appear. A third method consists of following the penetration of coloured liquid (usually methylene blue) in the pack.

Biotesting of the seam integrity is another possibility. This is a design failure test which can be used to look at the effect of design changes on integrity. If challenge with active cultures of bacteria results in growth in the bag, the bag is defective. The above methods are not very accurate and are time consuming, but they do not require high investment costs. These tests are also destructive, so they are only performed on a certain number of (statistically determined) samples.

For checking the strength of the seam, N_2 pressure can be used. N_2 is injected in the pouch until the seam ruptures (burst testing). In a second test, the pouch is filled with N_2 so that the pressure is just below the seam rupture point. When no leaks are present, the pressure within the bag should remain constant. Another method to check the seam is the use of electricity conduction. In case of leaks electricity is not conducted.

The most accurate, but also the most expensive leak detection method is the use of He in an on-line process. Holes as small as 3 to 4 μ can be detected. These holes are big enough for microorganisms to pass. First, the bag is exposed to a He-enriched atmosphere. Afterwards, the bag is brought into an environment without He. In case of leakage, He will escape from the package and will be detected by a He detector (a mass spectrometer, for instance). The use of He detection is still very limited in the food industry. The method is only justified for certain high value products, but can be useful for development of processes.

Loma Systems (U.K.) developed an on-line leak detection piece of equipment, based on a plug that exerts pressure on the package and that can reveal leaks by a minuscule change in position.

The use of a film with a non-skid layer put on top of the package is another possibility. In case of a leak in the package, the upper film can be moved since there is some space between the package and the vacuum-test film. These two methods can reveal holes as small as 0.5 mm. None of the above described methods seems to be accurate, quick and cheap at the same time.

Coextruded films

The extrusion process consists of rotating a metal screw inside a metal barrel to mix and melt the plastic, and then forcing the plastic through an opening, called a die, to form the desired shape. Frequently, this extruded shape is immediately "oriented" by stretching, uniaxially or biaxially, to give the material added strength and to improve other properties, in addition to reducing the thickness—all of which can result in a lower cost for the ultimate product. Polymers react differently to orientation, but some have their properties greatly enhanced by orientation and are thus very frequently used in the oriented form, e.g., PP, nylon, polyesters and PS. Resistance to temperature is highly improved by orientation. Although the type of polymers of different suppliers are the same, a great difference in usability can be found. For some applications it is desirable to have controlled shrinkage of the film, e.g., for vacuum packaging around foods such as hams. To produce film which will shrink by the desired amount, the operator must adjust the heat-setting or annealing temperatures, among other conditions.

Most of the films used in vacuum cooking are multilayer complexes of different plastics, produced by extrusion-lamination or gluing together. In some films EVOH is used to improve barrier properties. Plastics which are excellent in controlling water loss tend not to control O_2 gain well enough. No single plastic appears to do both, so the minimum number of plastics to control both is two—a water barrier and an O_2 barrier.

Barrier films, such as PA which give the multilaminate its high gas impermeability and mechanical strength, tend to be toxic and not suitable for food products. They are, therefore, sandwiched between layers of heat-resistant plastics approved for use with foods to provide a packaging film with the desired heat resistance and barrier qualities.

For vacuum cooking, pouches consisting of an outer PA layer and an inner PE layer are often used. The PA gives the pouch mechanical strength and impermeability to O_2 and other gases. PE is coextruded for its water and vapour impermeability. The pouches have a 100 to 150 μ thickness (the films for thermoforming of trays are thicker). The more PE, the longer the sealing time.

SAFETY REGULATIONS

Introduction

Food legislation in the EC is a good example of the changing opinion on the role of the state. In the beginning of the EC (1950s) member states and the commission hoped to regulate everything. Since the judgement of the European Court of Justice, known as the judgement "Cassis de Dijon" (20/02/79), all products legally made and marketed in a member state should be, in principle, freely admitted in all member states. Only two exceptions were allowed: public health and consumer protection. This judgement introduced the distinction clearly between obligatory and voluntary rules.

In 1985 the directives on food legislation (COM 85/603 final 08/11/1985) were published. For the first time "horizontal objectives," useful for all foodstuffs were published by the EC (Luxembourg). Only five aims have to be regulated: protection of public health, defence of consumer, fair competition, public inspections and environmental protection. But for some products (milk, meat, fish, etc.) the old approach is still used and some specific "vertical" rules are still decided. The directive on food hygiene was published in July, 1993. The prevention of microbial risks by applying HACCP is the cornerstone of this new regulation. One of the main topics of discussion is whether these directives can be applied to all kinds of businesses in the food sector; i.e., from small restaurants to food industry.

HACCP is very much related to quality assurance (ISO 900X). Although quality assurance is not an obligation, for most food companies it is a must in order to be allowed to deliver to the very strong buying combinations of the distribution sector. Also, the new European legislation on product liability places the producers in a weak position if they do not have good quality norms and a sound quality assurance system.

Horizontal Legislation

Labelling

From the point of view of safety, two council directives are important: 1) council directive on labelling, presentation and advertising of foodstuffs for sale to the ultimate consumer (79/112/EEC modified by

89/395/EEC); and 2) council directive on batch identification (89/396/EEC).

A producer has to put the following information on a label for safety reasons : 1) minimal durability date or limited date of consumption for very perishable products; 2) batch identification with a view to help a recall procedure; 3) storage and use conditions, including temperature; and, 4) directions for use if necessary, for example reheating conditions.

Additives-Contaminants-Packaging Materials

The council directive concerning food additives authorized for use in foodstuffs intended for human consumption is a framework directive (89/107/EEC). It defines an additive. The framework stipulates that a comprehensive directive of food additives will be a list of authorized food additives and a list of foodstuffs to which these additives may be added. Three drafts are in discussion: sweeteners, colours, and preservatives-antioxidants-miscellaneous additives. The council directive on contaminants is also a framework directive. This draft proposes to have a non-exhaustive list with limits in foodstuffs, limits of analytical detectability, sampling and analytical methods. Several council directives for specific products: cereals, animal foodstuffs, vegetal foodstuffs, etc., have been published before.

Official Controls

A framework directive of official control of foodstuffs was adopted in June 1989. Since 01/01/1993 the toll barriers have disappeared, and therefore it is important now to have the same principles and control procedures. These procedures include inspection, sampling and analysis, hygienic inspection of workers, and examination of auto control systems. In 1985, the criteria of sampling and analysis methods were defined, but there is still a lot of discussion of the norms that are often very different in many countries. A good example is the cooling requirement for sous vide products in different countries from the EC (Table 3.9).

Hygiene

The proposal of Council Directive on the hygiene of foodstuffs is a new step in hygiene legislation. It covers all stages of the life of

Table 3.9. Cooling requirements for sous vide cooked products in Europe (James, 1990).

Country	Temperature Range (°C)	Maximum Time Allowed (h)	Subsequent Temperature (°C)
Denmark	65 to 10	3	<5
France	70 to 10	2	0 to 3
West Germany	80 to 15	2	
	15 to 2	24	<2
Sweden	80 to 8	4	<3
U.K.	70 to 3	2	0 to 3

foodstuffs, not only production. It refers very strongly to the use of voluntary actions from manufacturers to assure the safety of products, e.g., food manufacturing codes, HACCP, and quality assurance.

Vertical Legislation

Meat Products

The basic directive was adopted in 1977. Since this date several modifications have been published, the last in February 1992. Annex A defines general conditions for the licensing of factories and general conditions of hygiene. To obtain this EC-license, many slaughterhouses and meat factories had to invest a lot of money, however hygienic conditions have improved considerably. Annex B focuses on specific conditions for meat products; including: 1) specific conditions for the licensing of factories; 2) specific conditions of hygiene for factories; 3) requirements for raw materials; 4) control of production by authorities; 5) packaging, wrapping and labelling; 6) healthiness stamp, 7) storage and transport; 8) specific conditions for pasteurized and sterilized products; and, 9) specific conditions for prepared meals and meat.

In most EC-countries there is a debate as to whether the vertical legislation of meat products applies to sous vide products. From the strict interpretation of this directive, the answer would be yes. In many countries most central production units are inspected by public health authorities and are not under veterinary control. The conflict between these inspections and the differing views on food hygiene makes it very difficult for a producer to know which regulations to follow.

Fish Product

This directive is in the line of the meat products, but maybe not as strict.

Zoonoses

This proposal contains: 1) an obligation for the member states to collect information on the effect of zoonotic agents on the public health and to inform the Commission; 2) an obligation to collect samples; 3) specific requirements to fight against zoonotic agents; and, 4) creation of reference laboratories to give necessary technical assistance.

Voluntary Actions for Quality

Variability in food manufacturing and composition is too wide, certainly at the European level, to have standards for every product. Food legislation can only define the essential characteristics in regards to security and health. Inspection services do not have to define all the detailed precautions that will lead to the desired result; this is the responsibility of the producers. The instruments to achieve the hygienic and quality norms are (see Table 3.10): 1) quality assurance following the ISO 900X guidelines; 2) HACCP; and, 3) Good Manufacturing Practices.

In 1991 ECFF (European Chilled Food Federation) was created by CFA (UK, France). Several other countries joined ECFF; Spain, Italy, Belgium, and Finland. The objective of ECFF was to develop a good manufacturing practice code: "These guidelines form a basis for designing safe manufacturing for chilled food (including "sous vide" food). The setting up and management of such operations will, however, require expert help to ensure that they are designed properly, with quality assurance procedures implemented to ensure their control." ECFF classified chilled foods, with an indication of the type of hygienic control and shelf life (Table 3.11). Categories 6, 7, and 10 apply to sous vide products.

This classification is a first step, but more research is needed to define safety criteria that will be product specific and that don't rely only on heat treatment. Also additional hurdles should be taken into account.

Table 3.10. Regulations and voluntary actions (Falconnet, 1993).

Quality	Compulsory Regulations	Voluntary Regulations	Voluntary Actions
Safety	Hygiene directives		Good manufacturing practices for safety, HACCP
	Contaminants, pesticides		Quality assurance
	Additives, flavours, extraction solvents		
	Materials in contact with foodstuffs, . . .		
Health			
Nutritional value		Nutritional labelling directive	
Ingredients			
Satisfaction			
Ethical aspects	Labelling directive	Claims	Name of foodstuffs
Ingredients			Brand, . . .
Organoleptic perception	Packaging waste	Biological products	Professional codes
Price, . . .		AOP-IGP	Standardization
Environmental aspects		Poultry, A.S., . . .	Certification of products
Origin			
Tradition			
Services			
Convenience			Definition of products
Packaging			
Availability, . . .			Trade policy
Consistency			Quality Assurance

RESEARCH AND TRAINING

Introduction

Many people have been convinced about the advantages of the sous vide technique. The main advantages are sensorial quality and possibilities to rationalize and improve work organization. But the main drawback of the technique is that it requires much higher skills and knowledge than

Table 3.11. ECFF classification of chilled foods.

Category and Hygiene Conditions Required		Status of Components	Further Preparation Required by Consumer for Safety	Shelf Life
1	GMP	Raw	None	Short
2	GMP	Raw	Heat process	Short
3	HCA*	Raw + heat processed (70°C, 2 min)	None	Short
4	HCA	Raw + heat processed (70°C, 2 min)	Heat process	Short
5	HRA**	Heat processed (70°C, 2 min) + decontaminated	None	Short
6	GMP	Heat processed in-pack (70°C, 2 min)	None	Short
7	GMP	Heat processed in-pack (70°C, 100 min or equivalent)	None	Med.
8	HRA	Oven pasteurised (70°C, 100 min or equivalent)	None	Med.
9	HRA	Oven pasteurised (90°C, 10 min or equivalent)	None	Long
10	GMP	Heat processed in-pack (90°C, 10 min or equivalent)	None	Long
11	HRA	Duo pasteurization (90° C, 10 min, then 70°C, 2 min in-pack	None	Long
12	GMP	a_w < 0.97, heat processed in-pack (70°C, 2 min)	None	Long

*HCA; high degree of care area.
**HRA; high risk area.

traditionally can be found in the food and catering industry. One case of food poisoning can affect everyone using the sous vide technique. For this reason, training on different levels is needed to improve safety and to define more meaningful safety criteria for this type of food.

Training

The "Ecole de Cuisine Georges Pralus" is very active in promoting "sous vide" and in training cooks. Different hotel schools in Belgium (2), Switzerland (2), Luxembourg (1), Netherlands (1) are affiliated with the school of Georges Pralus. Also in France, ISVAC (Institute for Sous Vide and Modified Atmosphere) is developing training programs. In the UK, Leeds Metropolitan University and Bournemouth University are offering courses on "Sous Vide Cook-Chill" as a part of the Leisure and Consumer Studies program. ALMA University of Leuven (Belgium) is offering more scientific background on the sous vide system. In 1993, the first European symposium on sous vide was held in Leuven, Belgium.

But in most cases, training is given by vendors of sous vide equipment. Training and demonstrations are an important part of their selling strategy. The quality of the training programs is very different. In order to stop the proliferation of courses on sous vide and minimize the potential risks of the application of sous vide, the content of these courses should be controlled and people who will use the sous vide technique should have a minimum training level, dependent on the scale of the operation. The codes of good manufacturing practices should also contain guidelines on skills and quality certification of training centres.

Research

A lot of research and development is kept secret by the companies that have developed their own recipes and technology. To improve the quality, safety and efficiency of sous vide, more fundamental and more systematic scientific knowledge is needed.

Table 3.12 shows that in the EC-research programs FLAIR and AAIR, many topics related to sous vide are covered.

Research and development in different European countries are very

Table 3.12. EC research projects related to sous vide.

Code	Description
FLAIR project AGRF-CT90	Improving the safety and quality of meat and meat products by modified atmospheres and assessment of novel methods
FAR project UP-2-515-FAR	Modified atmospheres: a new approach to improving safety of fresh fish products
FLAIR project AGRF-0048	Natural antimicrobial systems: new technologies for food safety, quality and environment protection
FLAIR project AGRF-0058	Development of computer aided process design procedures to improve quality and safety of products with a limited shelf life
FLAIR project AGRF-0017	Predictive modelling of microbial growth and survival in foods
AAIR project proj. ref. 921519	The microbial safety and quality of foods processed by the "sous vide" system as a method of commercial catering
FLAIR study	Sous vide cooking
AAIR project 125	Improvement of the safety and quality of refrigerated ready-to eat foods using novel mild preservation techniques

Table 3.13. Most important research topics.

1. Safety
   —kinetics of toxin formation of *Clostridium botulinum*
   —dynamic models for growth and inactivation in "sous vide" products
   —European code of good manufacturing practices
   —pasteurization values as a function of type of product and shelf life
2. Heat treatment
   —intelligent control systems for combined microwave and conventional heating/cooling
   —thermal efficiency of equipment
3. Sensorial and nutritional aspects
   —texture kinetics
4. Packaging materials
   —mechanical properties
   —recycling
5. Economic evaluation
   —total cost of "sous vide"in comparison with other catering systems
   —effect of "sous vide" on work organization, numbers and skills of personnel

complementary. There is a need for even more co-operation and exchange of research results (Martens, 1993). For this reason a second European symposium is being planned for 1995.

As part of a FLAIR study, Schellekens and Martens (1992) determined the most important research topics (Table 3.13).

CONCLUSIONS

The application of the sous vide technique is growing very fast in Europe. France and Belgium are leading the way; the Anglo-Saxon countries are sceptical for safety reasons but are following. However in these latter countries, as well as Germany, the interest to improve sensorial quality is growing. The application of sous vide is also hindered due to the fact that a lot of cooks and workers in restaurant and catering see "sous vide" as a job killer. But the fast rising cost for a meal in a restaurant is also a job killer. In the long term, the demographic evolution forces us to find efficient and qualitative solutions for the foodservice industry to serve the growing number of elderly people. Since, from a food standpoint, people are very conservative, the application of sous vide will be an evolution rather than a revolution.

REFERENCES

Allewijn, P. and Norder, E. 1992. "Recente ontwikkelingen in gasverpakken: luchtig verpakken," *J. Food Management*, 7:18–30.

Castle, L., Jickells, S. M., Gilbert, J. and Harrison, N. 1990. "Migration testing of plastics and microwave-active materials for high-temperature food-use applications," *Food Additives and Contaminants*, 7:779–796.

De Baerdemaeker, J. and Nicolaï, B. M. 1993. "Equipment considerations for sous vide cooking," *First European Sous Vide Cooking Symposium Proceedings*, March 25–26, 1993, Leuven, Belgium, pp. 100–116.

Falconnet, F. 1993, "European legislation and good manufacturing practice code for sous vide products," *First European Sous Vide Cooking Symposium Proceedings*, March 25–26, 1993, Leuven, Belgium, pp. 41–56.

Fiess, M. 1991. "Comment produire industriellement 150 recettes différentes par semaine? La 5ième gamme sans fausse notes," *J. RIA*, 464:38–39.

Fox, R., 1989. "Plastic packaging–The consumer preference of tomorrow," *J. Food Technology*, 43(12):84.

Grigaut, I. 1990. "Les fabricants en font un plat," *J. RIA*, 451:52–55.

IFT, 1989. "Microwave food processing." *J. Food Technology*, 43:117–126.

James, S. J. "Cooling systems for ready meals and cooked products," in *Process engineering in the food industry, Vol. 2. Convenience foods and quality assurance*. Proceedings of a Conference, "Engineering Innovation in the Food Industry. Its Role in Quality Assurance, "Univ. of Bath, 9–11 April 1990. R. W. Field and J. A. Howell, eds. Elsevier Science Publisher Ltd., London. pp. 88–97.

Keuning, R. 1990. "Food ingredients for the '90's," *Food for the 90's*. Chapter 8, Elsevier Science Publishers Ltd., London.

Lengkeek, G. 1990. "Modellering microwave in geïntegreerde proceslijn," *J. Food Management*, 11:61–62.

Light, N. and Walker, A. 1990. *Cook-Chill Catering: Technology and Management*. Elsevier Science Publishers Ltd., London.

Martens, T. 1976. Steriliseren in flexible verpakkingen. Thesis Faculty of Agricultural Sciences, Katholieke Universiteit Leuven, Belgium.

Martens, T. 1980. Mathematical model of heat processing in flat containers. Ph.D. thesis, Katholieke Universiteit Leuven, Belgium.

Martens, T. 1991, "Sous vide, an opportunity for small and medium sized enterprises," *J. Vita Magazine*, June 1992.

Marten, T., 1992. "Just in time, computer integrated manufacturing and sous vide," *FNK Symposium Proceedings*, November 4–5, 1992, Zürich, Switzerland.

Martens, T., 1993. "A European network of centres of competence in sous vide," *First European Sous Vide Cooking Symposium Proceedings*, March 25–26, 1993, Leuven, Belgium, pp. 125–129.

Pralus G. 1985. *Une Histoire d'Amour: La Cuisine Sous Vide*. Published by Georges Pralus, France.

Rozier, J., Carlier, V., Bolnot, F., Tassin, P., Rouve, G. and Gauthier, M. 1990. Plats cuisinés à l'avance et cuisson sous vide. Maîtrise de la qualité hygiénique. APRIA-CDIUPA.

Schellekens, W. and Martens, T. 1992. *"Sous Vide" Cooking. Part I: Scientific Literature Review. Part II: Feedback from Practice*. Leuven, ALMA Katholieke Universiteit Leuven, Belgium, EUR 15018 EN.

Sheard, M. A. and Rodger, C. 1993. "Optimum heat treatments for sous vide cook-chill products," *First European Sous Vide Cooking Symposium Proceedings*, March 25–26, 1993, Leuven, Belgium, pp. 117–126.

Southern, S. and Cakebread, D. 1991, "Microwave power," *J. Eurofood Trends and Technology*, 8:120–124.

Van Dijk, J. 1990. "De maaltijdenfabriek," *J. Voeding en Voorlichting*, 3:5–6.

CHAPTER 4

Microbiological Concerns Associated with MAP and Sous Vide Products

JOHN H. HANLIN—*Campbell Soup Company, U.S.A.*
GEORGE M. EVANCHO —*Campbell Soup Company, U.S.A.*
PETER J. SLADE —*Campbell Soup Company, U.S.A.*

INTRODUCTION

Sous vide was developed in the mid-1970s in France by Chef Georges Pralus as a means of improving the flavour, appearance and aroma of cooked food. The process, originally designed for use by chefs in restaurants, is now being applied to foods prepared in one location and shipped to another, very often by low- or non-skilled labour, many of whom may be poorly educated in even the rudiments of food safety, sanitation and hygiene. In its simplest form, sous vide consists of packaging foods under vacuum in hermetically sealed bags or pouches, followed by pasteurization, chilling and refrigerated storage. The term sous vide refers both to the food and the processing method by which the food is prepared. The process is generally as follows:

(1) Preparation of raw ingredients.
(2) Precooking (e.g., grilling, browning, etc.) of the ingredients, if appropriate.
(3) Placing food in high barrier (low moisture/oxygen transmission film), plastic bags or pouches.
(4) Applying a vacuum and heat sealing the bags or pouches.
(5) Cooking food by any number of means at a specific time/temperature to ensure pasteurization.
(6) Rapidly chilling the food within a specified time (generally less than 2 h) to a specific temperature (typically less than 3°C), but not freezing.
(7) Labeling with product name, date of production, use by date (shelf life should be determined by experiment), storage and preparation instructions.
(8) Storing under controlled refrigerated temperature (0 to 3°C) until prepared for consumption.

Modified atmosphere packaged (MAP) foods are prepared with or without a heat treatment step, any heat treatment being applied before the product is packaged (not to the packaged product as with sous vide products). MAP products may be packaged in alternative gaseous atmospheres, particularly elevated levels of CO_2 and/or N_2. Consequently, the safety concerns associated with sous vide products when compared to MAP products are similar in many respects, but different in others. By far, the most critical issue surrounding sous vide products is whether or not pathogens of concern might survive the heat treatment given the product. In MAP products the issue is rather one of good sanitation and hygiene in preparation of the food prior to packaging, and prevention of recontamination during packaging and gas flushing. General considerations regarding the safety of both types of product concern the ability of pathogenic microorganisms and/or their spores or toxins to: (a) survive any heat process given the product, (b) survive, germinate (in the case of spores), and proliferate in the gaseous atmosphere used in the process, and (c) survive and grow at the low, refrigeration temperatures used in storage and distribution of the product.

Food ingredients may be contaminated with a variety of pathogenic bacteria, but not all are of concern in MAP and sous vide products. The pathogens of greatest concern are those which can grow under low O_2 tension (anaerobes, facultative anaerobes and microaerophiles) and at low temperatures (psychrotrophs). Aerobes (which require free O_2 to grow) are of less concern since MAP products usually contain very low levels of O_2 while sous vide products are packaged under vacuum. Mesophiles (which do not grow below 10°C and have an optimum growth temperature of 30 to 45°C) are not of primary concern unless the food is subjected to gross temperature abuse.

This chapter aims to investigate the nature of the potential threat from bacterial pathogens, particularly how survival and growth of the organisms of concern relate to the microbial ecology of MAP and sous vide products. The psychrotrophic pathogens of major concern in these products are: nonproteolytic *Clostridium botulinum* types B, E and F, *Listeria monocytogenes*, *Yersinia enterocolitica* and other *Yersinia* spp., and *Aeromonas hydrophila* (Silliker and Wolfe, 1980; Hintlian and Hotchkiss, 1986; Farber, 1991; Lambert et al., 1991; Reddy et al., 1992). These will be considered in depth in this chapter. Other pathogens, such as *Bacillus cereus* and other *Bacillus* spp., *Campylobacter* spp., proteolytic *C. botulinum* types A, B and F, *Clostridium perfringens*, enteropathogenic *Escherichia coli* (EPEC), *Plesiomonas shigelloides*, *Salmonella* spp., *Shigella* spp., *Staphylococcus aureus*, *Vibrio cholerae*, *V. parahaemolyticus* and other *Vibrio* spp., will not be specifically addressed since these are generally considered not to present a threat in adequately processed and stored MAP and sous vide products.

From the above it follows that the most serious threat to the safety of MAP and sous vide products comes from those pathogens capable of surviving any heat process given the product, and which thrive in the gaseous environment within the package when held at the temperatures used to store and distribute the product. Hence, the topics covered in this chapter include: (a) the heat resistance of potential pathogens, (b) the O_2 relations of pathogens in these products, (c) the minimum growth temperatures of the organisms of concern, and (d) control of safety in MAP and sous vide products.

INFLUENCE OF HEAT PROCESSING ON SURVIVAL OF PATHOGENS

Inherent in any discussion on the growth of pathogens in sous vide products is the assumption that the product's packaging will remain intact and that the only organisms capable of growing are those that survive a thermal process. Knowledge of an organism's heat resistance is therefore of fundamental importance. Shapton and Shapton (1991) have summarized heat resistance data for many foodborne pathogens.

The heat resistance of bacterial spores or vegetative cells is generally measured by two parameters, the D-value and the z-value. D-value is the time (minutes) at temperature (*t*) to reduce the population by 90% or one $\log_{10}$. z-Value is the temperature change required to increase or decrease the D-value by a factor of 10.

This section will review the heat resistance of those pathogens of chief concern in sous vide products. The four pathogens of concern are: nonproteolytic *C. botulinum* types B, E and F, *L. monocytogenes*, *Y. enterocolitica*, and *A. hydrophila*. The assumption is made that if any of these pathogens (spore or vegetative cell) survive the heat process delivered to a sous vide product, they may grow and be a threat to human health.

Clostridium botulinum

The agent of botulism, *C. botulinum*, is perhaps the most notorious microbial foodborne pathogen. Spores of the organism, while not the most heat resistant known to food microbiologists, are targeted for destruction to ensure the safety of low-acid foods. Thermal inactivation of these spores thus provides the basis for successful heat-processing of low-acid canned foods (Pflug and Odlaug, 1978). Strains of *C. botulinum* are classified into four groups: Group I–proteolytic types A, B and F; Group II–nonproteolytic types B, E and F; Group III–types C and D; and, Group IV–type G.

The Group I strains grow poorly at 12°C and do not grow below 10°C. However, Group II (non-proteolytic) strains can grow below 10°C and are therefore of concern in MAP and sous vide products (Smith and Sugiyama, 1988). Group I strains of *C. botulinum* are primarily associated with soil in areas of comparatively low rainfall, whereas Group II strains are predominantly associated with water habitats, mud or fresh water sediment (Smith and Sugiyama, 1988).

Heat Resistance of Group II Spores of *Clostridium botulinum*

The heat resistance of spores of Group II strains is considerably lower than the classic values reported for Group I strains. Within Group II, the nonproteolytic type B strains are generally regarded as being more heat resistant than spores of type E strains. Scott and Bernard (1982) reported that nonproteolytic type B strains were more heat resistant than type E strains. They reported $D_{82.2°C}$ values of between 1.49 and 32.3 min for spores of nonproteolytic type B. The authors were careful to point out that the higher heat resistances were not reproducible between experiments. $D_{82.2°C}$ values for strains Minnesota E, Whitefish E and Saratoga E in the same pH 7.0 phosphate buffer were 0.52, 0.4 and 0.37 min, respectively.

The $D_{176°F\ (80°C)}$ values for spores of several strains of *C. botulinum* type E heated in ground whitefish chubs were calculated by Crisely et al. (1968). Spores were heated in the fish paste and recovered in a bacteriological recovery medium. The Alaska strain had a $D_{176°F}$ value of 4.3 min, while the Beluga (type E strain) had a $D_{176°F}$ value of 2.1 min. The $D_{185°F}$ values for nonproteolytic type B spores of *C. botulinum* were between 0.45 and 14.33 min. The $D_{185°F}$ value for *C. botulinum* type E was 0.2–0.32 min (Hackney et al., 1991).

Kautter's group at the Food and Drug Administration (FDA) investigated the thermal death time of *C. botulinum* type E spores heated in blue crab (Lynt et al., 1977). Sterilized blue crabmeat was seeded with spores (10^5-10^6 per g) and several 1 g samples were heated in thermal death time tubes at 165−185°F. The Beluga strain was the most heat resistant of the five strains tested. The reported $D_{180°F}$ value was 0.74 or 1.03 min if the upper 95% confidence interval was included.

Bucknavage et al. (1990) calculated the D-values of spores of the Beluga strain of *C. botulinum* from the slopes of the survival curves in oyster homogenates. The authors evaluated the effect of potassium sorbate and/or NaCl on spore survival in the heating menstruum and reported $D_{80°C}$ values of between 0.55−1.00 min. The heat resistance of five *C. botulinum* type E spore crops in oyster homogenates was determined by Chai and Liang (1992). The Type E Minnesota strain was the most heat resistant. D-values at 165, 167, 170, 175 and 180°F were 8.96, 5.28, 2.69, 1.03 and 0.43 min, respectively. Chai and Liang (1992) closed their discussion by expressing concern over the safety of oyster pasteurization. Their D-value determinations led them to conclude that the process would only lead to a 1.5-log inactivation of the Minnesota strain.

Gaze and Brown (1990) calculated D- and z-values for nonproteolytic *C. botulinum* type B and E spores heated in cod and carrot homogenate. These values are shown in Table 4.1.

Based on data from this and other published works, the Campden Food and Drink Research Association (CFDRA) recommends that the slowest heating point in a sous vide product be held at 90°C for 7 min (Gaze and Brown, 1990). This process will destroy six logs of the most heat resistant spores of Group II *C. botulinum.*

Table 4.1. Heat resistance data for nonproteolytic C. botulinum *type B and E heated in homogenates of cod and carrot (adapted from Gaze and Brown, 1990).*

Type/Strain/Substrate	$D_{90°C}$ Value (min)*	z-Value (C°)
Type B/ATCC 25765/Cod	1.10 (0.95)	8.64 (0.998)
Type E/ATCC 9564/Cod	0.79 (0.83)	8.26 (0.997)
Type B/ATCC 25765/Carrot	0.43 (0.94)	9.76 (0.995)
Type E/ATCC 9564/Carrot	0.48 (0.88)	9.84 (0.982)

*D-Value in minutes, followed by correlation coefficient in parentheses.

Heat Resistance of *Listeria monocytogenes*

L. monocytogenes is a small rod-shaped, gram-positive, facultatively anaerobic bacterium, motile at room temperature (Palumbo, 1986). There has been a considerable amount of information published on the heat resistance of *L. monocytogenes* over the past few years. The organism is particularly noteworthy because it appears to be more heat resistant than many other vegetative bacteria, but considerably less heat resistant than spores of Group II *C. botulinum*.

As with other bacteria, the heat resistance of *L. monocytogenes* is affected by a number of parameters including the heating menstruum, pH, NaCl, curing salts, cell growth temperature and sublethal heat shock (Beuchat et al., 1986; Linton et al., 1990; Mackey et al., 1990; Smith et al., 1991; Yen et al., 1992). Mackey et al. (1990) reported the heat resistance of a natural isolate of *L. monocytogenes* in raw beef; D-values at 60, 65 and 70°C were 3.8, 0.93 and 0.14 min, respectively. The z-value was 7.2°C. D-Values of the same strain in chicken breast at 60, 65 and 70°C were 8.7, 0.52 and 0.13 min. The z-value was reported as 6.3°C. Mackey and Bratchell (1989) summarized the heat resistance of *L. monocytogenes* as determined by a number of research groups in the form of a z-value curve (Figure 4.1).

Gaze et al. (1989) reported the heat resistance of *L. monocytogenes* strains Scott A and NCTC 11994 in homogenates of chicken, steak and carrot. A summary is shown in Table 4.2.

Huang et al. (1992) found that a heat treatment greater than the 4-D process recommended by the USDA (1.48 min at 65.5°C) was required to inactivate the large numbers of *L. monocytogenes* Scott A that developed in chicken gravy during refrigerated storage. After inoculation of cooled gravy (40°C) with 10^5 cfu/ml and storage at 7°C, the population increased by about 1.5 logs after 24 h. After 1, 3, 5, and 10 d at 7°C the gravy was heated at 50, 55, 60, and 65°C. D-values were highest at the early stages of chill storage (day 0); 195 min at 50°C and 0.48 min at 65°C. Clearly, the fact that large numbers of *L. monocytogenes*, enough not to be inactivated by a 4-D process, will develop in poorly refrigerated gravy is of major consequence to the safety of MAP and sous vide products.

Heat Resistance of *Yersinia enterocolitica*

Y. enterocolitica is a facultatively anaerobic, gram-negative, short

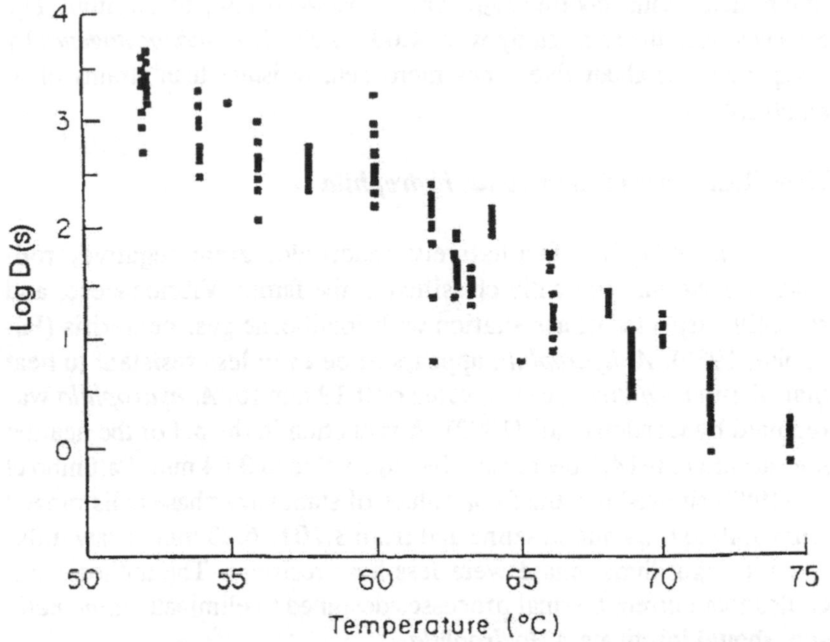

FIGURE 4.1 Published data on heat resistance of *Listeria monocytogenes* as compiled by Mackey and Bratchell (1989) (reproduced by kind permission of Blackwell Scientific Publications, Ltd).

rod-shaped bacterium currently classified as part of the family Enterobacteriaceae (Palumbo, 1986). The heat resistance of five strains of *Y. enterocolitica* in skim milk was reported by Hanna et al. (1977a). The ATCC cultures 23715 and 9610 were the most heat resistant. A $D_{55°C}$ value of approximately 2 min can be extrapolated from their published data. Sorquist (1989) also found *Y. enterocolitica* to be relatively sensi-

Table 4.2. $D_{70°C}$ and z-value for strains of L. monocytogenes *heated in homogenates of chicken, steak and carrot (adapted from Gaze et al., 1989).*

Strain	Menstrua	$D_{70°C}$ Value (min)	z-Value (C°)
Scott A	Chicken	0.16	6.72
Scott A	Steak	0.20	5.98
Scott A	Carrot	0.23	7.04
11994	Chicken	0.20	7.39
11994	Steak	0.14	5.98
11994	Carrot	0.27	6.70

tive to heat, with reported $D_{58°C}$ values between 1.4 and 1.8 min. The z-values from the same study were 4.00–4.52°C. *L. monocytogenes* by comparison, is about five times more heat resistant than strains of *Y. enterocolitica.*

Heat Resistance of *Aeromonas hydrophila*

A. hydrophila is a facultatively anaerobic, gram-negative, rod-shaped bacterium currently classified in the family Vibrionaceae, and recently suspected of association with foodborne gastroenteritis (Palumbo, 1986). *A. hydrophila* appears to be even less resistant to heat than *Y. enterocolitica.* A $D_{55°C}$ value of 0.19 min for *A. hydrophila* was reported by Condon et al. (1992). A reduction in the pH of the heating menstruum to pH 4.0 decreased the $D_{55°C}$ value to 0.04 min. Palumbo et al. (1987) showed that the $D_{48°C}$ values of stationary phase cells ranged from 3.49 to 6.64 min in saline and from 3.20 to 6.23 min in raw milk. Cells in logarithmic phase were less heat resistant. The authors concluded that current thermal processes designed to eliminate *Salmonella* spp. should inactivate *A. hydrophila*.

Accuracy of Heat Resistance Determinations

Accurate calculations of the heat resistance of all pathogens are critical for the establishment of safe thermal processes. For example, Harrison and Huang (1990), reported differences in the thermal death times for *L. monocytogenes* Scott A in crabmeat, depending upon the recovery medium used. Using a non-selective medium (trypticase soy agar) as the recovery medium, the D-values at 50, 55 and 60° were 40.4, 12 and 2.6 min, respectively. However when modified Vogel-Johnson was used as the recovery medium D-values were up to 50% lower. The reduced D-values were explained due to reduced recovery of heat-injured cells on the selective agar. The physiological condition of *L. monocytogenes* Scott A, the enumeration medium, and the growth environment have been found to greatly affect the measured heat resistance of log-phase cells (Linton et al., 1992). $D_{55°C}$ values for heat shocked cells were 2.1-fold higher than non-heat shocked cells when cells were enumerated on TSYE agar aerobically, and 2.2-fold higher for cells enumerated anaerobically (Linton et al., 1992). Variations in reported D-values are evidently related to differences in handling practices before, during and after exposure to heat. In particular, the effects of growth temperature and sub-lethal heat shock on the heat resistance of *L. monocytogenes* have received much attention recently. Knabel et al. (1990) showed that elevated growth temperature (i.e., 43°C compared to 37°C) increased the thermal resistance of *L. monocytogenes*. Furthermore, a simple heat shock at 43°C increased the thermal resistance of *L. monocytogenes*, but not to the same extent as a growth temperature of 43°C. Smith et al. (1991) showed that growth temperature significantly affected the heat resistance of *L. monocytogenes*. Cells grown at 37 or 42°C were more resistant to 52°C for 1 h than cells grown at 5, 10, 19 or 28°C.

Farber and Brown (1990) investigated the effect of prior heat shock on the thermal resistance of *L. monocytogenes* in meat. Twenty gram samples of a fermented sausage mix (66% pork/33% beef), inoculated with approximately 10^7 *L. monocytogenes* per g, were held at heat shock temperatures of 40, 44, 48 and 52°C for periods of time prior to thermal processing at 62 or 64°C. While a heat shock of 48°C for 30 min failed to increase D-values significantly, when heat shocked at 48°C for 2 h, the $D_{64°C}$ value increased from 3.3 min to 8.0 min. Bunning et al. (1990) showed that a prior heat shock induced thermotolerance in *L. monocytogenes*, but not significantly. Mackey and Derrick (1986, 1987) demonstrated that slow heating and prior mild heat shock (48°C for 30 min) could induce thermotolerance, affecting the heat resistance of *Salmonella* spp., under similar conditions.

If slow heating or sublethal heat shock induces physiological changes in bacterial cells that increase their resistance to heat, perhaps chilling them will decrease their heat resistance. Humphrey (1990) showed that the heat sensitivity of *S. enteritidis* phage type 4 in homogenized whole egg increased after storage at 4 or 8°C. The $D_{55°C}$ values for control cells, cells held at 8°C for 24 h and cells held at 4°C for 24 h were 8.0, 2.4 and 1.6 min respectively. Furthermore, there was an inverse correlation between the storage time at 4°C (0–24 h) and the $D_{55°C}$ value.

The work of Humphrey (1990) supported the conclusion of Mackey and Derrick (1987) that the phenomenon of "thermotolerance" is a general one and may have implications for the survival of pathogenic bacteria in foods given a marginal heat treatment. Mackey and Derrick (1987) recommended that the following steps be taken to limit the development of "thermotolerance" during pasteurization: 1) keep foods refrigerated prior to heating; 2) heat product rapidly to lethal temperatures: and, 3) if rapid heating is impossible, heat treatment should be sufficient to inactivate thermotolerant cells.

INFLUENCE OF GASEOUS ATMOSPHERE ON SURVIVAL AND GROWTH OF PATHOGENS IN MAP AND SOUS VIDE PRODUCTS

A critical factor in the safety and stability of O_2-reduced packaged foods is the O_2 relations of the pathogenic and spoilage microorganisms in the food. O_2 is required by many microorganisms, but not all (Shapton and Shapton, 1991). It is toxic to some. Microorganisms that require O_2 for energy-yielding metabolic processes are called aerobes, while those that do not utilize O_2 are called anaerobes. Organisms capable of using either respiratory or fermentation processes, depending on the availability of O_2 in the environment, are termed facultative.

Microorganisms display varying degrees of sensitivity to the oxidation/reduction (O/R) potential for their growth media (Jay, 1986). The O/R or "redox" potential of a substrate may be defined as the ease with which the substrate gains or loses electrons (Jay, 1986). Transfer of electrons between compounds creates a potential difference, measured in millivolts (mV). A highly oxidized substance has a positive potential, whilst that of a reduced substance is negative. The O/R potential of a system is expressed by the symbol Eh. Generally, aerobic microorganisms require positive Eh values (oxidized) for growth, whereas anaerobes require negative values (reduced). Some aerobic bacteria actually grow under slightly reduced conditions, and these are often referred to as microaerophiles (Jay, 1986). The O/R potential of a food is determined by: 1) the characteristic O/R potential of the food; 2) the "poising capacity," i.e., the resistance to change in potential of the food; 3) the O_2 tension of the atmosphere about the food; and, 4) the access which the atmosphere has to food (Jay, 1986).

In most cases a high Eh is caused by the presence of O_2 (Lund and Wyatt, 1984). The growth of anaerobes normally occurs at reduced Eh values, but the exclusion of O_2 may be necessary for some anaerobes (Jay, 1986). Inhibition of anaerobes at high Eh may be due to the presence of dissolved O2 rather than the Eh itself. For example, Walten and Hentges (1975) found that, when cultured in the presence of O_2, growth of *C. perfringens* was inhibited even when the medium had an Eh of −50 mV. However, growth occurred in a medium with an Eh of +325 mV in the absence of O_2. A high Eh caused by oxidizing agents other than O_2 may not be inhibitory (Lund and Wyatt, 1984).

Effect of Eh on the Growth of *Clostridium botulinum*

According to Smoot and Pierson (1979) the optimum Eh for growth of *C. botulinum* is −350 mV. Reported Eh maxima for initiation of growth of *C. botulinum* range from +30 to +250 mV (Ando and Iida, 1970; Huss et al. 1979; Morris, 1976: Smoot and Pierson, 1979). The metabolic activity of a high number of germinating spores may reduce the Eh and enhance the growth of vegetative cells (Lund and Wyatt, 1984). Spore germination may occur at Eh levels higher than those allowing growth (Ando and Iida, 1970), and decrease Eh (Smith, 1975). The presence of NaCl, and acidic conditions, lowers the maximum Eh level for initiation of growth (Smoot and Pierson, 1979). However, the Eh of most foods exposed to O_2 is usually low enough to permit growth of *C. botulinum* (Sperber, 1982). Huss et al. (1979) compared fresh fish with high Eh values, and viscera and spoiled fish with low Eh values, and found that toxin production by *C. botulinum* type E was not significantly different at either Eh. The storage temperature and spore load, however, were found to markedly influence the toxin titre.

Lund and Wyatt (1984) studied the effect of redox potential and different NaCl concentrations on the growth of *C. botulinum* type E. In medium at pH 6.8−7.0, containing 0.1% NaCl, the probability of growth of spores was not significantly lower up to an Eh of +60 mV (adjusted by the introduction of air) than under strictly anaerobic conditions, at Eh −400 mV. Between an Eh of +122 to 164 mV, the probability of growth was reduced by a factor of 10^5. However, these authors experienced difficulty in measuring redox potentials at Eh values between +100 and +200 mV. These levels, at pH 7.0, may be below those detectable with available platinum redox electrodes. Lund and Wyatt (1984) and Lund et al. (1984) thus recommended measurement of both redox potential and partial pressure of atmospheric O_2 in studies such as this. Indeed, the nature of most food products makes Eh very difficult to measure (Shapton and Shapton, 1991).

Effect of Modified Atmospheres (MAs) on the Growth of Psychrotrophic Pathogens

In most MAP products, redox potentials are usually not considered since, as found by Lund and Wyatt (1984), they are difficult to measure and effects on the microorganisms may be due to the presence or absence of O_2 and/or CO_2, and not the particular redox potential of the system. Gaseous atmospheres are usually modified by adjusting the O_2 concentration in the package by substitution with CO_2 and/or N_2 and are measured empirically. In MAP and sous vide products the constituent gases may well change with time depending upon: 1) the permeability of the packaging material used; 2) the indigenous changes in the atmosphere resulting from reactions in the food; and, 3) those changes produced by growth of surviving microorganisms associated with the food.

Residual O_2 may be consumed with concomitant production of CO_2 as a result of bacterial respiration (Lefevre, 1991). Added CO_2 has bacteriostatic and fungistatic properties, and retards the growth of most molds and aerobic bacteria (Day, 1992). The combined negative effects on various enzymatic and biochemical pathways result in an increase in the lag phase and generation time of susceptible microorganisms (Day, 1992). The inhibitory effects of CO_2 increase with decreasing temperature due primarily to the increased solubility of CO_2 at lower temperatures (Jay, 1986). The solubility of CO_2 diminishes with increased NaCl concentration (Lefevre, 1991). The reduction in pH caused by the production of carbonic acid from CO_2 is a minor inhibiting factor (Jay, 1986).

CO_2 is the most important gas used in MAP. It may kill, inhibit, have no effect on, or stimulate the growth of microorganisms (Day, 1992). The ability of CO_2 to inhibit microorganisms is a complex phenomenon, not only dependent on the type, numbers and age of microorganisms, but also on the concentration of CO_2, a_w, pH and storage temperature. Explanations of possible inhibitory mechanisms are given by Enfors and Molin (1978, 1981), Daniels et al. (1985), Genigeorgis (1985), and Lefevre (1991).

The effects of MAs on the major psychrotrophs of concern, namely nonproteolytic *C. botulinum* types B, E, and F, *L. monocytogenes, Y. enterocolitica* and *A. hydrophila*, have been reviewed by Genigeorgis (1985), Hintlian and Hotchkiss (1986), Farber (1991), Lambert et al. (1991), and Reddy et al. (1992). The effects on mesophilic foodborne pathogens have been briefly summarized by these authors, most thoroughly by Farber (1991). Day (1992) stated that high levels of CO_2 have inhibitory effects on *E. coli, Salmonella*, spp., and *S. aureus.*

Clostridium spp.

The effects of MAs on survival of *Clostridium* spp. is a contentious issue. *C. botulinum* and *C. perfringens* are not greatly affected by the presence of CO_2 and may flourish in the anaerobic conditions created by MAP. Silliker and Wolfe (1980) concluded that MAs neither increase nor decrease the *C. botulinum* hazard. Christiansen and Foster (1965) found that the rate of toxin production by *C. botulinum* type A in sliced bologna was the same whether the product was packaged in air or under vacuum. Ajmal (1968) reported that *C. botulinum* type E produced toxin in meat and fish products under both aerobic and anaerobic conditions. Baker et al. (1986) found that populations of *C. sporogenes* (which have properties similar to proteolytic *C. botulinum*) inoculated into minced chicken and held under 80% CO_2 at 2, 7 and 13°C, declined less in the elevated CO_2 atmosphere than in controls stored in air. They concluded that elevated CO_2 levels may protect clostridia.

The incidence of *C. botulinum* in finfish is high. The incidence of types A, B, E and F spores in one study was 21.7%, and in another, 66.7% (Baker et al., 1990a). Estimates of numbers of *C. botulinum* spores per 100 g of samples ranged from 9 to 240 in red snapper, and 3 to 120 in salmon (Baker et al., 1990a). The high concentration of nonprotein nitrogenous compounds in finfish make them particularly susceptible to microbial spoilage under conventional refrigerated storage (Mayer and Ward, 1991). The normal spoilage patterns will often indicate to the consumer that the product should not be consumed. However, if a shift in the spoilage microorganisms occurs as a result of MA packaging, the normal course of fish spoilage may not ensue. Indeed, studies have shown that under some conditions, *C. botulinum* may grow and produce toxin in vacuum or MAP fish prior to the appearance of the signs of spoilage (Post et al., 1985).

Salmon fillets inoculated with *C. botulinum* types A, B, and E spores and held at 4.4°C for 57 d in air or under MA (60% CO_2/25% O_2/15% N_2) did not produce detectable toxin (Stier et al. 1981). At 22.2°C, both air and MAP samples became toxic within 2 d. The authors concluded that MA did not enhance the growth of *C. botulinum* at 4.4°C, whilst at 22.2°C toxin production was preceded by, or coincided with spoilage. The major effect of the MA was repression of the gram-negative spoilage microorganisms.

Post et al. (1985) inoculated three types of fish with spores of *C. botulinum* type E, then stored samples under vacuum, in CO_2, N_2, or in MAs containing 65–90% CO_2 and 2–4% O_2 (balance N_2) at 4 to 26°C. In several instances, they found that toxin could be formed before the onset of spoilage. Although flounder spoiled before it became toxic, MAP cod and whiting became toxic prior to or simultaneously with organoleptic rejection. Interestingly, toxin formation was also evident in samples packed under 2 or 4% O_2 prior to rejection based on organoleptic changes. Under certain circumstances the addition of 6% O_2 has been found to be more inhibitory to toxin production by *C. botulinum* than an atmosphere devoid of O_2 (Garcia and Genigeorgis, 1987), but generally the inclusion of small amounts of O_2 in a gas mixture does not appear to provide more safety than elevated CO_2 atmospheres. The false sense of security inspired by the presence of O_2 must thus be resisted (Reddy et al., 1992). Eklund (1982) showed that salmon became toxic more readily when stored under normal atmosphere as compared to atmospheres containing 60 or 90% CO_2. The safety net was short-lived, however, because all samples became toxic after 10 d at 10°C. The CO_2 atmospheres used in these packages did not increase the safety of the finfish, but did delay the time to spoilage. Reddy et al. (1992) compared published data on the times to toxicity and the times to spoilage for MAP fish held at a variety of temperatures. Their review showed that spoilage coincided with toxin production at storage temperatures above 20°C. However between 4 and 12°C, there seems to be no clear pattern. Whereas some studies have shown that spoilage precedes toxicity in this temperature range, Post et al. (1985) clearly demonstrated that cod and whiting fillets may become toxic prior to spoilage.

Baker and Genigeorgis (1990) reasoned that if "sensory spoilage could not be regarded as an accurate warning signal for toxicity," then perhaps a model could be used to predict the lag time to toxicity (*LT*) and the probability (*P*) of growth of a spore. Based on this assumption, they designed and conducted a factorial experiment with multiple spore levels, fish types, tissue types and MAs. Their results showed that temperature was by far the most important factor controlling toxin production. Other factors played a minor role in affecting the *LT*. The earliest *LT* for all their experiments at 30, 20, 16, 12, 8 and 4°C were 0.5, 1, 2, 3, 3 and 18 d, respectively.

Following a 1987 outbreak of botulism thought to be associated with the consumption of coleslaw made from packaged shredded cabbage, the FDA investigated the potential for shredded cabbage to support the growth and toxin production of proteolytic *C. botulinum* under MAP (Solomon et al., 1990). In an atmosphere of 70% CO_2/30% N_2, inoculated cabbage (100–200 spores/g) became toxic within 4 d at 22–25°C despite its acceptability based on appearance, odour and texture (Solomon et al., 1990).

While manufacturers may intentionally modify the atmosphere of a packaged food, sometimes the product per se can modify its own environment. Sugiyama (1982) showed that fresh mushrooms sealed in plastic film-wrapped packages respired, used up much of the available O_2 and created an environment that could support toxin production by proteolytic strains of *C. botulinum*. Although this work showed that packaged mushrooms inoculated with high numbers of spores became toxic, Kautter et al. (1978) showed that none of 1,078 packages of uninoculated fresh mushrooms became toxic after storage at room temperature for 7 d.

Growth of *C. botulinum* in vacuum packages has also been investigated. Since a component of the sous vide process is evacuation of the package prior to delivery of the thermal process, it is important to consider the potential for surviving spores of *C. botulinum* to germinate, outgrow and produce toxin in the anaerobic environment of a sous vide packaged food. Brown and Gaze (1990) evaluated the potential for the growth of psychrotrophic *C. botulinum* in fillets and homogenates of chicken and cod. The authors simulated the sous vide process by heating packaged product to an equivalent center temperature of 70°C for 2 min. Type E strain (ATCC 9564) formed toxin within six weeks at 5°C, three weeks at 8°C, one week at 15°C and 5 d at 30°C. This study demonstrated that vacuum packaged foods with extended shelf life are a potential risk with respect to botulinum toxigenesis. While these products may be safe if held under proper refrigeration (i.e., <4°C), due to the possibility for storage temperature abuse, they must be viewed as potentially unsafe. Notermans et al. (1981) reported that both proteolytic and non-proteolytic spores of *C. botulinum* could survive the process typically given potatoes (95°C for 40 min), and then grow and form toxin under vacuum within six weeks at 10°C. No toxin was detected at 4°C. The potatoes did not exhibit signs of spoilage, and were organoleptically acceptable.

Listeria monocytogenes

Berrang et al. (1989b) investigated the effect of controlled atmosphere (CA) storage (CO_2 concentrations <10%) on fresh asparagus, broccoli, and cauliflower inoculated with *L. monocytogenes* and stored at 4 and 15°C. CA storage lengthened the acceptable shelf life of the products. Populations of *L. monocytogenes* increased, but CA-storage did not influence the rate of growth. *L. monocytogenes* grew well on shredded lettuce stored under MA (3% O_2/97% N_2) at 5 and 10°C

(Beuchat and Brackett, 1990). They concluded *L. monocytogenes* is capable of growing on lettuce subjected to commonly used packaging and distribution practices in the food industry. The organism also grew on beef in CO_2 packs held at 10°C, but not at 5°C (Gill and Reichel, 1989). In vacuum-packaged meat, *L. monocytogenes* grew after long lag periods at refrigerator temperatures, but at slower rates than the spoilage microorganisms. This was also the case in studies by Grau and Vanderlinde (1992) with vacuum-packaged processed meats. Kaya and Schmidt (1991) determined that on vacuum-packed beef with a pH of <5.8 inoculated with a population of 10^3 *L. monocytogenes*, there was no increase during nine weeks when held at 2 and 4°C, but at 7°C the count increased 100-fold after 3 to 7 d. At pH > 6.0 the listeriae increased 10-fold during the first week at 2 and 4°C, and within 3 d at 7°C, but by 1000-fold at 10°C. Similarly, in turkey loaf stored under vacuum at 3°C, no significant changes were observed in the numbers of *L. monocytogenes* held for up to 15 d (Ingham and Tautorus, 1991). In an anaerobic atmosphere (75% CO_2/25% N_2), *L. monocytogenes* did not survive well on minced raw chicken (Wimpfheimer et al. 1990). However, with addition of as little as 5% O_2 the organism grew, even at 4°C. These conditions inhibited the aerobic spoilage microorganisms of the meat. On precooked chicken nuggets, growth of *L. monocytogenes* was stimulated at 3°C in the presence of *Pseudomonas fluorescens* in air and in an atmosphere composed of 76% CO_2/13.3% N_2/10.7% O_2, but not in 80% CO_2/20% N_2/0% O_2 (Marshall et al., 1992). This growth stimulation did not occur at 7 or 11°C. At the lower temperatures (3 and 7°C), growth of *P. fluorescens* was not affected by the presence of *L. monocytogenes*, regardless of the atmosphere. However, at 11°C, early inhibition of the pseudomonad was observed in both air and the O_2-free atmosphere. Conversely, late growth of *L. monocytogenes* was inhibited by *P. fluorescens* at 11°C in the O_2-replete atmosphere.

Yersinia enterocolitica

Gill and Reichel (1989) found *Y. enterocolitica* grew at rates similar to, or faster than, those of the spoilage microorganisms in vacuum-packed beef and at temperatures between −2 to 10°C. In CO_2 packs, the organism only grew at 5°C or above and after a relatively prolonged lag period. However, Eklund and Jarmund (1983) found that, compared to growth in air, growth at 2, 6, and 20°C for 23 d under an atmosphere of 100% CO_2 was reduced by 100, 98, and 43%, respectively. Zee et al. (1984) examined the effect of different atmospheres on growth of *Y. enterocolitica* in trypticase soy broth at 25°C. While 10% CO_2 (balance argon) stimulated the growth of the organism as compared to growth in air, 40% CO_2 increased the lag phase, and 100% CO_2 increased the lag phase and decreased the growth rate during the logarithmic phase. In an atmosphere of 20% CO_2/80% O_2, and at temperatures of 1, 4, 10, and 15°C, Kleinlein and Untermann (1990) determined that *Y. enterocolitica* was markedly inhibited by the background microorganisms of minced beef (10^2 and 10^5 cfu/g), except in samples with low levels (10^2 cfu/g) incubated at the two highest temperaures.

Aeromonas hydrophila

Enfors et al. (1979) found that *A. hydrophila* could proliferate to levels of about 10^6/cm^2 in pork stored under N_2 for 10 d at 4°C, but not in pork stored in CO_2. In beef, the organisms grew at rates similar to or faster than those of the background microorganisms when packed under vacuum and stored at temperatures between −2 to 10°C (Gill and Reichel, 1989). However, in CO_2-packs, it did not grow at 5°C or lower. In surimi-type products (minced fish, salt-added, and low-salt surimi), Ingham and Potter (1988) found that MA-storage (36% CO_2/13% O_2/51% N_2) at both 4 and 13°C did not stimulate the growth of *A. hydrophila* when compared to growth in air. The organism competed well with a co-inoculated strain of *Pseudomonas fragi* on minced and low-salt surimi (both species were inhibited more in the high-salt product, *A. hydrophila* to a greater extent than *P. fragi*), when stored in both air and MA. The authors concluded that MA-storage of surimi-based products could not be used to significantly inhibit *Aeromonas*, and that potentially significant numbers of the organism may be present even in organoleptically acceptable product.

Slight growth of *A. hydrophila* occurred during air- and vacuum-storage at 2°C for 6 d on crayfish (Ingham, 1990). No growth occurred during MA-storage (80% CO_2/10% O_2) at 2°C. At a higher temperature (8°C), substantial growth of *A. hydrophila* occurred after 6 d. Growth was inhibited equally well in vacuum- and MA-storage, but only slightly better than by storage in air. The author concluded that vacuum- or MA-storage in conjunction with good refrigeration effectively inhibited growth of *A. hydrophila* on the product over a 6 d period. On the contrary, Berrang et al. (1989a) found that CA-storage of fresh vegetables at 4 or 15°C (for up to 21 d and 10 d, respectively) did not significantly affect the growth of *A. hydrophila* as compared to growth in air. Whilst CA-storage effectively lengthened the organoleptic acceptance time of the vegetables, it also permitted growth of *Aeromomas* to "perhaps significantly" high levels.

INFLUENCE OF STORAGE TEMPERATURE ON SURVIVAL AND GROWTH OF PATHOGENS IN MAP AND SOUS VIDE PRODUCTS

By convention, bacteria are differentiated into four major physiological groups based on their temperature ranges of growth, namely: thermophiles, mesophiles, psychrotrophs and psychrophiles (ICMSF, 1980). Most foodborne pathogens are mesophiles which prefer moderate temperatures with an optimum generally between 30 and 45°C, and minimum growth temperatures ranging from about 5 to 10°C (Table 4.3). Poorly refrigerated MAP and sous vide products, may be exposed to the latter range of temperatures (and sometimes even more severely abused at higher temperatures), so the importance of effective heat treatments and gaseous barriers to prevent growth of pathogens in these products cannot be overemphasized. Some thermophiles may grow at temperatures as high as 75 to 90°C with an optimum between 55 and 65°C. None present direct safety concerns in MAP products, since they are not known to be pathogenic and, in any case, will not grow below 35°C.

Psychrophiles are generally recognized as organisms having an optimum temperature for growth of about 15°C or slightly lower, a maximum of about 20°C or lower, and a minimum of 0°C or lower (ICMSF, 1980). Since no true psychrophiles are recognized foodborne pathogens, they do not present a threat to the safety of MAP and sous vide products. Some are spoilage organisms, and may therefore play a beneficial role in indicating that a MAP or sous vide product has been stored beyond its acceptable consumption period. Additionally, since psychrophiles are sensitive to temperatures greater than 20°C, they are rarely found in effectively pasteurized sous vide products.

Organisms capable of growth at temperatures as low as 0°C, but not meeting the optimal and maximal temperature requirements for psychrophiles are termed psychrotrophs (*psychros* = "cold"; and *trephein* = "to nourish upon" or "to develop"). Psychrotrophic microorganisms are most often defined as those that can grow, producing visible colonies (or turbidity), within 7 to 10 d at temperatures between 0 and 7°C (Jay, 1986). Psychrotrophic pathogens include nonproteolytic

*Table 4.3. Growth range temperatures of some pathogenic microorganisms (°C).**

Organism	Minimum	Optimum	Maximum
Listeria monocytogenes	0.0	28.0	45.0
Yersinia enterocolitica	3.0	32 to 34	44.0
Aspergillus flavus	3.0	25.0	44.0
A. parasiticus	3.0	25.0	44.0
*Bacillus cereus***	3 to 4	30.0	48.0
C. botulinum (non-proteolytic)	3.3	25 to 37	40 to 45
Vibrio parahaemolyticus	5.0	37.0	43.0
Pseudomonas aeruginosa	5.0	25 to 30	42.0
Salmonella spp.	5 (8 to 10 in product)	37.0	46.0
Staphylococcus aureus	7 (8 to 15 in product)	37.0	48.0
S. aureus toxin production	10.0	40 to 45	48.0
C. botulinum (proteolytic)	10 to 12	30 to 40	48.0
E. coli (EPEC)	<10	30 to 42	44.0
C. perfringens	15.0	43 to 45	52.0
Campylobacter spp.	30.0	42.0	45.0

*From Shapton and Shapton (1991) (reproduced by kind permission of Butterworth-Heinemann Ltd).

**Some *B. cereus* strains can grow at 3 to 4°C, 2 and 3 log increases occur in 7 and 10 d, respectively.

C. botulinum types B, E and F, *L. monocytogenes*, *Y. enterocolitica*, and *A. hydrophila* (Kraft, 1992). The effects of temperature on growth of *C. botulinum* have been considered quite extensively (Notermans et al., 1990: Conner et al., 1989; Lund and Notermans, 1993). Effects on the other psychrotrophs have not been considered in as much detail (Farber, 1991).

Some strains of other species, such as enteropathogenic *E. coli* (EPEC), *P. shigelloides* and *V. parahaemolyticus*, may be psychrotrophic under certain conditions (Kraft, 1992). A number of mesophilic pathogens are known to survive but not to grow at refrigeration temperatures ≤5°C (Kraft, 1992). These include strains of *Bacillus* spp., proteolytic *C. botulinum* types A, B and F, *C. perfringens, Salmonella* spp., *Shigella* spp., *S. aureus* and *V. cholerae*. Many of these, however, are known to grow at various temperatures slightly above 5°C, the so-called "refrigerator abuse" range of temperatures (Palumbo, 1986). Farber (1991) has considered the potential for other pathogens to grow in poorly refrigerated MAP products at temperatures from 5 to 12°C. Of these, the pathogen of most recent concern of which little is known with regard to survival and growth in these products is enterohemorrhagic *E. coli* (EHEC) serotype 0157:H7, the cause of hemorrhagic colitis and hemolytic uremic syndrome or "HUS" (Doyle and Padhye, 1989). A previous review (Kornacki and Marth, 1982) indicated that holding food at 5°C appeared adequate to prevent growth and toxin production by enterotoxigenic *E. coli* (ETEC). However, Olsvik and Kapperud (1982) showed that heat stable toxin (ST) could be produced by ETEC strains growing at 4°C in both broth and broth with cream. It has been known for some time that *E. coli* and other Enterobacteriaceae can proliferate in dairy products at low temperatures (Witter, 1961). Presently, very little is known about the survival and growth of enteropathogenic *E. coli*, (including EHEC) at low temperatures and/or in MAP and sous vide products. The involvement of *E. coli* 0157:H7 with colitis and HUS outbreaks associated with undercooked ground beef (Doyle and Padhye, 1989) suggests that these are areas for urgent investigation.

Nonproteolytic *Clostridium botulinum* Types B, E, and F

The pathogens of most concern in MAP products are nonproteolytic *C. botulinum* types B, E and F (Conner et al., 1989; Baker et al., 1990b). Nonproteolytic strains grow and produce toxin under certain conditions at refrigeration temperatures (Kim and Foegeding, 1993). Therefore proper refrigeration alone is not a complete safeguard against botulism (Sperber, 1982). *C. botulinum* type E can grow and produce toxin at temperatures as low as 3.3°C (Kautter, 1964). Schmidt et al. (1961) found that four strains of type E could grow and produce toxin in beef stew within 36 d at 3.3°C, but not in 104 d at 2.2°C. Similarly, in fresh oysters and clams, type E grew when stored at 4°C, but not at 2°C (Patel et al., 1978). Cann et al. (1965) observed growth and toxin production by type E *C. botulinum* in vacuum-packed herring after 15 d at 5°C. The initial inoculum was 10^2 spores/package. Nonproteolytic *C. botulinum* types B and F can also grow and produce toxin at temperatures as low as 3.3°C (Eklund, 1967a 1967b). Toxin was detected from two strains of type F and one type B strain at this temperature. Sperber (1982) determined that, at temperatures of 5.6, 4.4, or 3.3°C, the times for nonproteolytic *C. botulinum* type B to produce detectable toxin were 27, 33 and 129 d, respectively.

Listeria monocytogenes

Gray and Killinger (1966) reported growth of *L. monocytogenes* at temperatures as low as 3°C. An early recommendation for selective isolation of *L. monocytogenes* involved cold enrichment of the sample for several weeks at 4°C (Gray et al., 1948). More recently, *L. monocytogenes* has been observed to grow at temperatures around 1°C (Juntilla et al., 1988), some strains growing in a range as low as −0.1 to −0.4°C (Walker and Stringer, 1987). At these temperatures, the lag phase is greatly extended and may exceed 20 d (Walker and Stringer, 1987).

L. monocytogenes is capable of growth at 4 to 6°C in milk and lamb (Khan et al., 1973), and grows well in soft cheeses, even when refrigerated at 4°C (Kraft, 1992). However *L. monocytogenes* populations inoculated onto packaged catfish fillets stored at 4°C did not change over 16 d (Leung et al., 1992).

Gill and Reichel (1989) found that *L. monocytogenes* was incapable of multiplying on high-pH beef packaged under 100% CO_2 at temperatures of 5°C or less. However, the organism did grow in CO_2-packed meat stored at 10°C, as well as on vacuum-packed meat stored at 0, 2, 5, and 10°C. High residual nitrite or reduced a_w was found to reduce growth of the organism on vacuum-packaged processed meats stored at 0 to 5°C (Grau and Vanderlinde, 1992). When the storage temperature was increased to 15°C, the growth rate of *L. monocytogenes* increased more rapidly than that of commensal bacteria (lactic acid bacteria and *Brochothrix thermosphacta*). Numbers of *L. monocytogenes* inoculated onto uncured, vacuum-packed turkey loaf did not change during storage within 15 d at 3°C (Ingham and Tautorus, 1991). Likewise, Beuchat and Brackett (1990) found no significant changes in populations of *L. monocytogenes* on chlorine-treated, MAP (97% N_2/3% O_2) shredded lettuce during the first 8 d of storage at 5°C. Significant increases occurred between 8 and 15 d at 5°C, but after only 3 d at 10°C.

Yersinia enterocolitica

Y. enterocolitica and related organisms can be isolated from a number of foods, particularly those of animal origin, many of which have been implicated in outbreaks of food poisoning (Palumbo, 1986). *Y. enterocolitica*, and other members of the genus *Yersinia*, are capable of growth at 5°C (Bercovier and Mollaret, 1984). Leistner et al. (1975) examined the growth of *Y. enterocolitica* inoculated into pork and held at 0 to 1°C. The bacterial counts increased from 10^2-10^3 cfu/g to 10^8 cfu/g after 3 weeks. The organism grows readily in raw beef and pork held at 1 to 7°C even in the presence of the competitive microorganisms normally associated with these commodities (Hanna et al., 1977b). The bacteria used in this study were originally isolated from vacuum-packaged beef stored at 1 to 3°C for 21 to 35 d (Hanna et al., 1977c). In ground beef held under normal atmospheric conditions, Kleinlein and Untermann (1990) reported increases in *Y. enterocolitica* populations of 1-log and 3.5-logs after 14 d at 1 and 4°C, respectively. When packaged under 20% CO_2/80% O_2, growth was effectively halted at 1°C, but only slightly inhibited at 4°C. *Y. enterocolitica* can grow in CO_2-packaged beef with high pH stored at 5 and 10°C, but not at temperatures lower than 2°C (Gill and Reichel, 1989).

Y. enterocolitica is not readily isolated from raw milk (Stern et al., 1980). Whether this is due to competitive microorganisms, nutrient content or some other factor(s) in whole milk is presently unknown (Palumbo, 1986). *Y. enterocolitica* isolated from milk supplies in Northern Ireland produced detectable enterotoxin when inoculated into milk and milk media at elevated temperatures, but not in media incubated for 7 d at 4°C (Walker and Gilmour, 1990). Formation of toxin in refrigerated products was considered not to present a major health problem. Schiemann (1988) arrived at the same conclusion from work with a number of foods. Slurries of lettuce, beef, pork, chicken, pasteurized milk and fish inoculated with toxigenic *Y. enterocolitica* were negative for toxin after incubation for 4 d at 9.8°C. *Y. enterocolitica* inoculated into unsalted, commercially pasteurized liquid egg products grew at temperatures of 2, 6.7, and 12.8°C after 14 d, with a delayed (>4 d) growth response at the two lower temperatures (Erickson and Jenkins, 1992). Growth was inhibited in egg containing >5% NaCl at all three temperatures. The ability of *Y. enterocolitica* to produce enterotoxin in refrigerated foods and, if so, its significance are still questionable issues (Palumbo, 1986).

Aeromonas hydrophila

Early studies showed that *A. hydrophila* could grow at 1 to 5°C (Eddy, 1960). Rouf and Rigney (1971) determined that 6 of 13 strains of A. *hydrophila* were "psychrophiles," capable of good growth in 6 to 11 d at 0°C. Five strains described as mesophiles grew at 10°C, and one at 5°C. Palumbo et al. (1985b) determined that clinical isolates of *A. hydrophila* are capable of growth from 4 to 42°C. Five strains studied in close detail grew from 10^3/ml to over 10^8/ml in 14 d at 4°C. Other clinical isolates also grew well at this temperature. In studies with strains from clinical, raw meat, and ready-to-eat fleshfood sources, the food-derived strains were better adapted to growth at lower temperatures than those from clinical or raw meat sources (Hudson, 1992). In another study with various proteinaceous foods, the numbers of *A. hydrophila*/g or ml increased 10- to 1000-fold during one week of storage at 5°C (Palumbo et al., 1985a). The organism is capable of growing in a number of foods at low temperatures (Callister and Agger, 1987; Palumbo and Buchanan, 1988).

Gill and Reichel (1989) packaged samples of inoculated, high pH (>6.0) beef under vacuum or under CO_2 and stored them at −2, 0, 2, 5 or 10°C. In vacuum packs, *A. hydrophila* grew at all storage temperatures at rates similar to or faster than the spoilage microorganisms. In CO_2 packs the organism only grew at 10°C.

In investigations of food model systems using variable levels of NaCl, $NaNO_2$ and pH, combined effects were found to reduce the growth of *A. hydrophila* under anaerobic conditions at low temperatures, but individual variables did not affect the growth kinetics (Palumbo et al., 1992). The organism was most effectively inhibited with 100−200 mg/L $NaNO_2$, and 2.5−3.5% NaCl at pH 6.3 or lower, and temperatures of 5 to 12°C. In flaked crab, a population of 2×10^3 inoculated *A. hydrophila* grew at 10 and 15°C, but not at 0 or 5°C (Yoon and Matches, 1988). It was speculated that at the lower temperatures either "metabolic crowding" by competitive aerobic spoilage organisms occurred (APC of 10^9 cfu/g within 2 weeks at 5°C), or the strain used was inhibited at these temperatures. On packaged catfish fillets the *A. hydrophila* population increased from 10^3/cm^2 to 10^6/cm^2 after 16 d when stored at 4°C (Leung et al., 1992). A faster growth rate was observed on overwrapped packaged fillets. At 2°C, Ingham (1990) only observed slight growth on crayfish tailmeat stored in air or under vacuum for 6 d with more profuse growth at 8°C. No growth occurred during storage at 2°C under a commercial MA gas mix (80% CO_2/0% O_2).

CONTROL OF SAFETY IN MAP AND SOUS VIDE PRODUCTS

Rhodehamel (1992) identified at least four major safety concerns associated with sous vide products, namely: 1) sous vide products are formulated with little or no preservatives; 2) sous vide products receive minimal thermal processes and are not shelf-stable; 3) vacuum packaging provides an anaerobic environment, which extends shelf life by inhibiting normal aerobic spoilage microorganisms; and, 4) refrigerated conditions must be maintained absolutely during a sous vide product's shelf life.

The majority of MAP and sous vide products have few "barriers" to prevent the growth of pathogenic microorganisms, other than heat treatment (not always given with MAP products) and modification of the gaseous environment within the package (evacuation in the cases of sous vide and some MAP products, and introduction of various gases to other MAP products). Hence, controlling the temperature at which these products are stored, distributed and handled prior to consumption is extremely important. In the restaurant environment, and in the hands of a skilled chef, the pasteurization process and the temperature to which products are exposed can be monitored and controlled with relative ease, and such products present limited risk. However, when products must be shipped to remote locations and marketed to the retail trade, there is ample opportunity for temperature abuse to occur.

A food product temperature abuse study conducted by Audits International (1989) evaluated temperatures of refrigerated products in twenty-five major metropolitan areas in the U.S. The study showed that over 80% of the products were above 4.4°C, over 50% were above 7.2°C, over 25% were above 10°C, and almost 10% were above 12.8°C. Independent studies conducted by Campbell Soup Company gave similar results (Hutton et al., 1991). Product temperatures ranged from 2.2 to 12.8°C, with a mean temperature of 7.8°C and a sigma of 2.05°C. Further studies conducted by Audits International (1989) revealed temperature abuse in the home. Here, greater than 25% of products were held at temperatures higher than 7.2°C, while 10% were above 10°C. Studies conducted by van Garde and Woodburn (1987) showed that as many as 21% of home refrigerators were at temperatures above 10°C. Additionally, 60% of the individuals surveyed left perishable foods at room temperature for 2 h or longer, and 95% of those surveyed conduct at least one unsafe practice with respect to food safety (Woodburn and Vanderiet, 1985). Rhodehamel (1992) stated that "strict adherence to temperature control must be mandatory not only for the sous vide manufacturer, distributor and retailer, but also for the consumer."

The pasteurization process (time/temperature) applied to a food will determine, to a large extent, its safety and shelf life. The pasteurization conditions should be designed to achieve sufficient internal temperature to destroy all vegetative pathogens and reduce the level of nonpathogenic vegetative microorganisms to a level appropriate to ensure the desired shelf life of the product. The longer the desired shelf life, the greater the level of destruction of vegetative microorganisms required.

Destruction of vegetative pathogens can be accomplished by controlling the internal temperature of the food to a level which would ensure the destruction of any contaminating organisms of known heat resistance likely to be encountered in the food. Of the vegetative pathogens of greatest concern, *L. monocytogenes* is the most heat resistant. Therefore, any pasteurization process designed to eliminate *L. monocytogenes* should be adequate to sufficiently reduce the number of other pathogens. A target reduction of 10^6 would be appropriate to ensure the safety of sous vide products. The minimum pasteurization process applied to a food which can be relied upon to achieve a six-log reduction in the numbers of *L. monocytogenes* would achieve an equivalent lethality of 2 min at 70°C throughout the food (Gaze et al., 1989). A greater than 10^6 reduction of other pathogens should occur, since all but *S. senftenberg* 775W or salmonellae that have been heat shocked or those found in products with reduced water activity, are less heat resistant than *L. monocytogenes* (Shapton and Shapton, 1991).

An equivalent process of 2 min at 70°C, while adequate to render a food free of vegetative pathogens such as *Listeria*, is not adequate to eliminate bacterial spores, such as those of *C. botulinum*. Psychrotrophic strains of *C. botulinum* (nonproteolytic B and E), though more heat sensitive than proteolytic strains, are likely to survive the above minimum process. These same strains have been reported to grow at temperatures as low as 3.3°C (Eklund et al., 1967b). *C. botulinum* is the major microbiological hazard associated with sous vide processing because: 1) it can withstand mild heat treatments given some sous vide products; 2) normal spoilage microorganisms, and thus indicators of incipient spoilage are completely or partially inactivated; and, 3) vacuum packaging creates an anaerobic environment, which is conducive to growth (Smith et al., 1990).

With extended refrigerated shelf life of product at a temperature above

3°C, the potential for growth of psychrotrophic strains of *C. botulinum* is very real. Brown and Gaze (1990) and Brown et al. (1991) demonstrated that toxin could be produced in sous vide products in as little as 21 d at 8°C and in as little as 7 d at 15°C. Mesophilic strains of *C. botulinum*, which would also survive the pasteurization process, would not grow if the food were held below 10°C. However, under conditions of temperature abuse, these organisms would grow and produce toxin, making the food unfit for consumption. Details on the time to toxicity for *C. botulinum* were discussed earlier in this chapter.

Eliminating the potential hazard from psychrotrophic *C. botulinum* in products with extended shelf life requires the application of a thermal process adequate to destroy a likely number of contaminants. The hazard from psychrotrophic *C. botulinum* can be controlled by applying a process of at least 7 min at 90°C, which is adequate to achieve a six-log reduction in psychrotrophic *C. botulinum* spores (Gaze and Brown, 1990). The U.K. government has recommended building in a safety margin to allow for variations in heat resistance of strains in various foods and recently published guidelines (10 min at 90°C) for inactivation of spores of psychrotrophic *C. botulinum* (Anon., 1992). French legislation (Anon., 1974) requires that sous vide products with a 21 d shelf life be heated for 100 min at 70°C, and that products with a 42 d shelf life be heated for 1,000 min at 70°C. These processes are based on the thermal resistance of *Enterococcus faecalis*. A process of 1,000 min at 70°C is equivalent to 10 min at 90°C (Anon., 1992).

Eliminating the potential hazard from mesophilic *C. botulinum* under conditions of temperature abuse would be impossible, short of giving the product a thermal process with an F_0 of 3 min. A process with this lethality goes well beyond the pasteurization requirements for sous vide, and a product subjected to this level of heating would undoubtedly suffer significant quality loss. This hazard must therefore be controlled through proper temperature control, from rapid cooling after processing to maintenance of low temperatures (0 to 3°C) during distribution and handling by the food service operator or the consumer.

To prevent growth and toxin production in refrigerated foods with a shelf life of 10 d or longer, the U.K. Advisory Committee on the Microbiological Safety of Food has recommended the following, in addition to refrigerated storage, be used singly or in combination: 1) heat treatment of 90°C for 10 min or equivalent lethality; 2) a pH of ≤5.0 throughout all components of complex foods; 3) a minimum salt level of 3.5% in the aqueous phase of all components of complex foods; 4) an a_w of ≤ 0.97 throughout all components of complex foods; and, 5) combinations of heat and preservative factors which have been shown consistently to prevent growth and toxin production by psychrotrophic *C. botulinum* at temperatures up to 10°C (Anon., 1992).

Role of HACCP in MAP and Sous Vide Products

As with any other food processing operation, the quality and safety of MAP and sous vide products depend heavily on the quality of the original raw materials, and the care that goes into their preparation, handling, and processing as dictated by Good Manufacturing Practices (GMPs), and implementation of Hazard Analysis Critical Control Point (HACCP) procedures (NACMCF, 1992). The potential for growth of pathogens which survive the pasteurization process given sous vide products can be controlled through the application of a HACCP program. HACCP is a preventative system in which safety is designed into the food formulation and the process by which it is produced. It is an excellent risk reduction tool and can be applied to reduce the risks associated with sous vide products. It is a proactive process which looks critically at all ingredients, as well as production, distribution, marketing, preparation and consumption.

Applying HACCP to the production of sous vide products will identify the pasteurization process as a critical control point (CCP) with the time and temperature defined in terms of critical limits (e.g., 10 min at 90°C). The method and frequency of monitoring will be specified, as well as the actions to be taken by the operator should either critical limit not be achieved. All measurements and actions will be documented to verify the HACCP plan has been adhered to. Other CCPs will be part of the plan, such as cooling rate after pasteurization, final product temperature, package integrity, labeling (preparation instructions, use-by date, etc.), all of which will be treated similarly to the pasteurization process.

Similarly with MAP products, the gas mix within a package will be a CCP. All CCPs identified in the production process are under the direct control of production personnel. If the product is to be produced, stored, prepared and consumed on site (e.g., in a restaurant), the number of critical factors are limited and relatively easy to control. If, however, product is to be shipped and prepared at a remote site, the number of CCPs increases but, more importantly, many of them fall outside the direct control of the producer. The lack of direct control over CCPs contributes significantly to the safety concerns with sous vide products.

Control of a manufacturing process to produce safe, high quality sous vide products, does not ensure that the same level of safety and quality will be delivered to the consumer. The best HACCP plans are for naught if control of temperature is lost during distribution and handling by the foodservice operator or the consumer. Training and education are important elements in implementing HACCP in a production facility. Production workers must understand what actions are necessary to control the safety of the foods they produce and the consequences of their actions if critical limits are not adhered to. Training and educating personnel outside the direct employ of the producer cannot be relied upon to assure the safety of sous vide products. For those fortunate enough to own their own refrigerated distribution system, exerting control over critical factors of CCPs should be no different than controlling CCPs during manufacture. Distributing sous vide products through an independent system can present numerous problems all beyond direct control.

When direct control of CCPs is impossible or impractical, the incorporation of multiple "barriers" or "hurdles" (Scott, 1989) into the product formula (when and where possible) to retard or prevent the growth of bacteria, including pathogens, is recommended. The gaseous environment and temperature of storage of the product are two important extrinsic parameters applied in the preservation of the food. Heat treatments given to some MAP and sous vide products may be thought of as "processing factors," i.e., procedures that modify them in such a way so as to bring about a change in their microbial ecology (Gould, 1992). A number of intrinsic properties of the food (e.g., pH, a_w, anti-microbial constituents, etc.) must also be considered along with these parameters when safety and storage stability of a particular product are considered. The use of barriers or hurdles to control bacterial growth involves the incorporation of several subinhibitory factors (e.g., reduced pH and a_w), none of which alone is entirely inhibitory but which, in combination, will help to inhibit bacterial growth. Hence, the "net effects" take account of the interactive effects of a number of factors, a process commonly referred to as "combination preservation" (Gould, 1992). The effectiveness of barriers must be verified by conducting appropriate challenge studies (e.g., inoculated packs), and the critical limits of each barrier, once established must be controlled. In addition, the inclusion of a consumer readable time/temperature abuse indicator on each package, which would alert the consumer if the product has been temperature abused and might present a hazard, is recommended. Abuse indicators must be evaluated under anticipated normal and abuse conditions to ensure they provide accurate information. Changes in the indicator must be correlated with signs of bacterial spoilage or the growth of pathogens.

Absolute safety of sous vide products can never be guaranteed, but by recognizing hazards and taking steps to reduce them, sous vide products can be produced, distributed and marketed with acceptable and manageable risks.

REFERENCES

Ajmal, M. 1968. "Growth and toxin production of *Clostridium botulinum* type E," *J. Appl. Bacteriol.*, 31:120–123.

Ando, Y. and Iida, H. 1970. "Factors affecting the germination of spores of *Clostridium botulinum* type E," *Jpn. J. Microbiol.*, 14:361–370.

Anon. 1974. Réglementation des conditions d'hygiène relatives à la préparation, la conservation, la distribution et la vente des plats cuisinés à l'avance. Arrêté du 26 Juin 1974. République Francaise Ministére de l'Agriculture.

Anon. 1992. Report on Vacuum Packaging and Associated Processes. Advisory Committee on the Microbiological Safety of Food. HMSO, London.

Audits International. 1989. Audits International/Monthly, April, 1989. Audits International, Highland Park, IL.

Baker, D.A. and Genigeorgis, C. 1990. "Predicting the safe storage of fish under modified atmosphere with respect to *Clostridium botulinum* toxigenesis by modeling length of the lag phase of growth," *J. Food Prot.*, 53:131–140.

Baker, D. A., Genigeorgis, C. and Garcia, G. 1990a. "Prevalence of *Clostridium botulinum* in seafood and significance of multiple incubation temperatures for determination of its presence and type in fresh retail fish," *J. Food Prot.*, 53:668–673.

Baker, D. A., Genigeorgis, C., Glover, J. and Razavilar, V. 1990b. "Growth and toxigenesis of *C. botulinum* type E in fishes packaged under modified atmospheres," *Int. J. Food Microbiol.*, 10:269–290.

Baker, R. C., Qureshi, R. A. and Hotchkiss, J. J. 1986. "Effect of an elevated level of carbon dioxide-containing atmosphere on the growth of spoilage and pathogenic bacteria at 2, 7, and 13°C," *Poultry Sci.* 65:729–737.

Bercovier, H. and Mollaret, H. H. 1984. Chapter XIV, *Yersinia.* In *Bergey's Manual of Systematic Bacteriology, Vol. 1*, N. R. Krieg and J. G. Holt, eds., Williams and Williams, Baltimore, MD, pp. 498–506.

Berrang, M. E., Brackett, R. E. and Beuchat, L. R. 1989a. "Growth of *Aeromonas hydrophila* on fresh vegetables stored under a controlled atmosphere," *Appl. Environ. Microbiol.*, 55:2167–2171.

Berrang, M. E., Brackett, R. E. and Beuchat, L. R. 1989b. Growth of *Listeria monocytogenes* on fresh vegetables stored under controlled atmosphere," *J. Food Prot.*, 52:702–705.

Beuchat, L. R. and Brackett, R. E. 1990. "Survival and growth of *Listeria monocytogenes* on lettuce as influenced by shredding, chlorine treatment, modified atmosphere packaging and temperature," *J. Food Sci.*, 55:755–758, 870.

Beuchat, L. R., Brackett, R. E., Hao, D. Y. -Y. and Conner, D. E. 1986. "Growth and thermal inactivation of *Listeria monocytogenes* in cabbage and cabbage juice," *Can. J. Microbiol.*, 32:791–795.

Brown, G. D. and Gaze, J. E. 1990. "Determination of the growth potential of *Clostridium botulinum* types E and nonproteolytic B in sous vide products at low temperatures," Technical Memorandum No. 593. Campden Food and Drink Research Association, Chipping Campden, Glos., U.K.

Brown, G. D., Gaze, J. E. and Gaskell, D. E. 1991. "Growth of *Clostridium botulinum* non-proteolytic type B and type E in sous vide products stored at 2–15°C," Technical Memorandum No. 635. Campden Food and Drink Research Association, Chipping Campden, Glos., U.K.

Bucknavage, M. W., Pierson, M. D., Hackney, C. R. and Bishop, J. R. 1990. "Thermal inactivation of *Clostridium botulinum* type E spores in oyster homogenates at minimal processing temperatures," *J. Food Sci*, 55:372–373, 429.

Bunning, V. K., Crawford, R. G., Tierney, J. T. and Peeler, J. T. 1990. "Thermotolerance of *Listeria monocytogenes* and *Salmonella typhimurium* after sublethal heat shock," *Appl. Environ. Microbiol.*, 56:3216–3219.

Callister, S. M. and Agger, W. A. 1987. "Enumeration and characterization of *Aeromonas hydrophila* and *Aeromonas caviae* isolated from grocery store produce," *Appl. Environ. Microbiol.*, 53:249–253.

Cann, D. C., Wilson, B. B., Hobbs, G. and Shewan, J. M. 1965. "The growth and toxin production of *Clostridium botulinum* in certain vacuum packed fish," *J. Appl. Bacteriol.*, 28:431–436.

Chai, T.-J. Liang, K. T. 1992. "Thermal resistance of spores from five type E *Clostridium botulinum* strains in eastern oyster homogenates," *J. Food Prot.*, 55:18–22.

Christiansen, L. N. and Foster, E. M. 1965. "Effect of vacuum packaging on growth of *Clostridium botulinum* and *Staphylococcus aureus* in cured meats," *Appl. Microbiol.*, 13:1023–1029.

Condon, S., Garcia, M. L., Otero, A. and Sala, F. J. 1992. "Effect of culture age, preincubation at low temperature, and pH on the thermal resistance of *Aeromonas hydrophila*," *J. Appl. Bacteriol.*, 72:322–326.

Conner, D. E., Scott, V. N., Bernard, D. T. and Kautter, D. A. 1989. "Potential *Clostridium botulinum* hazards associated with extended shelf-life refrigerated foods: A review," *J. Food Safety*, 10:131–153.

Crisely, F. D., Peeler, J. T., Angelotti, R. and Hall, H. E., 1968. "Thermal resistance of spores of five strains of *Clostridium botulinum* type E in ground whitefish chubs," *J. Food Sci.*, 33:411–416.

Daniels, J. A., Krishnamurthi, R. and Rizvi, S. S. H. 1985. "A review of effects of carbon dioxide on microbial growth and food quality," *J. Food Prot.*, 48:532–537.

Day, B. P. F. 1992. "Guidelines for the good manufacturing and handling of modified atmosphere packed food products," Technical Manual No. 34. Campden Food and Drink Research Association, Chipping Campden, Glos., U.K.

Doyle, M. P. and Padhye, V. V. 1989. *Escherichia coli*, In *Foodborne Bacterial Pathogens*, M. P. Doyle, ed., Marcel Dekker, Inc., New York, NY, pp. 235–281.

Eddy, B. P. 1960. "Cephalotrichous, fermentative Gram-negative bacteria: The genus *Aeromonas*," *J. Appl. Bacteriol*, 23:216–249.

Eklund, M. W. 1982. "Significance of *Clostridium botulinum* in fishery products preserved short of sterilization," *Food Technol.*, 36(12):107–112, 115.

Eklund, M. W. and Jarmund, T. 1983. "Microculture model studies on the effect of various gas atmospheres on microbial growth at different temperatures," *J. Appl. Bacteriol.*, 55:119–125.

Eklund, M. W., Poysky, F. T. and Wieler, D. T. 1967a. "Characteristics of *Clostridium botulinum* type F isolated from the Pacific coast of the United States," *Appl. Microbiol.*, 15:1316–1323.

Eklund, M. W., Wieler, D. I. and Poysky, F. T. 1967b. "Outgrowth and toxin production of nonproteolytic type B *Clostridium botulinum* at 3.3 to 5.6°C," *J. Bacteriol.*, 93:1461–1462.

Enfors, S.-O. and Molin, G. 1978. "Mechanism of the inhibition of spore germination by inert gases and carbon dioxide," *Spores*, 7:80–84.

Enfors, S.-O. and Molin, G. 1981. The effect of different gases on the activity of microorganisms, In *Psychrotrophic Microorganisms in Spoilage and Pathogenicity*, T. A. Roberts, G. Hobbs, J. H. B. Christian, and N. Skovgaard, eds., Academic Press, New York, NY, pp. 335–343.

Enfors, S.-O., Molin, G. and Ternstrom, A. 1979. "Effect of packaging under carbon dioxide, nitrogen, or air on the microbial flora of pork stored at 4°C," *J. Appl. Bacteriol.*, 47:197–208.

Erickson, J. P. and Jenkins, P. 1992. "Behavior of psychrotrophic pathogens *Listeria monocytogenes, Yersinia enterocolitica*, and *Aeromonas hydrophila* in commercially pasteurized eggs held at 2, 6.7 and 12.8°C," *J. Food Prot.*, 55:8–12.

Farber, J. M. 1991. "Microbiological aspects of modified atmosphere packaging technology–a review," *J. Food Prot.*, 54:58–70.

Farber, J. M. and Brown, B. E. 1990. "Effect of prior heat shock on heat resistance of *Listeria monocytogenes* in meat," *Appl. Environ. Microbiol.* 56:1584–1587.

Garcia, G. W. and Genigeorgis, C. 1987. "Quantitative evaluation of *Clostridium botulinum* non-proteolytic types B, E and F growth risk in fresh salmon tissue homogenates stored under modified atmospheres," *J. Food Prot.*, 50:390–397.

Gaze, J. E. and Brown, G. D. 1990. "Determination of the heat resistance of a strain of non-proteolytic *Clostridium botulinum* type B and a strain of type E, heated in cod and carrot homogenate over a temperature range 70–92°C," Technical Memorandum No. 592. Campden Food and Drink Research Association, Chipping Campden, Glos., U.K.

Gaze, J. E., Brown, G. D., Gaskell, D. E. and Banks, J. G. 1989. "Heat resistance of *Listeria monocytogenes* in non-dairy food menstrua," Technical Memorandum No. 523. Campden Food and Drink Research Association, Chipping Campden, Glos., U.K.

Genigeorgis, C. A. 1985. "Microbial and safety implications of the use of modified atmospheres to extend the storage life of fresh meat and fish," *Int. J. Food Microbiol.*, 1:237–251.

Gill, C. O. and Reichel, M. P. 1989. "Growth of the cold-tolerant pathogens *Yersinia enterocolitica, Aeromonas hydrophila* and *Listeria monocytogenes* on high-pH beef packaged under vacuum or carbon dioxide," *Food Microbiol.*, 6:223–230.

Gould, G. W. 1992. "Ecosystem approaches to food preservation," *J. Appl. Bacteriol. Suppl.*, 73:58S–68S.

Grau, F. H. and Vanderlinde, P. B. 1992. "Occurrence, numbers, and growth of *Listeria monocytogenes* on some vacuum-packaged processed meats," *J. Food Prot.*, 55:4–7.

Gray, M. L. and Killinger, A. H. 1966. "*Listeria monocytogenes* and listeric infections," *Bacteriol. Rev.*, 30:309–382.

Gray, M. L., Stafseth, J., Thorp, F., Sholl, L. B. and Riley, W. F. 1948. "A new technique for isolating *Listerellae* from the bovine brain," *J. Bacteriol.*, 55:471–476.

Hackney, C. R., T. E. Rippen, and D. R. Ward, 1991. "Principles of pasteurization and minimally processed seafoods," In *Microbiology of Marine Food Products*, D. R. Ward and C. R. Hackney, eds., Van Nostrand Reinhold, New York, NY.

Hanna, M. O., Stewart, J. C., Carpenter, Z. L. and Vanderzant, C. 1977a. "Heat resistance of *Yersinia enterocolitica* in skim milk," *J. Food Sci.*, 42:1134, 1136.

Hanna, M. O., Stewart, J. C., Zink, D. L., Carpenter, Z. L. and Vanderzant, C. 1977b. "Development of *Yersinia enterocolitica* on raw and cooked beef and pork at different temperatures," *J. Food Sci.*, 42:1180–1184.

Hanna, M. O., Zink, D. L., Carpenter, Z. L. and Vanderzant, C. 1977c. "*Yersinia enterocolitica*-like organisms from vacuum-packaged beef and lamb," *J. Food Sci.*, 41:1254–1256.

Harrison, M. A. and Huang, Y.-W. 1990. "Thermal death times for *Listeria monocytogenes* (Scott A) in crabmeat," *J. Food Prot.*, 53:878–880.

Hintlian, C. B. and Hotchkiss, J. H. 1986. "The safety of modified atmosphere packaging: A review," *Food Technol.*, 40(12):70–76.

Huang, I-P. D., Yousef, A. E., Marth, E. H. and Matthews, M. E. 1992. "Thermal inactivation of *Listeria monocytogenes* in chicken gravy," *J. Food Prot.*, 55:492–496.

Hudson, J. A. 1992. "Variation in growth kinetics and phenotype of *Aeromonas* spp. from clinical, meat processing and fleshfood sources," *Int. J. Food Microbiol.*, 16:131–139.

Humphrey, T. J. 1990. "Heat resistance of *Salmonella enteritidis* phage type 4: The influence of storage temperature before heating," *J. Appl. Bacteriol.*, 69:493–497.

Huss, H. H., Schaeffer I., Petersen, E. R. and Cann, D. C. 1979. "Toxin production by *Clostridium botulinum* type E in fresh herring in relation to the measured oxidation reduction potential (Eh)," *Nord. Vet.-Med.*, 31:81–86.

Hutton, M. T., Chehak, P. A. and Hanlin, J. H. 1991. "Inhibition of botulinum toxin production by *Pediococcus acidilactici* in temperature abused refrigerated foods," *J. Food Safety*, 11:225–267.

ICMSF (International Commission on Microbiological Specifications for Foods). 1980. *Microbial Ecology of Foods. Vol. 1. Factors Affecting Life and Death of Microorganisms.* Academic Press, New York, NY.

Ingham, S. C. 1990. "Growth of *Aeromonas hydrophila* and *Plesiomonas shigelloides* on cooked crayfish tails during cold storage under air, vacuum and a modified atmosphere," *J. Food Prot.*, 53:665–667.

Ingham, S. C. and Potter, N. N. 1988. "Growth of *Aeromonas hydrophila* and *Pseudomonas fragi* on mince and surimis made from Atlantic pollock and stored under air or modified atmosphere," *J. Food Prot.*, 51:966–970.

Ingham, S. C. and Tautorus, C. L. 1991. "Survival of *Salmonella typhimurium, Listeria monocytogenes* and indicator bacteria on cooked uncured turkey loaf stored under vacuum at 3°C," *J. Food Safety*, 11:285–292.

Jay, J. M. 1986. *Modern Food Microbiology*, 3rd edition. Van Nostrand Reinhold, New York, NY.

Juntilla, J. R., Niemala, S. I. and Hirn, J. 1988. "Minimum growth temperature of *Listeria monocytogenes* and non-haemolytic *Listeria*," *J. Appl. Bacteriol.*, 65:321–327.

Kautter, D. A. 1964. "*Clostridium botulinum* type E in smoked fish," *J. Food Sci.*, 29:843–849.

Kautter, D. A., Lilly, T. and Lynt, R. 1978. "Evaluation of the botulinum hazard in fresh mushrooms wrapped in commercial polyvinylchloride film," *J. Food Prot.*, 41:120–121.

Kaya, M. and Schmidt, U. 1991. "Behaviour of *Listeria monocytogenes* on vacuum-packed beef," *Fleischwirtsch.*, 71:424–426.

Khan, M. A., Palmas, C. V., Seaman, A. and Woodbine, W. 1973. "Survival versus growth of a facultative psychrotroph," *J. Sci. Food Agr.*, 24:491–495.

Kim, J. and P. M. Foegeding. 1993. Principles of control, *Clostridium botulinum: Ecology and Control in Foods*, A. H. W. Hauschild and K. L. Dodds, eds., Marcel Dekker, New York, NY, pp. 121–176.

Kleinlein, N. and Untermann, F. 1990. "Growth of pathogenic *Yersinia enterocolitica* strains in minced meat with and without protective gas with consideration of the competitive background flora," *Int. J. Food Microbiol.*, 10:65–72.

Knabel, S. J., Walker, H. W., Hartman, P. A. and Mendonca, A. F. 1990. "Effects of growth temperature and strictly anaerobic recovery on the survival of *Listeria monocytogenes* during pasteurization," *Appl. Environ. Microbiol.*, 56: 370–376.

Kornacki, J. L. and Marth, E. H. 1982. "Foodborne illness caused by *Escherichia coli*: A review," *J. Food Prot.*, 45:1051–1067.

Kraft, A. A. 1992. *Psychrotrophic Bacteria in Foods: Disease and Spoilage*. CRC Press. Inc., Boca Raton, FL.

Lambert, A. D., Smith, J. P. and Dodds, K. L. 1991. "Shelf life extension and microbiological safety of fresh meat–A review," *Food Microbiol.*, 8:267–297.

Lefevre, D. 1991. The effects of modified atmospheres upon microorganisms, In *Conference Proceedings: International Conference on Modified Atmosphere Packaging*. Campden Food and Drink Research Association, Chipping Campden, Glos., U. K.

Leistner, L., Hechelmann, H., Kashiwazaki, M. and Albertz, R. 1975. "*Yersinia enterocolitica* in faeces and meat from pigs, sheep and birds," *Fleischwirtsch.*, 55:1599–1601.

Leung, C.-K, Huang, Y.-W. and Harrison, M. A. 1992. "Fate of *Listeria monocytogenes* and *Aeromonas hydrophila* on packaged channel catfish fillets stored at 4°C," *J. Food Prot.*, 55:728–730.

Linton, R. H., Pierson, M. D. and Bishop, J. R. 1990. "Increase in heat resistance of *Listeria monocytogenes* Scott A by sublethal heat shock," *J. Food Prot.*, 53:924–927.

Linton, R. H., Webster, J-B., Pierson, M. D., Bishop, J. R. and Hackney, C. R. 1992. "The effect of sublethal heat shock and growth atmosphere on the heat resistance of *Listeria monocytogenes* Scott A," *J. Food Prot.*, 55:84–87.

Lund, B. M. and Notermans, S. H. W. 1993. "Potential hazards associated with REPFEDS, "In *Clostridium botulinum: Ecology and Control in Foods*, A. H. W. Hauschild and K. L. Dodds, eds., Marcel Dekker, New York, NY, pp. 279–303.

Lund, B. M. and Wyatt, G. M. 1984. "The effect of redox potential, and its interaction with sodium chloride concentration, on the probability of growth of *Clostridium botulinum* type E from spore inocula," *Food Microbiol.*, 1:49–65.

Lund, B. M., Knox, M. R., Wyatt, G. M. 1984. "The effect of oxygen and redox potential on growth of *Clostridium botulinum* type E from a spore inoculum," *Food Microbiol.*, 1:277–287.

Lynt, R. K., Solomon, H. M., Lilly, T. and Kautter, D.A. 1977. "Thermal death time of *Clostridium botulinum* type E in meat of the blue crab," *J. Food Sci.*, 42:1022–1025, 1037.

Mackey, B. M. and Bratchell, N. 1989. "A review: The heat resistance of *Listeria monocytogenes*," *Lett. Appl. Microbiol.*, 9:89–94.

Mackey, B. M. and Derrick, C. M. 1986. "Elevation of the heat resistance of *Salmonella typhimurium* by sublethal heat shock," *J. Appl. Bacteriol.*, 61:389–393.

Mackey, B. M. and Derrick, C. M. 1987. "The effect of prior heat shock on the thermoresistance of *Salmonella thompson* in foods," *Lett. Appl. Microbiol.*, 5:115–118.

Mackey, B. M., Pritchet, C., Norris, A. and Mead, G. C. 1990. "Heat resistance of *Listeria*: Strain differences and effects of meat type and curing salts," *Lett. Appl. Microbiol.*, 10:251–255.

Marshall, D. L., Andrews, L. S., Wells, J. H. and Farr, A.J. 1992. "Influence of modified atmosphere packaging on the competitive growth of *Listeria monocytogenes* and *Pseudomonas fluorescens* on precooked chicken," *Food Microbiol.*, 9:303–309.

Mayer, B. K. and D. R. Ward. 1991. Microbiology of finfish and finfish processing, In *Microbiology of Marine Food Products*, D. R. Ward and C. R. Hackney, eds., Van Nostrand Reinhold, New York, NY, pp. 3–17.

Morris, J. G. 1976. "Oxygen and the obligate anaerobe," *J. Appl. Bacteriol.*, 40:229–244.

NACMCF (National Advisory Committee on Microbiological Criteria for Foods). 1992. "Hazard analysis and critical control point system," *Int. J. Food Microbiol.*, 16:1–23.

Notermans, S., Dufrenne, J. and Keybets, M. J. H. 1981. "Vacuum-packed cooked potatoes–Toxin production by *Clostridium botulinum* and shelf life," *J. Food Prot.*, 44:572–575.

Notermans, S., Dufrenne, J. and Lund, B. M. 1990. "Botulism risk of refrigerated, processed foods of extended durability," *J. Food Prot.*, 53:1020–1024.

Olsvik O. and Kapperud, G. 1982. "Enterotoxin production in milk at 22 and 4°C by *Escherichia coli* and *Yersinia enterocolitica*," *Appl. Environ. Microbiol.*, 43:997–1000.

Palumbo, S. A. 1986. "Is refrigeration enough to restrain foodborne pathogens?" *J. Food Prot.*, 49:1003–1009.

Palumbo, S. A. and Buchanan, R. L. 1988. "Factors affecting growth or survival of *Aeromonas hydrophila* in foods," *J. Food Safety*, 9:37–51.

Palumbo, S. A., Maxino, A. F., Williams, A. C., Buchanan, R. L. and Thayer, D. W. 1985a. "Starch ampicillin agar for the quantitative detection of *Aeromonas hydrophila*," *Appl. Environ. Microbiol.*, 50:1027–1030.

Palumbo, S. A., Morgan, D. R. and Buchanan, R. L. 1985b. "The influence of temperature, NaCl on the growth of *Aeromonas hydrophila*," *J. Food Sci.*, 50:1417–1421.

Palumbo, S. A., Williams, A. C., Buchanan, R. L. and Phillips, J. G. 1987. "Thermal resistance of *Aeromonas hydrophila*," *J. Food Prot.*, 50:761–764.

Palumbo, S. A., Williams, A. C., Buchanan, R. L. and Phillips, J. G. 1992. "Model for the anaerobic growth of *Aeromonas hydrophila* K144," *J. Food Prot.*, 55:260–265.

Patel, H. R., Patel, P. C. and York, G. K. 1978. "Growth and toxicogenesis of *Clostridium botulinum* type E on marine mollusks at low temperatures" *J. Food Sci. Technol. India*, 15:231.

Pflug, I. J. and Odlaug, T. E. 1978. "A review of z and F values to ensure the safety of low-acid canned food," *Food Technol.*, 32(6):63–70.

Post, L. S., Lee, D. A., Solberg, M., Furang, D., Specchio, J. and Graham, C. 1985. "Development of botulinal toxin and sensory deterioration during storage of vacuum and modified atmosphere packaged fish fillets," *J. Food Sci.*, 50:990–996.

Reddy, N. R., Armstrong, D. J., Rhodehamel, E. J. and Kautter, D. A. 1992. "Shelf-life extension and safety concerns about fresh fishery products packaged under modified atmospheres: A review," *J. Food Safety*, 12:87–118.

Rhodehamel, E. J. 1992. "FDA's concerns with sous vide processing," *Food Technol.*, 46(12):73–76.

Rouf, M. A. and Rigney, M. M. 1971. "Growth temperatures and temperature characteristics of *Aeromonas*," *Appl. Microbiol.*, 22:503–506.

Schiemann, D. A. 1988. "Examination of enterotoxin production at low temperature by *Yersinia* spp. in culture media and foods," *J. Food Prot.*, 51:571–573.

Schmidt, C. F., Lechowich, R. V. and Folinazzo, J. F. 1961. "Growth and toxin production by type E *Clostridium botulinum* below 40°F," *J. Food Sci.*, 26:626–630.

Scott, V. N. 1989. "Interaction of factors to control microbial spoilage of refrigerated foods," *J. Food Prot.*, 52:431–435.

Scott, V. N. and Bernard, D. T. 1982. "Heat resistance of spores of non-proteolytic type B *Clostridium botulinum*," *J. Food Prot.*, 45:909–912.

Shapton, D. A. and Shapton, N. F. 1991. ***Principles and Practices for the Safe Processing of Foods***. Butterworth Heinemann, Oxford, U.K.

Silliker, J. H. and Wolfe, S. K. 1980. "Microbiological safety considerations in controlled-atmosphere storage of meats," *Food Technol.*, 34(3):59–63.

Smith, M. V. 1975. The effect of oxidation-reduction potential on the outgrowth and chemical inhibition of *Clostridium botulinum* type E spores. Ph.D. Thesis, Virginia Polytechnic Institute, Blacksburg, VA.

Smith L. DS. and Sugiyama, H. 1988. ***Botulism: The Organism, Its Toxins, the Disease***. Thomas, Springfield, IL.

Smith, J. L., Marmer, B. S., and Benedict, R. C. 1991. "Influence of growth temperature on injury and death of *Listeria monocytogenes* Scott A during a mild heat treatment," *J. Food Prot.*, 54:166–169.

Smith, J. P., Toupin, C., Gagnon, B., Voyer, R., Fiset, P. P. and Simpson, M. V. 1990. "Hazard analysis critical control point approach (HACCP) to ensure the microbiological safety of sous-vide processed meat/pasta product," *Food Microbiol.*, 7:177–198.

Smoot, L. A. and Pierson, M. D. 1979. "Effect of oxidation-reduction potential on the outgrowth and chemical inhibition of *Clostridium botulinum* 10755A spores," *J. Food Sci.*, 44:700–704.

Solomon, H. M., Kautter, D. A., Lilly, T. and Rhodehamel, E. J. 1990. "Outgrowth of *Clostridium botulinum* in shredded cabbage at room temperature under a modified atmosphere," *J. Food Prot.*, 53:831–833.

Sorquist, S. 1989. "Heat resistance of *Campylobacter* and *Yersinia* strains by three methods," *J. Appl. Bacteriol.*, 67:543–550.

Sperber, W. H. 1982. "Requirements of *Clostridium botulinum* for growth and toxin production," *Food Technol.*, 36(12):89–94.

Stern, N. J., Pierson, M. D. and Kotula, A. W. 1980. "Growth and competitive nature of *Yersinia enterocolitica* in whole milk," *J. Food Sci.*, 45:972–974.

Stier, R. F., Bell, L., Ito, K. A., Shafer, B. D., Brown, L. A., Seeger, M. L., Allen, B. H., Porcuna, M. N. and Lerke, P. A. 1981. "Effect of modified atmosphere storage on *Clostridium botulinum* toxigenesis and the spoilage microflora of salmon fillets," *J. Food Sci.*, 46:1639–1642.

Sugiyama, H. 1982. "Botulism hazards from nonprocessed foods," *Food Technol.*, 36(12):113–115.

Van Garde, S. J. and M. J. Woodburn. 1987. "Food discard practices of households," *J. Am. Diet. Assoc.*, 87(3):322–329.

Walker, S. J. and Gilmour, A. 1990. "Production of enterotoxin by *Yersinia* species isolated from milk," *J. Food Prot.*, 53:751–754.

Walker, S. J. and Stringer, M. F. 1987. "Growth of *Listeria monocytogeneses* and *Aeromonas hydrophila* at chill temperatures," *J. Appl. Bacteriol.*, 63:R20.

Walten, W. C. and Hentges, D. J. 1975. "Differential effects of oxygen and oxidation–reduction potential on the multiplication of three species of anaerobic intestinal bacteria," *Appl. Microbiol.*, 30:781–785.

Wimpfheimer, L., Altman, N. S. and Hotchkiss, J. H. 1990. "Growth of *Listeria monocytogenes* Scott A, serotype 4 and competitive spoilage organisms in raw chicken packaged under modified atmospheres and in air," *Int. J. Food Microbiol.*, 11:205–214.

Witter, L. D. 1961. "Psychrophilic bacteria–Review," *J. Dairy Sci.* 44:983–1015.

Woodburn, M. J. and Vanderiet, S. 1985. "Safe food: Care labeling for perishable foods," *Home Econ. Res. J.*, 14:3–10.

Yen, L. C., Sofos, J. N. and Schmidt, G. R. 1992. "Thermal destruction of *Listeria monocytogenes* in ground pork with water, sodium chloride and other curing agents," *Lebensm.-Wiss. u. -Technol.* 25:61–65.

Yoon, I. H. and Matches, J. R. 1988. "Growth of pathogenic bacteria in imitation crab," *J. Food Sci.*, 53:688–690.

Zee, J. A., Bouchard, C., Simard, R. E., Pichard, B. and Holley, R. A. 1984. "Effect of N_2, CO, CO_2 on the growth of bacteria from meat products under modified atmospheres," *Microbiol. Aliments Nutri.*, 2:351–370.

CHAPTER 5

MAP and CAP of Fresh, Red Meats, Poultry and Offals

C. O. GILL—*Agriculture and Agri-Food Canada*

INTRODUCTION

THE storage life of fresh meats can be extended by sealing products into packages that contain an atmosphere differing from air in the concentrations of N_2, O_2 and CO_2. During the course of storage, the atmosphere first established in the pack may alter, through interactions between the product and the atmosphere, and/or by the exchange of gases across the plastic film used for forming the pack. Packages in which such atmospheric changes occur arc properly referred to as Modified Atmosphere Packagings (MAP). This type of packaging can be distinguished from Controlled Atmosphere Packagings (CAP), in which the atmosphere that is first established remains unaltered in composition throughout the life of the package (Gill, 1990).

The MA and CA packagings are variously appropriate for different types of meat and for differing commercial circumstances. Their proper usages can therefore be appreciated only with understanding of the critical differences in deteriorative processes between various fresh meats, the preservative capabilities of the various packaging systems, and the various commercial circumstances within which preservative packagings may be required to function.

DETERIORATIVE PROCESSES IN FRESH MEATS

Microbial Spoilage

All fresh meats are contaminated with spoilage bacteria that largely derive, either directly or indirectly, from the hides of slaughtered animals (Grau, 1986). Those bacteria are deposited on meat surfaces during the dressing and breaking down of carcasses (Nottingham, 1982).

The deep tissues of most meats will remain sterile until well after the onset of microbial spoilage (Gill and Penney, 1977; Gill, 1979). The exceptions are offals, such as liver, that have an open structure at the cellular level. In those, spoilage bacteria can penetrate into the spaces between cells, but any such bacteria in the deep tissues will be of little consequence because of the persistently greater numbers of the surface bacteria (Gill and DeLacy, 1982).

All fresh meats provide a rich source of nutrients to support microbial growth, and interspecific inhibitions do not occur until the numbers of bacteria are large (Gill and Newton, 1977; Newton and Gill, 1978). Under those circumstances, a population of spoilage bacteria is dominated by the species that grow most rapidly in the environment provided by the meat (Gill, 1986a).

When O_2 is present in concentrations non-limiting for growth, the spoilage population of bacteria will be composed of strictly aerobic species; acinetobacteria, moraxellae and pseudomonads. The spoilage process will be dominated by the pseudomonad fraction, because these organisms produce malodorous, putrid compounds when they exhaust the available glucose at a meat surface and commence the metabolism of amino acids, whereas the metabolic byproducts of the other species are relatively innocuous (Gill, 1976; Gill and Newton, 1978; Shaw and Latty, 1982).

When the absence of O_2 prevents the growth of the strictly aerobic species, the bacterial population will usually be dominated by the anaerobic, but aerotolerant, lactobacilli. These will be the only species present in significant numbers when meat held under an anaerobic atmosphere provides an environment of pH less than 5.8 (Grau, 1983a; Egan, 1983). Unlike the pseudomonads, the lactobacilli do not precipitate spoilage while they are increasing in numbers. Instead, they will impart mild, acid-dairy flavours to meat only some time after they have attained their maximum numbers (Dainty et al., 1979; Seideman and Durland, 1983).

If, however, the meat provides an environment of pH greater than 5.8, then facultatively anaerobic species, notably psychrotrophic enterobacteria, *Brochothrix thermosphacta* and *Alteromonas putrefaciens*, will also be able to grow on meat held under an anaerobic atmosphere (Grau, 1980, 1981; Taylor and Shaw, 1977). Although all these types of organisms grow more slowly than the lactobacilli, they will often persist in a population of spoilage bacteria when their initial numbers are very much greater than those of the lactobacilli (Gill, 1986a). Like the pseudomonads, the enterobacteria will precipitate putrid spoilage when the spoilage bacteria exhaust the available glucose at the meat surface (Patterson and Gibbs, 1977). That occurs when the mixed population attains its maximum numbers. In contrast, *B. thermosphacta* produces a strong, sour-aromatic odour and *A. putrefaciens* produces putrid, sulphurous odours, that increase in intensity with the numbers of these organisms (Blickstad, 1983; Grau,1983b). Thus, the anaerobic spoilage process in meat of higher pH is often dominated by the activities of the enterobacteria, or of *B. thermosphacta* or *A. putrefaciens* when those latter organisms are present.

The time before microbial spoilage develops will always be highly dependent on the extent to which the product is contaminated with spoilage organisms before storage commences, and the temperature of the product during storage. Obviously, the higher the initial numbers of spoilage bacteria, the shorter the time required for them to grow to the numbers sufficient to cause spoilage (Figure 5.1). Also, the time before

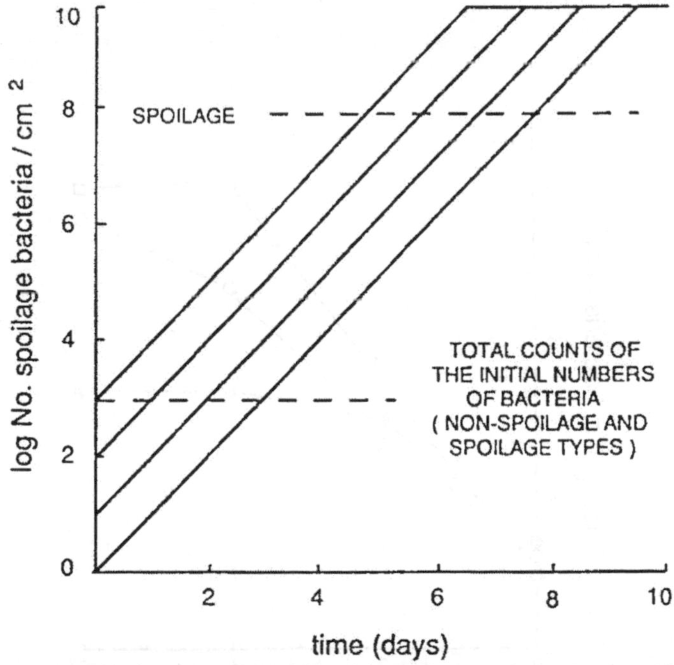

FIGURE 5.1 The effect of the initial numbers of spoilage bacteria on the time of spoilage. Growth of the spoilage bacteria will be detected from counts of the total numbers of bacteria only when the spoilage bacteria become predominant in the bacterial population. The figure represents the behavior of aerobic spoilage bacteria growing on meat stored at 2°C.

the growth of the spoilage organisms becomes apparent as an increase in the total bacterial counts will decrease with increasing fractions of spoilage organisms in the initial population. With the populations of bacteria that grow anaerobically, the effect of initial numbers on the time to spoilage can be further exaggerated. When initial numbers are very low, the growth rate advantage of the lactobacilli over the facultative anaerobes allows the former species to overgrow the latter on high-pH product. When the lactobacilli approach their maximum numbers, they impose an extensive inhibition on the growth of their competitors (Klaenhammer, 1988; Schillinger and Lucke, 1989, 1991). Thus, the facultative anaerobes of high spoilage potential can be maintained at numbers too low for their activities to dominate the spoilage process (Figure 5.2). Conversely, when the numbers of enterobacteria are relatively high on meat of pH < 5.8, those organisms may initiate slow growth after an extended lag phase to cause putrid spoilage of the meat, although forming only about 1% of a population otherwise dominated by lactobacilli (Gill and Penney, 1988).

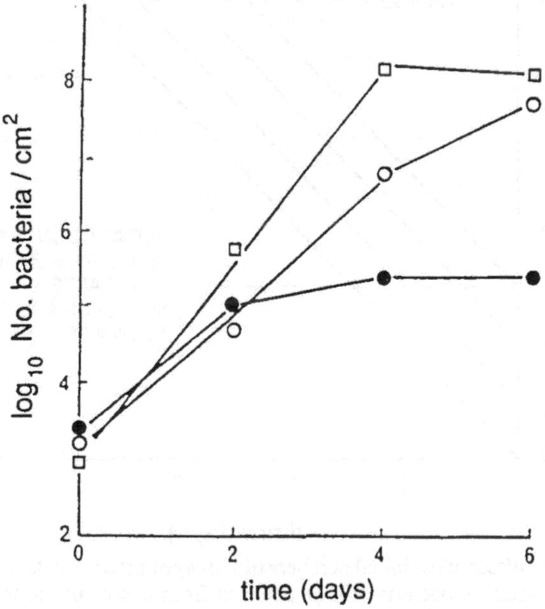

FIGURE 5.2 Growth of bacteria inoculated onto sterile meat slices and stored at 10°C under anaerobic conditions. An *Enterobacter* sp. alone (○); and the *Enterobacter* sp. (●) growing in mixed culture with a *Lactobacilli* sp. (□). [Reprinted from Newton and Gill (1978), with permission from Blackwell Scientific Publications Ltd.]

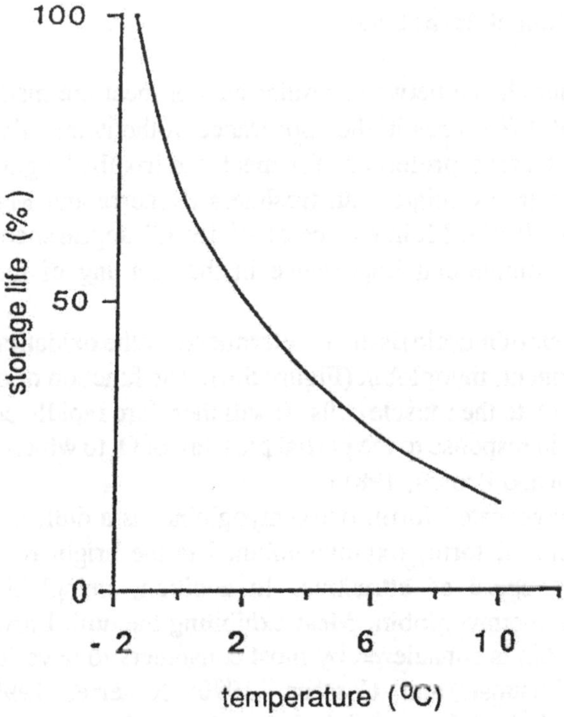

FIGURE 5.3 The effect of the storage temperature on the storage life of chilled meat. The proportional loss of storage life with increasing storage temperature is broadly similar for all atmospheres under which meat may be stored.

The rate at which bacteria grow is highly temperature dependent. The minimum temperature at which packaged meat can be held indefinitely without the meat freezing is −1.5°C, whereas the minimum temperature for growth of the psychrotrophic bacteria involved in meat spoilage is about −3°C (Gill, 1988b; Lowry and Gill, 1984). Therefore, microbial spoilage of fresh meat cannot be prevented by chilling alone, but the maximum storage life will be obtained at the lowest possible storage temperature. The rates of bacterial growth increase rapidly with small increases in temperature above the minimum for storage. The proportional loss of storage life for incremental increases in the storage temperature are broadly similar, whatever the nature of the spoilage bacteria (Gill and Jones, 1992; Pooni and Mead, 1984). Thus, at 0, 2, and 5°C, the storage life of product will respectively be about 70, 50 and 30% of the storage life obtained at −1.5°C (Figure 5.3). Any storage life ascribed to a fresh meat product will therefore be highly ambiguous, unless the storage temperature is also specified.

Deterioration of Meat Colour

Consumer choice between similar cuts of meat are made largely on the basis of differences in the appearance of the items. There is thus a general and strong preference for meat that has the bright red colours that consumers associate with freshness (Renerre and Mazuel, 1985; Young et al. 1988). Maintenance of a "fresh" appearance is therefore of critical commercial importance in the retailing of meat (Lefens, 1987).

The colour of muscle tissue is determined by the oxidation state of the muscle pigment, myoglobin (Figure 5.4). The function of myoglobin is to transfer O_2 to the muscle cells. It will therefore rapidly combine with or lose O_2, in response to the partial pressure of O_2 to which it is exposed (Livingston and Brown, 1981).

The deoxygenated form, deoxymyoglobin, is a dull, purple colour. The oxygenated form, oxymyoglobin, has the bright red colour that consumers regard as attractive. In addition, myoglobin is slowly oxidized to metmyoglobin. Meat exhibiting the dull, brown colour of metmyoglobin is considered by most consumers to have lost desirable quality (Faustman and Cassens, 1990; Renerre, 1990). Deoxymyoglobin is more susceptible to oxidation than oxymyoglobin, so metmyoglobin forms most rapidly at low O_2 concentrations (Ledward, 1970). That oxidation is only slowly reversed by enzyme-mediated processes, termed collectively metmyoglobin reducing activity (Ledward, 1985; Echevarne et al., 1990). The metmyoglobin reducing

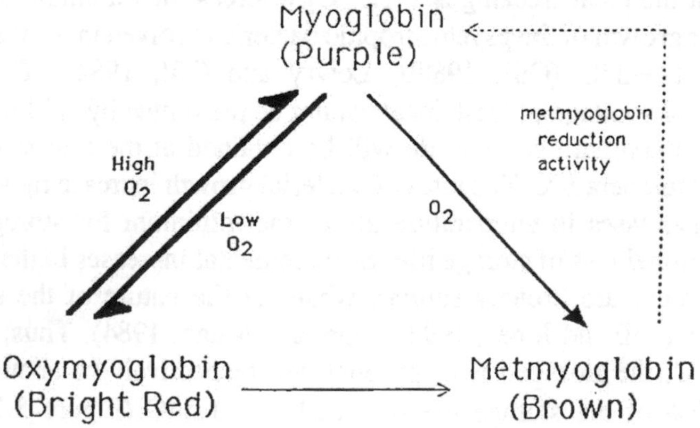

FIGURE 5.4 The reactions of the muscle pigment, myoglobin, with O_2.

capacity of any muscle tissue is limited, and once that capacity is exceeded the oxidation of myoglobin is essentially irreversible.

The display life of meat is usually limited by the time required for the metmyoglobin forming in the surface layers of the muscle tissue to reach such proportions of the total pigment that the meat appears first dull, then frankly brown. The rate at which metmyoglobin forms varies by factors of up to five between different types of muscle, and increases with increasing temperature for all muscles (O'Keeffe and Hood, 1982). The increase in the rate of loss of colour stability with increasing temperature is proportionally greater for muscles with high intrinsic stability than in those with low intrinsic stability (Hood, 1980). The time available for display will decline in proportion to the amount of metmyoglobin formed at the tissue surfaces during storage before display. Even if metmyoglobin formation has been prevented by rigorous exclusion of O_2 during pre-display storage, aged muscle tissue will have reduced ability to resist discolouration once it is exposed to O_2 (O'Keeffe and Hood, 1980–81a, 1980–81b). After about 12 weeks storage at the optimum storage temperature, the colour stability of muscles without preformed metmyoglobin will have reached a minimum value (Moore and Gill, 1987). That minimum value is apparently similar for all muscles.

Exposed marrow in bone-in cuts also loses colour stability with prolonged storage. During display, exposed and aged bone-marrow surfaces will darken relatively rapidly to brown, then black, as haemoglobin from the marrow oxidizes and accumulates at the exposed surfaces. Such darkened bone is unattractive to consumers (Hatch and Stadelman, 1972).

The appearance of fat tissue can also deteriorate with prolonged storage. The fat tissue can be stained with myoglobin from muscle tissue exudate and, after lengthy storage times, can be coloured by pigment imbibed from the muscle tissue even when the fat is not bathed in exudate. While that staining can give the fat an attractive pink appearance, oxidation of the pigment, either slowly in pack or rapidly after exposure to air, will cause the fat to appear dull and discoloured.

Deterioration of Texture, Flavour and Appearance Unrelated to Microbial Activities

Non-microbial deterioration of texture during the storage of raw meats in preservative packagings is not of significant concern for many types of product. In fact, the usual tenderization of muscle tissue during storage is considered a very desirable textural change (Morissey and Fox, 1981). However, the tenderization process involves the progressive degradation of muscle proteins (Dransfield et al., 1980–81). That can ultimately lead to a loss of the fibrous texture of meat (Radouco-Thomas et al., 1959). The food then has a meal-like mouthfeel which does not meet with consumer expectations.

Peptides and free amino acids are released during the breakdown of proteins (Rhodes and Lea, 1961). Those peptides can impart a liver-like flavour to the meat. Some consumers find such a flavour objectionable at low intensities when the flavour is not usual for a particular product (Gill, 1988b). Few consumers would seem to accept an intense liver flavour. Increasing concentrations of the amino acid tyrosine impart an undesirable, bitter flavour to meat. Tyrosine has a very low solubility in water, and will consequently crystallize out at low concentrations. In some offals, proteolysis can become so advanced that visible tyrosine crystals form at product surfaces (Lawrie, 1963). As the crystals are usually assumed to be microbial colonies, product of such compromised appearance is not acceptable to consumers.

Additionally, odour or flavour changes from oxidative deterioration may occur with some types of meat in some types of packaging (Ordonez and Ledward, 1977).

PACKAGING MATERIALS AND EQUIPMENT

As a preservative, packaging will only be effective while an appropriate in-pack atmosphere is maintained, and thus a packaging must necessarily be composed of a material that has limited capabilities of transmitting gases. Moreover, the pack must be sealed, rather than closed by clipping or crimping, to avoid any direct exchange of gases between the pack atmosphere and the air.

In general discussion, the suitability of films for constructing preservative packagings is usually considered with reference to the O_2 transmission rates of the films measured under specified conditions of humidity and temperature. For such discussion it is necessary to appreciate that not only will the absolute rates of O_2 transmission be very different when films are used at refrigeration temperatures but also that, at those temperatures, the relative magnitudes of specified O_2 transmissions may well not be conserved (Lambden et al., 1985).

Given those limitations to discussing the suitability of films in terms of their nominal O_2 transmission rates at room temperatures, then films with O_2 transmission rates much greater than 500 cc $O_2/m_2/24$ h/atm will be unsuited for any type of preservative packaging of meat. Generally, films with nominal O_2 permeabilities between 10 and 100 cc $O_2/m_2/24$ h/atm are used for vacuum and MA packages (Jenkins and Harrington, 1991; Brown, 1992). However, the laminates used for CAP will preferably incorporate a layer or layers of aluminum, to give materials that are essentially gas-impermeable (Kelly, 1989).

A wide range of equipment, designed for various commercial functions, is available for the preservative packaging of meat. Despite that diversity of form and function, it is, for the purposes of this discussion, possible to distinguish three types of machines. Those machine types are differentiated by the extent to which air is displaced from packagings during usual commercial operation of the equipment. The three groups can be referred to as low, high, and very high air-displacement types.

The first group would encompass those machines where air is swept from the packagings by a stream of gas, without any vacuum being applied, and machines that evacuate and gas a flexible pouch through a retractable snorkel(s) while the pouch is exposed to the ambient air pressure. With either type of machine, the atmosphere established in the pack is likely to contain a large fraction of residual air, in the first type because large volumes of gas would be required to dilute the air to a negligible fraction, and in the second because exhausting too much gas from the pack will lead to compaction or crushing of the product, and possible pouch puncture.

In the high air-displacement types of machine, the packaging is enclosed by a hood, with simultaneous evacuation of both the packaging and hood, commonly to a final vacuum of between 100 and 20 torr. Gassing may involve filling the hood with the intended atmosphere, as in the formation of lidded trays containing a modified atmosphere, or filling a pouch with simultaneous breaking of the hood vacuum to the air. In either case, the final atmosphere is likely to contain a moderate fraction of residual air. It should, however, be recognized that with some high air-displacement equipment, and the snorkel type of low airdisplacement equipment, very high displacements of air can be achieved by repeated evacuation and gassing cycles, but that such extended packaging cycles are not usual in commercial practice.

In very high air-displacement systems, a pouch is evacuated through a snorkel, to which the pouch mouth is compressed by sealing pads, with simultaneous evacuation of a hood covering the pouch. A high vacuum, about 1 torr, is drawn in both the hood and the pouch. The pouch is then filled with gas while the hood is broken to the air. When operated properly such systems can give consistent residual air levels of less than 500 ppm (Gill, 1990).

TYPES OF PRESERVATIVE PACKAGING

Four types of preservative packaging for raw meats can be distinguished (Gill and Molin, 1991); vacuum packs, high O_2 modified atmosphere packs (High-O_2 MAP), low O_2 modified atmosphere packs (Low-O_2 MAP), and controlled atmosphere packs (CAP). Those four types of packs differ considerably in their preservative capabilities.

Vacuum Packs

Vacuum packs are commonly formed from films with O_2 transmission rates about 50 cc $O_2/m_2/24$ h/atm which can be shrunk to the packaged product. However, non-shrink films, and films of either higher (>100 cc/m²/24 h/atm) or lower (ca 10 cc/m_2/24 h/atm) gas permeabilities are also used.

In forming a vacuum pack, the film must be applied closely to all surfaces of the product or to opposed film (Rizvi, 1981). When that is done, the atmosphere remaining in the pack is negligible. The level of vacuum established in the pack before it is sealed is then irrelevant, apart from the pressure difference that is required to form the film to the product. A higher pressure difference will be required to form stiff films to products of irregular shape than is needed to form highly flexible films to smoothly curved, more regular objects.

Residual O_2, and O_2 transmitted through the film, is scavenged by the product. In a fully effective vacuum pack, the scavenging reactions prevent the O_2 tension at the meat surface from rising to levels where there is rapid formation of metmyoglobin or aerobic growth of microorganisms. Thus, the meat colour is preserved and, when the product is uniformly of pH < 5.8, the bacterial population that develops is composed of the relatively innocuous lactobacilli (Egan, 1984).

However, fully effective packs cannot be formed with all products. If the shape of the product is such that the packaging film will bridge to leave vacuities, then an atmosphere containing increasing concentrations of O_2 will develop within such voids. Any such pockets of atmosphere will precipitate metmyoglobin formation in their vicinity, and permit the growth of potent spoilage bacteria (O'Keeffe and Hood, 1982; Grau, 1981).

If the product has a pH > 5.8, then potent spoilage bacteria, such as enterobacteria, *B. thermosphacta* or *A. putrefaciens*, will grow to cause early spoilage (Gill and Penney, 1986). An area of high pH may form only a portion of the product surface, as, for example, a layer of covering fat over muscle. Fat is of neutral pH and, if not bathed in muscle exudate of pH < 5.8, will provide a high-pH environment for microbial growth and early spoilage. Bacterial growth on fat may be accelerated by ineffective O_2 scavenging at fat surfaces (Grau, 1983a).

Also, muscle tissue exudate will tend to progressively accumulate in a few pockets. The dissolved myoglobin, and other proteins in that exudate, are not protected from oxidation by components of the muscle tissue. Consequently, the exudate becomes turbid and darkens as storage is prolonged. The appearance of the product can then be compromised by accumulation of a dark precipitate on meat surfaces (Jeremiah et al., 1992a).

High-O_2 MAP

Displayed red meats are usually rendered unsaleable by discolouration of the muscle tissue surfaces, rather than by microbial spoilage. High concentrations of O_2 will retard that deterioration, by increasing the fraction of oxidation-resistant oxymyoglobin in the pigment at the tissue surface. Moreover, the increased oxymyoglobin concentration will intensify the desirable red colour of the meat (Renerre, 1989; Shay and Egan, 1990).

Although a high O_2 concentration will delay colour deterioration, it will have no effect on the rates of growth of the aerobic spoilage bacteria. The growth of the aerobic bacteria can, however, be slowed by moderate concentrations of CO_2. When the CO_2 content of the atmosphere exceeds 20%, the rate of growth of the population of aerobic bacteria is approximately halved (Figure 5.5). The rate of growth increases with decreasing concentration below that value, but higher concentrations cause little further decrease of the growth rate (Gill and Tan, 1980).

Thus, an atmosphere of 80% O_2: 20% CO_2 would be optimal for

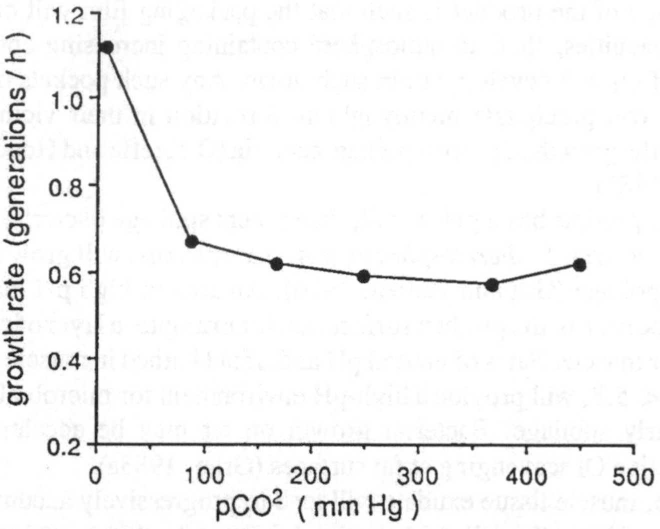

FIGURE 5.5 The effect of the partial pressure of CO_2 on the rate of growth of *Pseudomonas fluorescens* in nutrient both at 30°C. [Reprinted from Gill and Tan (1979), with permission from the American Society for Microbiology.]

stabilizing the colour and microbiological condition of red meat, but in practice such an atmosphere cannot readily be maintained within a display pack. The difficulty arises because O_2 is converted to CO_2 by both the meat and the bacteria; and because CO_2 is highly soluble in meat, and lost more rapidly than O_2 through the films used for lidding display trays (Gill, 1988c). Consequently, the initial concentration of CO_2 must be greater than 20%to allow for the dissolution of that gas in the meat. As storage progresses, the O_2 concentration and the total gas volume will tend to fall. Some decrease in the O_2 partial pressure will not be critically important, but excessive reduction of the atmosphere volume will lead to pack collapse. Collapsed retail packs will be essentially unsaleable.

To counter the unavoidable changes in the pack atmosphere, it is necessary to have a volume of atmosphere which is large in comparison to the volume of the product. A ratio of atmosphere to product of 3 to 1 is recommended for fresh meat (Holland, 1980). In addition, the O_2 concentration of the input gas is reduced from the possible 80% maximum, the CO_2 concentration is increased above 20%, and some N_2 is added as an inert filler. Then, the input gas will have a composition of about 60% O_2: 25% CO_2: 15% N_2 (Taylor, 1985). In a pack of sufficient volume in relation to the product, the CO_2 content of such an atmosphere will remain relatively constant after an initial period of stabilization, while the O_2 concentration declines and the N_2 concentration increases (Nortjé and Shaw, 1989).

The necessarily large volume of a High-O_2 MAP display pack in relation to the amount of product it contains imposes costs, from the increased amounts of packaging materials and the space that must be used for each saleable meat item. To avoid those costs, many MAP display packs are, in commercial practice, constructed without a large volume of atmosphere in relation to the amount of product. Such packs give product of an initial improved appearance, with some modest extension of the display life, provided that they are delivered to retail outlets on the same day that the product was packaged. If the distribution processes require that the packaged product be stored and/or transported for more than a day, the display performance of that type of package is likely to be unsatisfactory. To ameliorate the cost disadvantage without large sacrifice of product stability, recourse has been made to master-packaging a number of conventional tray-plus-overwrap display packs within a single pouch filled with a High-O_2 atmosphere. The unit costs of packaging and of the relatively costly display space are thus reduced, while the consumer is presented with a familiar type of packaging (Scholtz et al., 1992).

Although High-O_2 MAP master packaging is in use for distribution of centrally prepared retail packs (Lefens, 1987), the system has limited application, because product colour will deteriorate, albeit at a reduced rate, while the product is in store, and the preservative atmosphere is lost once the master pack is opened for display of the somewhat compromised product.

Low-O_2 MAP

Low-O_2 MAP is aimed at exploiting the inhibitory effects of CO_2 on the spoilage bacteria without any particular regard for the preservation of meat colour. Low-O_2 MAP is typically used for bulk-packed product, with a non-hooded snorkel machine being used to establish the pack atmosphere. It is used to draw only a low vacuum in the pouch, so substantial and variable volumes of air persist when the pouch is gassed.

Although pouches may be gassed with CO_2 alone, the dissolution of that gas in the product will require the pouch to be initially over-inflated if the pouch is not to collapse tightly around the product. Therefore, N_2

is often included in the input gas. Moreover, there is a widespread belief that O_2 in a pack atmosphere will preclude the growth of *Clostridium botulinum*. In fact, anaerobic niches where botulinum organisms can grow if any are present and if other factors of the environment are favourable, will occur in any package of raw meat irrespective of the surrounding atmosphere (Lambert et al., 1991). None the less, some O_2 is often included in the gas mixture used for establishing Low-O_2 MAP.

As a result of those various factors, the atmosphere initially established in Low-O_2 MAP can vary widely, with CO_2 from 50 to 90%, N_2 from 10 to 40%, and O_2 from 1 to 10%. During storage, O_2 concentrations are likely to decrease, so conditions within a pack of initially low O_2 content may finally preclude the growth of strictly aerobic organisms. In other packs, the O_2 content may remain sufficient for the growth of strict aerobes to be uninhibited, except by the effects of CO_2. Consequently, the delaying of microbial spoilage by Low-O_2 MAP can range from the modest time obtained with High-O_2 MAP to approaching the very long times attainable with CAP.

Obviously, Low-O_2 MAP cannot be used when the meat colour will be unacceptably degraded by exposure of the product to low concentrations of O_2.

CAP

A truly controlled atmosphere is obtained when product is sealed in a gas-impermeable pouch filled with an O_2-free atmosphere. If CO_2 is a major, or the sole component of the input atmosphere, then the quantity of added gas must be adjusted to assure that the intended atmosphere persists after dissolution of the gas in the product.

CO_2 is highly soluble in both muscle and fat tissues. The final concentration of CO_2 in either tissue will be proportional to the concentration of the gas in the equilibrated atmosphere. The solubility of CO_2 in muscle tissue decreases with decreasing tissue pH, and with increasing temperature (Figure 5.6). The solubility of the gas in fat tissue is also greatly affected by the tissue composition, being greater in the softer fats, such as that which is usual in pork, than in the harder fats, such as is commonly encountered in beef. However, within the temperature range used for the storage of fresh meat (−1.5 to 5°C), the solubility of CO_2 in fat increases with increasing temperature (Figure 5.7). With increasing temperatures beyond that range, the solubility of CO_2 attains a maximum value, then declines. The temperature at which CO_2 solubility is maximal is lower for hard than for soft fats, being about 8°C for beef tallow but

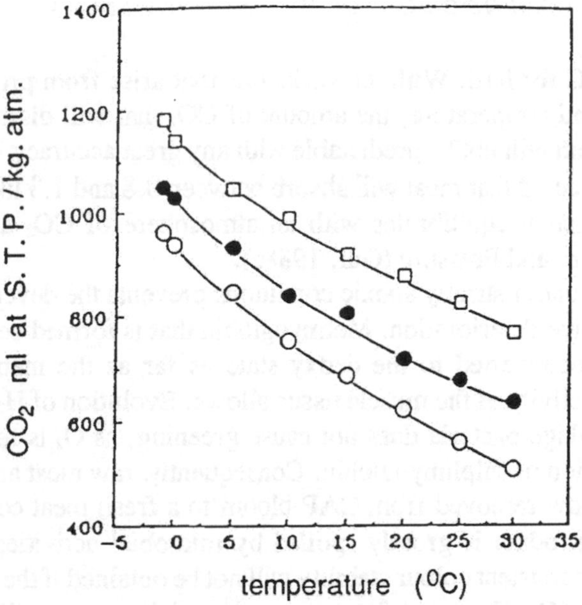

FIGURE 5.6 The effects of temperature and pH on the solubility of CO_2 in muscle tissue. Solubility of CO_2 in beef, pH 5.42 (○); pork; pH 5.75 (●), lamb, pH 6.20 (□). [Reprinted from (1988c), with permission from Elsevier Applied Science Publishers Ltd.]

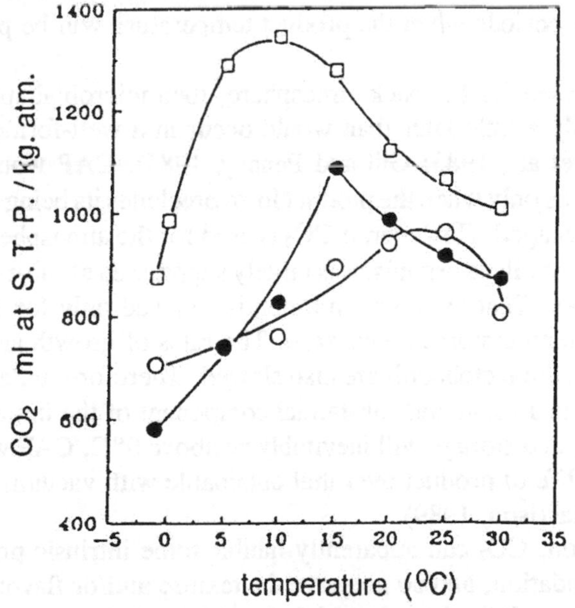

FIGURE 5.7 The effect of temperature on the solubility of CO_2 in fat tissues. Solubility of CO_2 in beef fat (○), pork fat (●), and lamb fat (□). [Reprinted from Gill (1988c), with permission from Elsevier Applied Science Publishers Ltd.]

about 25°C for lard. With the variations that arise from product composition and temperature, the amount of CO_2 that will dissolve in any product item will not be predictable with any great accuracy, although it can be expected that meat will absorb between 0.8 and 1.3 litres of CO_2 per kg when it equilibrates with an atmosphere of CO_2 at Standard Temperature and Pressure (Gill, 1988c).

Storage under strictly anoxic conditions prevents the development of any oxidative deterioration. Metmyoglobin that is formed before packaging is reconverted to the deoxy state as far as the metmyoglobin reducing activity of the muscle tissue allows. Evolution of H_2S by some of the spoilage bacteria does not cause greening, as O_2 is required for the formation of sulphmyoglobin. Consequently, raw meat and exposed bone marrow removed from CAP bloom to a fresh meat colour, even when the product is grossly spoiled by microbial activities. This indefinitely persistent colour stability will not be obtained if the packaging can transmit O_2. However, for some commercial purposes, films of very low O_2 permeability (<2 cc $O_2/m^2/24$ h/atm) can be satisfactorily substituted for gas-impermeable foil laminate or metalized films. Such substitution is possible, for example, when the product appearance is not rapidly degraded by low O_2 concentration, or when there is assurance that the product will not have to tolerate extended, uncertain storage periods, or periods when the product temperature will be poorly controlled.

If N_2 is used for the pack atmosphere, then microbial spoilage will develop only a little later than would occur in a well-formed vacuum pack (Lee et al., 1983; Gill and Penney, 1985). CAP would then be advantageous only when the product form precludes its being effectively vacuum packaged. However, if CO_2 is used for the atmosphere, growth of all potent spoilage organisms is largely suppressed at sub-zero storage temperatures. That extreme inhibition is relieved only for *B. thermosphacta* at temperatures above zero. The rates of growth and spoilage activities of the lactobacilli are also slowed. Therefore, unless *B. thermosphacta* is a usual and substantial component of the initial bacterial population, and storage will inevitably be above 0°C, CAP will greatly extend the life of product over that obtainable with vacuum packaging (Gill and Harrison, 1989).

In addition, CO_2 can apparently inhibit some intrinsic processes of tissue degradation, and so preserve the texture and/or flavour of some products beyond the times at which those degradative processes would usually render the products unacceptable (Gill, 1988d).

THE USE OF PRESERVATIVE PACKAGINGS WITH COMMERCIAL PRODUCT

Venison

In European markets it is considered undesirable to wash game meats, so visibly contaminated tissue must be excised from game carcasses. The long, black guard hairs from deer pelts are highly visible on dressed carcasses. Therefore, venison packers serving European markets have developed dressing techniques that aim at eliminating direct and indirect contacts between the carcass and the pelt, to avoid the excision of tissue to which hair adheres. In consequence, those packers place very few bacteria on carcass surfaces during dressing. Total counts then generally average $<10^2$ cfu/cm^2, and spoilage organisms are all but undetectable (Seman et al. 1988; Gill, 1989a). As the carcasses are not washed, their surfaces rapidly dry during chilling, to preclude any bacterial growth. With generally good attention to cutting room hygiene, the product is of excellent microbiological condition when it is presented for packaging.

Venison is essentially free of fat cover. Although the meat obtained from stags and castrates can be of high pH when the animals are severely stressed before they are slaughtered (MacDougall et al., 1979), the muscle tissue is generally of pH < 5.8.

With those characteristics, the meat is highly stable in well-formed vacuum packages. In commercial practice, when temperatures are maintained near the optimum for storage (-1.5°C), a storage life of about 18 weeks is obtained for vacuum-packaged primal cuts (Seman et al., 1988). Preservatively packaged retail cuts are as yet but little traded. However, retail cuts of venison would also be highly stable in vacuum pack, and possibly acceptable to consumers, given that the traditional preference for bright red meat may not be decisive in the selection of a meat with which many consumers are unfamiliar. Venison is also highly stable under CO_2 in CAP, the storage life reported for venison in that packaging being greater than 30 weeks at -1°C (Gill, 1990). Use of High-O_2 and Low-O_2 MAPs with venison have not been reported, but the colour stability of venison appears to be comparable with that of beef (MacDougall et al., 1979; Seman et al., 1988). Hence, High-O_2 MAP should be useful for extending the storage life of venison, but Low-O_2 MAP could be expected to rapidly degrade the appearance of the meat.

The long storage life achieved for vacuum-packaged venison of initially excellent microbiological condition cannot be expected when the dressing procedures that are used result in carcasses being heavily contaminated with bacteria (Smith et al., 1974; Sumner et al., 1977).

Beef

The average values of the total bacterial counts on beef generally exceed 10^3 cfu/cm^2, (Roberts et al., 1984; Anderson et al., 1987) with spoilage organisms forming 1 to 10% of the initial population (Newton et al., 1978). Much of the muscle tissue is of pH < 5.8 (McCollum and Hendrickson, 1977). Therefore, bone out primal cuts, such as striploins, will store in vacuum pack for about 12 weeks when they are held at -1.5 ± 0.5°C during transportation in refrigerated sea freight containers (Gill et al., 1988). However, more localized trading often involves irregularly shaped cuts with extensive fat cover. In addition, some carcasses yield Dark, Firm, Dry (DFD), high pH meat (Tarrant, 1981). On both these types of product, a population of potent spoilage bacteria will develop in vacuum packs to spoil the product within 8 weeks at -1°C.

In current practice, much beef traded overland is loaded to transport while still relatively warm. Refrigerated road trailers lack the capacity to rapidly reduce the temperatures within stacks of product, so much packaged beef experiences average temperatures of between 2 and 5°C during storage and distribution, despite refrigerated facilities being operated at 0°C or below. At those temperatures, the vacuum packaged high-pH beef cuts will have a storage life of little more than 2 weeks (Gill and Jones, 1992). That is, at best, barely adequate for many commercial purposes.

Given that the commercial circumstances may preclude the substantial improvement of product hygiene and/or much improved control over product temperature, then a packaging for beef primal cuts with preservative capabilities superior to that of the vacuum pack would seem to be required. The only available option is CAP using a CO_2 atmosphere; CAP with an N_2 atmosphere would give little increase in storage life over the vacuum pack, High-O_2 MAP has preservative capabilities that are less than those of the vacuum pack, and Low-O_2 MAP cannot be used with beef because the meat colour would be unacceptably degraded.

High-pH beef of average microbiological quality will have a shelf life of about 6 weeks at 5°C, or about 20 weeks at -1.5°C, when stored in CAP under CO_2 (Gill and Penney, 1988). Storage stability of that order should be adequate for most overland trades, with sufficient leeway to allow for uncertain temperature control during the multistage distribution of product.

Although beef is distributed to retail outlets predominantly as vacuum packaged primal cuts, many retailers would greatly prefer to receive the product as display packed consumer cuts (Farris et al., 1991). Substantial attempts have been made to investigate the latter type of trading using vacuum packs or High-O_2 MAP (Allen, 1989). Vacuum packaging has largely failed in that trade, because few consumers will accept the dull, purple colour of anoxic muscle despite the product being unspoiled. High-O_2 MAP has met with considerable success in some restricted markets (Renerre, 1989) but only limited success in general, because the storage life, of about 3 weeks under optimum conditions, is generally inadequate to cope with the vagaries of temperature control encountered in retail distribution. Thus, CAP master packaging remains the only possibility for advancing the central cutting of beef.

As vacuum packaging has proved adequate for preserving the product in commercial circumstances, CAP master packaging under a N_2 atmosphere should give a storage life sufficient for the convenient distribution and storage of beef before its display. The display stability of the product should be similar to that of freshly cut vacuum packaged beef. However, the commercial practicability of such a postulated trade remains to be demonstrated.

Lamb

The hygienic condition of many lamb carcasses is still rather poor, with average total bacterial counts $>10^4$ cfu/cm^2 and 10% or more of the population being spoilage types (Shaw et al., 1980; Ellerbroek et al., 1992). However, those seeking to sea-freight fresh lamb to distant markets have adopted dressing procedures comparable to those used for deer carcasses, with comparable hygienic benefits (Gill, 1986b).

Many of the muscles of lamb carcasses have a usual ultimate pH $\geq$ 5.8 (Chrystall and Haggard, 1975; Chrystall and Devine, 1985). Moreover, there is a layer of cover fat over most superficial muscle surfaces. Thus, the meat invariably behaves as a high pH product in anoxic packagings.

Lamb primal cuts of usual hygienic quality will store for no more than 6 weeks when vacuum packaged (Jeremiah et al., 1972). Vacuum packaged lamb cuts of high initial hygienic quality can remain unspoiled for 12 weeks when their temperature during sea freighting is maintained at -1.5 ± 0.5°C (Gill and Penney, 1985; Warburton and Gill, 1989). If microbial spoilage does not develop about that time, the product will still become unacceptable shortly after, because the muscle tissue loses its fibrous texture to give a porridge-like mouthfeel to the cooked product (Gill, 1988b).

As with beef, High-O_2 MAP has been used for packaging consumer cuts of lamb, with limited commercial success. Again, Low-O_2 MAP cannot be used with lamb, because the meat discolours at low O_2 concentrations.

CAP with a CO_2 atmosphere is increasingly being used for sea freighting lamb primals, and whole lamb carcasses, to distant overseas markets (Anon., 1992). With that type of packaging, and optimum storage temperatures, microbial spoilage can be delayed beyond 20 weeks when the product is prepared to a high hygienic standard. Moreover, the high concentration of CO_2 inhibits the textural deterioration that would otherwise ruin the eating qualities of the product after that time in storage (Gill, 1989b).

Pork

The initial microbiological condition of pork is poor. Although the average total bacterial numbers of 10^3 to 10^4 cfu/cm^2 are not much greater than those usually found on lamb, most of the contaminants are spoilage types. That unfortunate situation arises because, although the carcasses are pasteurised during scalding, they are recontaminated with spoilage bacteria during their passage through dehairing equipment. Many of those spoilage bacteria survive the subsequent singeing and polishing of carcasses. As the skin is not usually removed until the carcasses are broken down, and is not removed at all from some cuts, there is every opportunity for bacteria from the skin to be spread over all meat surfaces (Gill and Bryant, 1992).

Normal pork muscle is of pH < 5.8. Muscle of Pale, Soft, Exudative (PSE) quality tends to be of abnormally low pH, <5.5. High-pH, DFD muscle will also occur in a fraction of carcasses (Pearson, 1987). The superficial muscles are covered by a relatively thick layer of subcutaneous fat.

When cuts, such as bone-out loins, are stripped of subcutaneous fat, and are of normal or PSE character, the vacuum-packaged product will be spoiled by lactobacilli within 8 weeks at the optimum storage temperature (Shay and Egan, 1986). Vacuum-packaged cuts of DFD muscle, with extensive fat cover and/or with retained skin will spoil within 5 weeks from the activities of enterobacteria or *B. thermosphacta* (Gill and Harrison, 1989).

High-O_2 MAP has been used for the distribution of display ready consumer cuts, with the attainment of a product storage life comparable to that of similarly packaged beef (Lefens, 1987). Adequate microbiological stability is apparently obtained when the skin and most fat is stripped from the cuts, and there is careful attention to cutting room hygiene. Although the numbers of aerobic spoilage bacteria would still then be relatively high, the high glucose content of pork muscle, 1000 g/g vs 100 g/g in beef (Fischer and Augustini, 1977), ensures that the bacterial population must approach its maximum numbers of $10^9/cm^2$ before the onset of putrid spoilage (Gill, 1976).

Low-O_2 MAP is widely used for bulk packaging of pork primal cuts. That is possible because, in pork, the muscle pigment is resistant to oxidation at low partial pressures of O_2. Moreover, the myoglobin content of pork muscle is low, so the formation of metmyoglobin causes the muscle to appear dull and grey, rather than to become the dark brown that consumers find so objectionable in beef and lamb (Cross et al., 1986).

Why pork myoglobin resists oxidation does not seem to have been examined. However, it can be suggested that the metmyoglobin reducing capacity of pork muscle is generally sufficient to maintain the small content of myoglobin in its native state for prolonged periods, whereas that capacity is soon exhausted in heavily pigmented muscle from other species. Some support for that suggestion is offered by the behaviour of the myoglobin in PSE pork, which is as sensitive to oxidation at low partial pressures of O_2 as the myoglobin in other red meats (Jeremiah et al., 1992b). The PSE condition is the result of protein denaturation consequent on a rapid fall in pH while the muscle tissue is still warm (Mitchell and Heffron, 1982). That denaturation could well involve enzymes that function in the metmyoglobin reducing activity, and elimination of all metmyoglobin reducing capacity would remove the major constraint on pigment oxidation.

As PSE pork is a substantial and continuing problem for the pork industry, it might be thought commercially unsound to use a packaging that must, at an early time, further degrade the already compromised appearance of PSE product. However, the pigment in freshly cut PSE meat will oxidize rapidly once it is exposed to air, so its earlier, in-pack oxidation is, in commercial circumstances, of little consequence.

CAP with a CO_2 atmosphere can, at the optimum storage temperature, extend the storage life of skin-on pork to about 18 weeks, even when the initial hygienic condition of the product is very poor (Gill and Harrison, 1989). Unfortunately, pork is often initially contaminated with high numbers of *B. thermosphacta*. Growth of that organism at sub-zero temperatures is totally inhibited in an atmosphere of CO_2, but at temperatures only a little above zero, it will proliferate to spoil the meat. Thus, the extension of the storage life is not proportionally maintained at storage temperatures above zero. That accelerated collapse of preservative capability with increasing temperature would make CAP an uncertain system for preserving pork in many trades, unless the packaged product was of the type that consistently behaves as a low pH meat. Then, some consistent gain in storage life over vacuum-packaged product could be achieved.

Poultry

Like pork, poultry is usually contaminated initially with a large bacterial population that is predominantly composed of spoilage organisms. In poultry, those bacteria derive from the plucking equipment, and the chilling water in which carcasses are immersed after they have been eviscerated (Mulder et al., 1978; Schmitt et al., 1988). Again like pork, the skin provides a high-pH environment for the bacteria. In addition, the leg muscles are invariably of high-pH (McMeekin, 1977).

Good vacuum packages cannot usually be formed with raw poultry. The packaging film cannot be closely applied to the body cavity surfaces with either whole or half carcasses, and some bridging is difficult to avoid with collections of smaller portions. With poorly formed packs and a high-pH product, the storage life of poultry in vacuum packs is short. Putrid spoilage develops within 2 weeks at the optimum storage temperature (Barnes et al., 1979; Jones et al., 1982). The poor package formation may allow strict aerobes as well as facultative anaerobes to appear in the gram-negative fraction of the population. *B. thermosphacta* is not usually important in the spoilage of raw poultry, because it is not commonly a major component of the initial bacterial population.

High-O_2 MAP is used for display packs of poultry portions. The major effect being sought is some slowing in the rate of development of the spoilage bacteria. A high O_2 atmosphere will impart some brightening of the product, but that is not a major commercial consideration. In fact, little is gained for poultry by using the high O_2 concentrations required to stabilize red meat colour.

Low-O_2 MAP is extensively used with bulk packaged poultry. The colour of the product is not greatly degraded by low O_2 concentrations, but in commercial practice pack atmospheres will usually contain substantial fractions of O_2. Packaging is usually performed rapidly, using snorkel-type, low air-displacement equipment. A high vacuum is not drawn on packs, as that would crush the product, and the packs are not flushed before sealing. Thus, there is a large O_2 fraction in the atmosphere, from residual air, even when O_2 is not added with the input gas. The storage life of poultry in Low-O_2 MAP is then, in practice, much the same as that in High-O_2 MAP, about 2 weeks at the optimum storage temperature, despite a longer life being attainable when a microaerobic atmosphere is established (Hotchkiss et al., 1985).

With CAP under a CO_2 atmosphere, a storage life of about 12 weeks is attainable for poultry (Gill et al., 1990). In the effective absence of *B. thermosphacta*, the proportional loss of storage life with increasing temperature is similar to that obtained with other packagings. Despite superior preservative capabilities, CAP does not as yet seem to have been used commercially for packaging raw poultry.

Offals

Offals comprise a heterogeneous group of products, with a range of tissue types. Their varied behaviour in preservative packagings must therefore be expected. For the purposes of this discussion, consideration will be given only to those offal types for which some detailed information is available; namely hearts, livers, kidneys and sweetbreads (beef thymus gland).

The microbiological condition of offals at the time they are packaged is often poor (Patterson and Gibbs, 1979). That heavy contamination with spoilage bacteria is generally the result of poor practices during offal collection (Hinson, 1968a,b). When removed from the carcass, offals such as those previously noted will commonly carry microbial loads that are less, or at least no more, than those on the carcasses from which the offals are withdrawn. Unfortunately, it is a usual practice to collect warm offals into bulk containers. Such product masses will remain at warm temperatures for long periods even when they are held in refrigerated facilities. More often than not, the bulked product remains on unrefrigerated slaughter floors until such time as it is packed (Oblinger, 1983).

Temperatures of 30°C or more will persist within those tissue masses. The conditions within a mass of soft tissue will be anaerobic, because of the consumption of O_2 by the tissues. Offal tissues do not develop the low pH of skeletal muscle, their pH tending to remain at values >6.0. Under those circumstances, both spoilage and pathogenic types of facultatively anaerobic bacteria proliferate rapidly. At the time of packaging, offals will, therefore, often carry a varied and large bacterial population that is enriched for enterobacteria and lactobacilli as compared with the carcass meat (Gill, 1988b).

In addition to the microbiological deterioration, autolytic activities will proceed rapidly at warm temperatures in biochemically highly active tissues such as liver. Autolysis will lead to loss of texture, and the development of progressively stronger liver-like and bitter flavours (Stiffler et al., 1985). Thus, the condition of offals, both hygienically and organoleptically, is often seriously compromised by mishandling in the immediate post-slaughter period. Obviously, consistent early cooling of product should be assured in offal collecting systems, to avoid the rapid product deterioration that will arise from inadequate control of product temperatures.

Given that product is collected in a manner that presents offals for packaging that are cool, and of similar hygienic condition to carcass meat, then hearts and sweetbreads in vacuum packs will spoil as other high-pH meats. That is, they will develop a mixed bacterial population of lactobacilli and enterobacteria, with putrid spoilage becoming evident after about 8 weeks storage at −1.5°C.

Liver and kidney will also develop a mixed bacterial population, which will persist on kidney. However, fermentation of the large amounts of glucose available in liver will cause the pH of that tissue to fall below 5.8 (Gill and DeLacy, 1982). When that occurs, growth of the enterobacteria is inhibited, and the final bacterial population is composed solely of lactic acid bacteria. These bacteria impart to vacuum-packaged liver an acid odour and sour flavour, after 6 weeks at −1.5°C. By that time, the tissue is discoloured by grey patches, has a stale odour, and strong sour, bitter and liver flavours (Gill and Jeremiah, 1991).

There is no fall in the pH of kidney, similar to that occurring in vacuum-packaged liver. Moreover, putrid spoilage of vacuum-packaged kidney is delayed beyond 8 weeks, despite the presence of enterobacteria in the bacterial population. However, white granules of tyrosine can render the product's appearance unacceptable after about 6 weeks storage at −1.5°C. Tyrosine, an amino acid of very low solubility, is released during the autolytic breakdown of protein, and precipitates at tissue surfaces. This protein breakdown product also imparts the bitter flavour to aged liver (Lawrie, 1963). Thus, autolytic deterioration tends to render vacuum-packaged liver and kidney unacceptable at much the same time, or before, microbial spoilage becomes evident.

Fresh heart is dark red in colour, washed sweetbreads are white (fat tissue) and gray (glandular tissue), while fresh liver and kidney show a range of brown and red colours. Darkening and browning of heart does not appear to detract from its acceptability, sweetbreads are not discoloured by O_2, while liver and kidney become a uniform dark red-brown after their storage under O_2. The more uniform colour of aged liver and kidney is in fact more acceptable to consumers than the varied colours displayed by fresh product. Thus, there is little point in trying to preserve offal colours by packaging them under high O_2 atmospheres. Low O_2 atmospheres, though little used with offals, would seem more appropriate for delaying aerobic spoilage. In fact, biochemically highly active tissue like liver will tend to produce a microaerobic and CO_2 rich atmosphere in any closed container or pouch (Gill and Penney, 1984). In those circumstances, the mechanical establishment of a low O_2 atmosphere is hardly warranted.

As with other high pH meats, CAP with a CO_2 atmosphere can greatly delay the putrid spoilage of hearts and sweetbreads. Lactic spoilage of liver is also somewhat retarded, but autolytic deterioration is unaffected. Thus, CAP of liver would achieve little. However, the autolytic deterioration of kidney does seem to be inhibited by high CO_2 concentrations, so CAP packaging extends the storage life of that product to about 12 weeks at −1.5°C (Gill and Jeremiah, 1991).

CONCLUDING REMARKS

Selecting a preservative packaging that is appropriate for a particular fresh meat product in a particular trade requires knowledge of the principal deteriorative processes to which the product is subject, the general hygienic condition of the product when it is presented for packaging, and the temperature history that the product can be expected to experience during its storage and distribution. Unfortunately, many of those seeking to extend the storage life of fresh meats are uncertain about all three factors. Consequently, there is a tendency to assume that the storage life obtained with the most stable products, when they are stored at a near-optimum temperature, will be achievable with other products in circumstances where good control of product temperature is not assured. Such unwarranted optimism has inevitably led to extensive disappointment, and further misunderstanding of what is realistically achievable when preservative packagings are properly applied within a total, well-controlled system for the production, packaging and distribution of product.

That misunderstanding undoubtedly continues to be a major factor slowing the use of diverse preservative packagings with raw meats. At present, only vacuum packaging and Low-O_2 MAP are widely used, although in many circumstances they give only a modest extension of product storage life. Both of the packagings are commonly misapplied, at least in part because one or both of the packaging systems are conveniently at hand in many meat plants.

However, increasing commercial pressures for central preparation of display ready product, and for augmented and diversified international trading of fresh meats, are directing meat packers to seek real understanding of the limited, but outstanding, commercial successes that have been obtained with High-O_2 MAP and CAP systems. If that interest is sustained, some wider, diversified and more effective usages of preservative packaging with fresh meats may yet develop in the relatively near future.

REFERENCES

Allen, J. W. 1989. "Centrally prepackaged fresh red meat: success or failure?" *Nat. Provision.*, 201(2):8–19.

Anderson, M. E., Huff, H. E., Naumann, H. D., Marshall, R. T., Damare, J. M., Pratt, M. and Johnston, R. 1987. "Evaluation of an automated beef carcass washing and sanitizing system under production conditions," *J. Food Prot.*, 50:562–566.

Anon. 1992. "Carbon dioxide packs are taking off," *Food Technol. N.Z.* 27(1):9–11.

Barnes, E. M., Impey, C. S. and Griffiths, N. M. 1979. "The spoilage flora and shelf life of duck carcasses stored at 2° or −1°C in oxygen permeable or oxygen impermeable film," *Brit. Poult. Sci.*, 20:491–500.

Benning, C. J. 1983. Films, laminates and their markets, In *Plastic Films for Packaging*, Technomic Publishing, Lancaster, PA, pp. 111–133.

Blickstad, E. 1983. "Growth and end product formation of two psychrotrophic *Lactobacillus* spp. and *Brochothrix thermosphacta* ATCC 115009 at different pH values and temperatures," *Appl. Environ. Microbiol.*, 46:1345–1350.

Brown, W. E. 1992. Food preservation, In *Plastics in Food Packaging*, Dekker, New York, pp. 66–102.

Chrystall, B. B. and Devine, C. E. 1985. Electrical stimulation, its early development in New Zealand, In *Adv. Meat Res., Vol. 1*, A. M. Pearson and T. R. Dutson, eds., AVI Publishing Co., Westport, CT, pp. 73–120.

Chrystall, B. B. and Haggard, C. J. 1975. "Accelerated conditioning of lamb," *N.Z. J. Agric.*, 130:7–12.

Cross, H. R., Durland, P. R. and Seideman, S. C. 1986. Sensory qualities of meat, In *Muscle as a Food*, P. J. Bechrel, ed., Academic Press, Orlando, FL, pp. 279–320.

Dainty, R. M., Shaw, B. G., Harding, C. O. and Michanie, S. 1979. The spoilage of vacuum packaged beef by cold tolerant bacteria, In *Cold Tolerant Microbes in Spoilage and the Environment*, A. D. Russell, and R. Fuller, eds., Academic Press, London, pp. 83–100.

Dransfield, E., Jones, R. C. D. and MacFie, N. J. H. 1980–81. "Tenderizing in *M. longissimus dorsi* of beef, veal, rabbit, lamb and pork," *Meat Sci.*, 5:139–147.

Echevarne, C., Renerre, M. and Labas, R. 1990. "Metmyoglobin reductase activity in bovine muscle," *Meat Sci.*, 27:161–172.

Egan, A. F. 1983. "Lactic acid bacteria of meat and meat products," *Antonie van Leeuwenhoek*, 49:327–336.

Egan, A. F. 1984. "Microbiology and storage life of chilled fresh meats," *Proc. 30th Europe. Meet. Meat Res. Work.*, Bristol, U. K., pp. 221–224.

Ellerbroek, L., Wegener, J. and Arndt, G. 1992. "Oberflachen Keimgehalt von Schafschlachttierkorpen," *Fleischwirtsch.*, 72:498–503.

Farris, D. E., Dietrich, R. A., and Ward, J. B. 1991. "Reducing the cost of marketing beef," *Meat Process.* (Feb.):60–62.

Faustman, C. and Cassens, R. G. 1990. "The biochemical basis for discolouration in fresh meat: A review," *J. Muscle Foods*, 1:217–243.

Fischer, K. and Augustini, C. 1977. "Stadien der postmortalen Glycogenolyse bei unterschiedelichen pH-Werten in Schweinefleish," *Fleischwirtsch.*, 57:1191–1193.

Gill, C. O. 1976. "Substrate limitation of bacterial growth at meat surfaces," *J. Appl. Bacteriol.*, 41:401–410.

Gill, C. O. 1979. "Intrinsic bacteria in meat," *J. Appl. Bacteriol.*, 47:367–378.

Gill, C. O. 1986a. The control of microbial spoilage of fresh meats, In *Advances in Meat Research, Vol. 2*, A. M. Pearson and T. R. Dutson, eds., AVI Publishing Co., Westport, CT, pp. 49–88.

Gill, C. O. 1986b. "Quality control procedures for chilled lamb packaged to give an extended storage life," *Meat Ind. Res. Inst. N.Z.*, RM 165.

Gill, C. O. 1988a. "CO_2 packaging–the technical background, *Proc. 25th Meat Ind. Res. Conf.*, Meat Ind. Res. Inst. N. Z., Hamilton, pp. 181–185.

Gill, C. O. 1988b. Microbiology of edible meat by-products, In *Adv. Meat Res., Vol. 5*, A. M. Pearson, and T. R. Dutson, Elsevier Applied Science, Barking, Essex, pp. 47–82.

Gill, C. O. 1988c. "The solubility of carbon dioxide in meat," *Meat Sci.*, 22:65–71.

Gill, C. O. 1988d. "Packaging meat under carbon dioxide: the Captech system," *Proc. 34th Int. Cong. Meat Sci. Technol.*, Industry Day, Brisbane, Australia, pp. 76–77.

Gill, C. O. 1989a. "The use of CO_2 packaging for venison," *Meat Ind. Res. Inst. N.Z.*, RM 184.

Gill, C. O. 1989b. "Packaging meat for prolonged chilled storage: The Captech process," *Brit. Food J.*, 91:11–15.

Gill, C. O. 1990. "Controlled atmosphere packaging of chilled meat," *Food Control*, 1:74–78.

Gill, C. O. and Bryant, J. 1992. "The contamination of pork with spoilage bacteria during commercial dressing, chilling and cutting of pig carcasses," *Int. J. Food Microbiol.*, 16:51–62.

Gill, C. O. and DeLacy, K. M. 1982. "Microbiol spoilage of whole sheep livers," *Appl. Environ. Microbiol.*, 43:1262–1266.

Gill, C. O. and Harrison, J. C. L. 1989. "The storage life of chilled pork packaged under vacuum or carbon dioxide," *Food Microbiol.*, 26:313–324.

Gill, C. O., Harrison, J. C. L., and Penney, N. 1990. "The storage life of chicken carcasses packaged under carbon dioxide," *Int. J. Food Microbiol.*, *11:151*–158.

Gill, C. O. and Jeremiah, L. E. 1991. "The storage life of non-muscle offals packaged under vacuum or carbon dioxide," *Food Microbiol.*, 8:339–353.

Gill, C. O. and Jones, S. D. M. 1992. "The storage efficiency of a commercial process for the storage and distribution of vacuum packaged beef," *J. Food Prot.*, 55:880–888.

Gill, C. O. and Molin, G. 1991. Modified atmospheres and vacuum packaging, In *Food Preservatives*, N. J. Russell, and G. W. Gould, eds., Blackie, Glasgow, pp. 172–199.

Gill, C. O. and Newton, K. G. 1977. "The development of aerobic spoilage flora on meat stored at chill temperatures," *J. Appl. Bacteriol.* 43:189–195.

Gill, C. O. and Newton, K. G. 1978. "The ecology of bacterial spoilage of fresh meat at chill temperatures," *Meat Sci.*, 2:207–217.

Gill, C. O. and Penney, N. 1977. "Penetration of bacteria into meat," *Appl. Environ. Microbiol.*, 33:1284–1286.

Gill, C. O. and Penney, N. 1984. "The shelf life of chilled sheep livers packed in closed tubs," *Meat Sci.* 11:73–77.

Gill, C. O. and Penney, N. 1985. "Modification of in-pack conditions to extend the storage life of vacuum packaged lamb," *Meat Sci.*, 14:43–60.

Gill, C. O. and Penney, N. 1986. "Packaging conditions for extended storage of chilled, dark, firm, dry beef," *Meat Sci.*, 8:41–53.

Gill, C. O. and Penney, N. 1988. "The effect of the initial gas volume to meat weight ratio on the storage life of chilled beef packed under carbon dioxide," *Meat Sci.*, 22:53–63.

Gill, C. O., Phillips, D. M. and Harrison, J. C. L. 1988. Product temperature criteria for shipment of chilled meats to distant markets, In *Refrigeration for Food and People*, International Institute of Refrigeration, Paris, France, pp. 40–47.

Gill, C. O. and Tan, K. H. 1979. "Effect of carbon dioxide on growth of *Pseudomonas fluorescens*," *Appl. Environ. Microbiol.*, 38:237–240.

Gill, C. O. and Tan, K. H. 1980. "Effect of carbon dioxide on growth of meat spoilage bacteria," *Appl. Environ. Microbiol.*, 39:317–319.

Grau, F. H. 1980. "Inhibition of the anaerobic growth of *Brochothrix thermosphacta* by lactic acid," *Appl. Environ. Microbiol.*, 40:433–436.

Grau, F. H. 1981. "Role of pH, lactate and anaerobiosis in controlling the growth of some fermentative Gram-negative bacteria on beef," *Appl. Environ. Microbiol.*, 42:1043–1050.

Grau, F. H. 1983a. "Microbial growth on fat and lean surfaces of vacuum packaged chilled beef," *J. Food Sci.*, 48:326–329.

Grau, F. H. 1983b. "End products of glucose fermentation by *Brochrothrix thermosphacta*," *Appl. Environ. Microbiol.*, 45:84–90.

Grau, F. H. 1986. Microbial ecology of meat and poultry, In *Adv. Meat Res.*, *Vol. 2.*, A. M. Pearson, and T. R. Dutson, eds., AVI Publishing Co., Westport, CT, pp. 1–48.

Hatch, V. and Stadelman, W. J. 1972. "Bone darkening in frozen chicken broilers and ducklings," *J. Food Sci.*, 37:850–852.

Hinson, L. E. 1968a. "Why off-condition offal," *Nat. Provision.* 159(21):14–18.

Hinson, L. E. 1968b. "Why off-condition offal; part 2," *Nat. Provision.*, 159(22):24–27.

Holland, G. C. 1980. "Modified atmospheres for fresh meat distribution," *Proc. 33rd Meat Ind. Res. Conf.*, Chicago, IL, pp. 21–39.

Hood, D. E. 1980. "Factors affecting the rate of metmyoglobin accumulation in pre-packaged meat," *Meat Sci.*, 4:247–265.

Hotchkiss, I. H., Baker, R. C. and Qureshi, R. A. 1985. "Elevated carbon dioxide atmospheres for packaging poultry. II. Effects of chicken quarters and bulk packages," *Poult. Sci.*, 64:333–340.

Jenkins, W. A. and Harrington, J. P. 1991. Fresh meat and poultry, In *Packaging Foods with Plastics*, Technomic Publishing, Lancaster, PA., pp. 109–122.

Jeremiah, L. E., Smith, G. C. and Carpenter, Z. L. 1972. "Vacuum packaging of lamb: Effects of storage, storage time and storage temperature," *J. Food Sci.*, 37:457–462.

Jermiah, L. E., Gill, C. O. and Penney, N. 1992a. "The effect on pork storage life of oxygen contamination in nominally anoxic packagings," *J. Muscle Foods*, 3:263–281.

Jeremiah, L. E., Penney, N. and Gill, C. O. 1992b. "The effects of prolonged storage under vacuum or CO_2 on the flavour and texture profiles of chilled pork," *Food Res. Internat.*, 25:9–19.

Jones, J. M., Mead, G. C., Griffiths, N. M. and Adams, B. W. 1982. "Influence of packaging on microbiological, chemical and sensory changes in chill-stored turkey portions," *Brit. Poult. Sci.*, 23:25–40.

Kelly, R. S. A. 1989. "High barrier metallized laminates for food packaging," In *Plastic Film Technology, Vol. 1.* K. M. Finlayson, ed., Technomic Publishing, Lancaster, PA, pp. 146–152.

Klaenhammer, T. R. 1988. "Bacteriocins of lactic acid bacteria," *Biochimie*, 70:337–349.

Lambden, A. E., Chadwick, D. and Gill, C. O. 1985. "Oxygen permeability at sub-zero temperatures of plastic films used for vacuum packaging meat," *J. Food Technol.*, 20:781–783.

Lambert, A. D., Smith, J. P. and Dodds, K. L. 1991. "Effect of initial O_2 and CO_2 and low-dose irradiation on toxin production by *Clostridium botulinum* in MAP fresh pork, *J. Food Prot.*, 54:939–944.

Lawrie, R. A. 1963. "A note on tyrosine production in frozen stored livers," *J. Sci. Food Agric.*, 14:65–66.

Ledward, D. A. 1970. "Metmyoglobin formation in beef stored in carbon dioxide-enriched and oxygen-depleted atmospheres," *J. Food Sci.*, 35:33–37.

Ledward, D. A. 1985. "Post-slaughter influences on the formation of metmyoglobin in beef muscle," *Meat Sci.* 15:149–171.

Lee, B. H., Simard, R. E., LaLeye, L. C. and Holley, R. A. 1983. "Microflora, sensory and exudate changes of vacuum- or nitrogen-packaged veal chucks under different storage conditions," *J. Food Sci.*, 48:1537–1542, 1563.

Lefens, M. 1987. "Wilson's tendercuts. One company's commitment to the future," *Meat Processing*, 26(6):58–63, 87.

Livingston, D. J. and Brown, W. D. 1981. "The chemistry of myoglobin and its reactions," *Food Technol.*, 35(5):244–252.

Lowry, P. D. and Gill, C. O. 1984. "Mould growth on meat at freezing temperatures," *Int. J. Refrig.*, 7:133–136.

MacDougall, D. B., Shaw, B. G., Nute, G. R. and Rhodes, D. N. 1979. "Effect of pre-slaughter handling on the quality and microbiology of venison from farmed red deer," *J. Sci. Food Agric.*, 30:1160–1167.

McCollum, P. D. and Hendrickson, R. L. 1977. "The effect of electrical stimulation on the rate of post-mortem glycolysis in some bovine muscles," *J. Food Qual.*, 1:15–22.

McMeekin, T. A. 1977. "Spoilage association of chicken leg muscle," *Appl. Environ. Microbiol.*, 33:1244–1246.

Mitchell, G. and Heffron, J. J. A. 1982. "Porcine stress syndrome," *Adv. Food Res.*, 28:167–230.

Moore, V. J. and Gill, C. O. 1987. "The pH and display life of chilled lamb after prolonged storage under vacuum or CO_2." *N. Z. J. Agric. Res.*, 30: 449–452.

Morissey, P. A. and Fox, P. E. 1981. "Tenderization of meat: A review," *Irish J. Food Sci. Technol.*, 5:33–47.

Mulder, R. W. A. W., Dorresteijn, L. W. J. and Vander Broek, J. 1978. "Cross-contamination during the scalding and plucking of broilers," *Brit. Poult. Sci.*, 19:61–70.

Newton, K. G. and Gill, C. O. 1978. "The development of anaerobic spoilage flora on meat stored at chill temperatures," *J. Appl. Bacteriol.*, 44:91–95.

Newton, K. G., Harrison, J. C. L. and Wauters, A. M. 1978. "Sources of psychrotrophic bacteria on meat at the abbatoir," *J. Appl. Bacteriol.*, 45:75–82.

Nortjé, G. L. and Shaw, B. G. 1989. "The effect of aging treatment on the microbiology and storage characteristics of beef in modified atmosphere packs containing 25% CO_2 plus 75% O_2," *Meat Sci.*, 25:43–58.

Nottingham, P. M. 1982. Microbiology of carcass meats, In *Meat Microbiology*, M. H. Brown, ed., Applied Science Publishers, London, pp. 13–66.

Oblinger, J. L. 1983. "Factors affecting the microbiological quality of variety meat," *Proc. Meat Ind. Res. Conf.*, Chicago, IL, pp. 63–76.

O'Keeffe, M. and Hood, D. E. 1980–81a. "Anoxic storage of fresh beef 1: Nitrogen and carbon dioxide storage atmospheres," *Meat Sci.*, 5:27–39.

O'Keeffe, M. and Hood, D. E. 1980–81b. "Anoxic storage of fresh beef 2: Colour stability and weight loss," *Meat Sci.*, 5:267–281.

O'Keeffe, M. and Hood, D. E. 1982. "Biochemical factors influencing metmyoglobin formation on beef from muscles of differing color stability," *Meat Sci.*, 7:204–228.

Ordonez, J. A. and Ledward, D. A. 1977. "Lipid and myoglobin oxidation in pork stored in oxygen and carbon dioxide-enriched atmospheres," *Meat Sci.*, 41–49.

Patterson, J. J. and Gibbs, P. A. 1977. "Incidence and spoilage potential of isolates from vacuum packaged meat of high pH value," *J. Appl. Bacteriol.*, 43:25–38.

Patterson, J. J. and Gibbs, P. A. 1979. "Vacuum packaging of bovine edible offals," *Meat Sci.*, 3:209–222.

Pearson, A. M. 1987. Muscle function and post-mortem changes, In *The Science of Meat and Meat Products*, 3rd edn., J. F. Price, and B. S. Schweigert, ed., Food and Nutrition Press, Westport, CT, pp. 155–192.

Pooni, G. S. and Mead, G. C. 1984. "Prospective use of temperature function integration for predicting the shelf-life of non-frozen poultry-meat products," *Food Microbiol.*, 1:67–78.

Radouco-Thomas, C., Lataste-Dorolle, C., Zender, R., Busset, R., Meyer, M. M. and Mouton, R. F. 1959. "The anti-autolytic effect of epinephrine in skeletal muscle: Non-additive process for preservation of meat," *Food Res.*, 24:453–682.

Renerre, M. 1989. "Retail storage and distribution of meats in modified atmospheres," *Fleischwirtschaft Internat.* (1):51–53.

Renerre, M. 1990. "Factors involved in the discolouration of beef meat," *Int. J. Food Sci. Technol.*, 25:613–630.

Renerre, M. and Mazuel, J. P. 1985. "Relations entre methodes demesurees instrumentals et sensorielles de la couleur de la viande," *Sci. Aliments*, 5:541–557.

Rhodes, D. N. and Lea, C. M. 1961. "Enzymic changes in lamb's liver during storage 1," *J. Sci. Food Agric.*, 12:211–227.

Rizvi, S. S. H. 1981. "Requirements for foods packaged in polymeric films," *CRC Crit. Rev. Food Sci. Nutr.*, 14:111–134.

Roberts, T. A., Hudson, W. R., Whelehan, O. P., Simonsen, B., Olgaard, K., Labots, M., Snijders, J. M. A., VanHoof, J., Debevere, J., Dempster, J. F., Devereux, J., Leistner, L., Gehra, H., Gledel, J. and Fournaud, J. 1984. "Numbers and distribution of bacteria on some beef carcasses at selected abattoirs in some member states of the European communities," *Meat Sci.*, 11:191–205.

Schillinger, U. and Lucke, F. K. 1989. "Antibacterial activity of *Lactobacillus sake* isolated from meat," *Appl. Environ. Microbiol.*, 55:1901–1906.

Schillinger, U. and Lucke, F. K. 1991. "Lactic acid bacteria as protective cultures in meat products," *Fleischwirtschaft Internat.* (1):3–10.

Schmitt, R. E., Gallo, L. and Schmidt-Lorenz, W. 1988. "Microbial spoilage of refrigerated fresh broilers. IV. Effect of slaughtering procedures on the microbial association of poultry carcasses," *Lebensm.-Wiss. u-Technol.*, 21:234–238.

Scholtz, E. M., Jordaan, E., Kruger, J., Nortjé, G. L. and Naudé, R. T. 1992. "The influence of different centralized pre-packaging systems on the shelf-life of fresh pork," *Meat Sci.*, 32:11–29.

Seideman, S. C. and Durland, P. R. 1983. "Vacuum packaging of fresh meat: A review," *J. Food Qual.*, 6:29–47.

Seman, D. L., Drew, K. P., Clarken, P. A. and Littlejohn, R. P. 1988. "Influence of packaging methods and length of chilled storage on microflora, tenderness and colour stability of venison loins," *Meat Sci.*, 22:276–282.

Shaw, B. G. and Latty, J. B. 1982. "A numerical taxonomy study of *Pseudomonas* strains from spoiled meat," *J. Appl. Bacteriol.*, 52:219–228.

Shaw, B. G., Harding, C. D. and Taylor, A. A. 1980. "The microbiology and storage stability of vacuum-packed lamb," *J. Food Technol.*, 15:397–405.

Shay, B. J. and Egan, A. F. 1986. "Studies of possible techniques for extending the storage life of chilled pork," *Food Technol. Australia*, 38:144–146.

Shay, B. J. and Egan, A. F. 1990. "Extending retail storage life of beef and lamb by modified atmosphere packaging," *Food Australia*, 42:399–400, 404.

Smith, F. C., Field, R. A. and Adams, J. C. 1974. "Microbiology of Wyoming big game meat," *J. Milk Food Technol.* 37:129–131.

Stiffler, D. M., Savell, J. W., Griffin, D. B., Gawlik, M. F., Johnson, D. D., Smith, G. C., and Vanderzant, C. 1985. "Methods of chilling and packaging beef, pork, and lamb variety meats for transoceanic shipment: physical and sensory characteristics," *J. Food Prot.*, 48:754–764.

Sumner, J. L., Perry, I. R. and Reay, C. A. 1977. "Microbiology of New Zealand farmed venison," *J. Sci. Food Agric.*, 28:1105–1108.

Tarrant, P. V. 1981. The occurrence, causes and economic consequences of dark cutting beef–A survey of current information, In *The Problem of Dark-Cutting Beef*, D. E. Hood and P. V. Tarrant, eds., Martius Ninjhoff, The Hague, pp. 3–36.

Taylor, A. A. 1985. Packaging fresh meat, In *Developments in Meat Science, Vol. 3*, R. A. Lawrie, ed., Elsevier Applied Science Publishers, Barking, Essex, pp. 89–114.

Taylor, A. A. and Shaw, B. G. 1977. "The effect of meat pH and packaging permeability on putrefaction and greening in vacuum packaged beef," *J. Food Technol.*, 12:515–521.

Warburton, D. J. and Gill, C. O. 1989. "New Zealand effort leads to extended meat shelf life," *Nat. Provision.* (Aug.): 33–36.

Young, L. L., Reviere, R. D. and Cole A. B. 1988. "Fresh red meats: A place to apply modified atmospheres," *Food Technol.*, 42(9):65–69.

CHAPTER 6

MAP of Cooked Meat and Poultry Products

JOSEPH H. HOTCHKISS–*Cornell University, U.S.A.*
SCOTT W. LANGSTON–*Cornell University, U.S.A.*

INTRODUCTION

THE rapid loss of quality of refrigerated fresh poultry and cooked meats presents a challenge to the food manufacturer. Properly processed fresh poultry has a shelf life of approximately 10–12 d at 7°C (Labuza, 1982). Deterioration primarily results from microbial growth which causes slime and off-odours. Particularly important to spoilage are the genera of *Pseudomonas*, *Achromobacter*, *Flavobacterium*, and *Micrococcus* along with coliform bacteria, and several yeasts (Frazier and Westhoff, 1988). In addition to spoilage, growth of pathogenic organisms such as *Salmonella*, *Campylobacter* and others make safety a concern. Poultry has been frequently implicated as a vehicle for food poisoning outbreaks from these and other pathogenic microorganisms (Jay, 1992). The shelf life and safety of cooked red meats and poultry is determined not only by environmental storage conditions but also the numbers and type of contaminating microorganisms.

The conventional method used to inhibit the rapid spoilage of poultry has been to ship packaged or unpackaged carcasses or cut-up parts packed in shaved ice. Shaved ice has the advantage of insuring chilling and preventing moisture loss. The disadvantage is the necessity of making and shipping large amounts of ice. Disposal of the contaminated ice at the receiving point also creates a problem. For these reasons, combined with the desire to achieve longer shelf life, there has been a search for alternatives to ice packing. Two alternatives have proven successful in the marketplace. The first is the super-chilled or "deep chilled," (i.e., crust frozen) poultry. In this process, fresh poultry is frozen to a depth of a few millimeters in a low temperature blast tunnel and stored at temperatures of 0 to −2°C until temperature equilibrium is achieved in the carcass. Shelf life is approximately 18 d depending on how long the low temperature is maintained. The second alternative to deep chilling is modified atmosphere packaging (MAP). A shelf life of 18 or more d has been reported for MAP of fresh poultry.

MAP has also been widely used to extend the shelf life of processed poultry products such as breaded and fully cooked chicken, fully cooked whole chickens, and cured poultry products (Guise, 1985). MAP has been used to a lesser extent to extend the shelf life of other cooked meats such as beef or pork. This is in part due to less consumer demand for these products as well as problems with warmed-over-flavour (WOF) with such products. Red meats develop a strong oxidized flavour (termed "WOF") when reheated. This can be offset to some degree by MAP, but WOF is still a problem in most MAP systems.

Extensive reviews of MAP science and technology for meats can be found in Wolfe (1980), Seideman and Durland (1984), Genigeorgis (1985), Hintlian and Hotchkiss (1986), Brody (1989), and Lawlis and Fuller (1990). General microbiological aspects of MAP (Farber, 1991), as well as preservation (Cunningham, 1987) and MAP of poultry products (Hotchkiss, 1989), and the shelf life of MAP meats (Lambert et al., 1991a) have been reviewed.

CURRENT MAP TECHNOLOGY

There are three basic requirements for successful application of MAP to poultry and cooked meat products. First, is that the gas or gases surrounding the product contain at least 20% (v/v) CO_2 compared to the normal 0.03% found in air. The second is that the product and modified atmosphere (MA) be contained in a package which prevents or inhibits the exchange of the gases with the exterior environment. Lastly, the storage temperature must be controlled to insure the effectiveness of the gas mixture at controlling microbial growth. The desired effect for poultry is a reduction of microbial spoilage and pathogenic potential, reduction in oxidation, slowing of catabolic enzymatic processes, prevention of irreversible pigment change, and reduction of weight loss due to evaporation.

Several different gas mixtures have been investigated for MAP of meats but nearly all successful tests use some combination of CO_2, N_2, and in a few cases O_2. Of these gases, CO_2 is the most important because it is the most inhibitory to the growth of spoilage microorganisms. Gram-negative, psychrotrophic, aerobic rods, such as those in the family Pseudomonadaceae, are significantly inhibited by elevated CO_2 atmospheres (Tan and Gill, 1982; Dixon and Kell, 1989). This inhibition is not a result of reducing the O_2 content of the atmosphere. Mixtures containing 20% CO_2 and 20% O_2 are also effective at inhibiting gram-negative organisms (Eyles et al., 1993; Hotchkiss et al., 1994). Facultative bacteria such as *Enterobacter* are less affected (Erichsen and Molin, 1981). The slower growing lactic acid bacteria are not affected, and can dominate a system in which pseudomonads have been inhibited by CO_2. The lactic acid bacteria become the predominant microorganisms in MAP poultry (Baker et al., 1986). CO_2 can both increase the length of the lag growth phase and decrease the rate of growth of spoilage organisms in poultry (Wimpfheimer et al., 1990).

MAP has been commercially applied to poultry products either as a method to increase shelf life only during wholesale distribution and storage, or as a method to increase shelf life in wholesale distribution and retail display. In the former method, fresh poultry products are often prepackaged in low barrier consumer-sized wrapped retail trays which are then packaged together in 9 to 23 kg groups and placed in large bags inside corrugated cartons. The air in the bags is evacuated and replaced with a MA prior to sealing. The low barrier nature of the wrapped retail trays allows the MA to permeate into the packages. The advantage is shelf life extension at a relatively low cost. However, the effect of the MA is lost when the individual packages are removed for retail display. This system has been termed "masterpack."

MAP is also used on individual pre-cooked poultry and to a lesser extent cooked red meat products. In most cases, pre-cooked products are packaged in high barrier consumer-sized packages in which the air has been removed by vacuum or gas flushing and replaced with a MA which does not contain O2. This has the advantage of continuing the bio-static effect of the MA until the package is opened by the consumer or the MA has dissipated. The disadvantage is the cost of the high barrier container. For this reason, this method is most widely used for high value-added processed products.

SHELF LIFE EXTENSION BY MAP

Effect of CO_2

There is a substantial literature dating back more than fifty years showing that MAP alone or combined with other technologies can inhibit microbiological growth in meats. Early work by Coyne (1933) showed that 25% CO_2 inhibited the growth of *Achromobacter, Bacillus, Flavobacterium, Micrococcus,* and *Pseudomonas*. Haines (1933) found that the lag and log phases of *Pseudomonas* and *Achromobacter* growth were increased by as much as 50% in an atmosphere of 15–20% CO_2 at 20°C.

Work on shelf life extension of fresh poultry dates back more than 25 years. Wabeck et al. (1968) reported that 20% CO_2 (balance air) in continuously flushed systems resulted in an increased shelf life of whole chicken carcasses while 10% was less effective. Gardner et al. (1977) found that ice packed chicken had a shelf life of 14 d compared to 23 d for carcasses packed in MAP containing 50–80% CO_2. Sander and Soo (1978) established that MA increased shelf life of fresh poultry when compared to both ice packing and vacuum packaging. Jurdi et al. (1980) found 100% CO_2 to be slightly more effective than 30% CO_2 or 100% N_2. Hotchkiss et al. (1985) demonstrated a 2.5-fold increase in the shelf life of chicken quarters packaged in MAP compared to refrigerated storage in air. Sensory studies of raw and cooked chicken indicated not only an improvement in microbial condition but also of organoleptic attributes. Kraft (1984) concluded that MAP using CO_2 was more effective than vacuum packaging for poultry and both were more effective than conventional tray packaging. Anderson et al. (1985) compared vacuum packaged, CO_2 and N_2-flushed packaged broiler drumsticks and concluded that MAP may help extend shelf life. Mead et al. (1986) compared the quality of duck meat packaged by stretch wrapping or gas flushing and found an improved microbiological condition for the gas flushed product, but a more rapid deterioration in appearance. Studer et al. (1988) found that simply sealing chicken carcasses in polyethylene bags increased shelf life by 30% compared to unbagged carcasses, most likely due to the build up of CO_2 inside the bags. Gill et al. (1990) also confirmed that CO_2 extends the shelf life of poultry. These and other data published over the last three decades indicates that MAP can extend shelf life by two- to four-fold depending on the initial quality of the product, storage temperature, and package construction and design.

The optimum concentration of CO_2 for maximum shelf life has not been clearly established. As pointed out above, the literature indicates that generally, higher amounts of initial CO_2 give longer shelf life. We tested levels of CO_2 ranging from 0 to 100% CO_2 (balance air) on the growth of indigenous microorganisms in ground poultry in gas flushed and sealed glass jars and confirmed that higher CO_2 resulted in lower microbial growth (Baker et al., 1985; Hotchkiss et al., 1985). This inhibition was not due to changes in surface pH. Zeitoun and Debevere

(1992) concluded that the optimum gas mixture for extending shelf life of fresh poultry was 90% CO_2:10% N_2.

Using total aerobic plate counts to monitor the effectiveness of MAs containing CO_2 only indicates a portion of the effect of MA on poultry. In addition to a reduction in numbers, Bailey et al. (1979) observed a shift from gram-negative to gram-positive microorganisms in chicken stored in CO_2-containing atmospheres compared to air controls. In similar tests we observed that *Pseudomonas* represented approximately 50% of all genera present in refrigerated fresh poultry. After storage in air, *Pseudomonas* comprised >98% of all genera present while in CO_2 stored samples, less than 10% of the total population was pseudomonads and approximately 90% were lactobacilli (Baker et al., 1985). This shift in the microbial population is no doubt, in part responsible for the increase in shelf life. In general, gram-positive organisms have less spoilage potential than gram-negatives.

Most of the studies cited above used a one-time change in atmosphere in sealed plastic film packaging. In these cases, it is not clear if the increase in shelf life seen with higher CO_2 concentrations was due solely to the presence of the elevated CO_2 or due simply to the fact that higher gas concentration means that some amount of CO_2 would be expected to be in the package longer, simply because there was more to begin with. In other words, a continuous and unchanging supply of 20% CO_2 might be as effective as a one-time exchange of 100% CO_2 if the latter gas were allowed to permeate from the container and be replaced with air.

The effectiveness of MA on the shelf life of cooked meat products has been investigated but to a lesser extent than fresh poultry. Hintlian and Hotchkiss (1987) found that the microbiological and sensory quality of cooked beef could be maintained in a high CO_2 atmosphere but that the generation of oxidized flavor (WOF) was still a problem when the product was reheated. Similar conclusions were more recently published by Penney et al. (1993). They found that an anaerobic CO_2 MA extended shelf life even when compared to vacuum packaging in high barrier films. As expected, the type of packaging system greatly influenced the microorganisms present in the product.

Effect of O_2

Several reports have noted that high concentrations of CO_2 can accelerate the deterioration of the bright red oxymyoglobin pigments of some meats to brown metmyoglobin. The inclusion of O_2 in the MAP gas mixture can have beneficial effects on poultry and meats by increasing oxymyoglobin levels. Deterioration of colour is not a problem for most fresh poultry products although a "greying" of poultry meat has been noted (Baker et al., 1985). Of course, for cooked meats metmyoglobin (brown) formation occurs due to the cooking process. For these products, flavour deterioration due to oxidation is the major problem and so addition of O_2 would be undesirable. High O_2 levels may enhance oxidative rancidity in these products.

The addition of O_2 could also inhibit growth of anaerobic pathogens (Hintlian and Hotchkiss, 1986) although under some circumstances the presence of O_2 has been shown to increase the potential for toxigenesis (Lambert et al., 1991b). Presumably, growth of aerobic organisms in the sealed package rapidly decreases the O_2 content, thus allowing for more rapid toxigenesis by the anaerobic pathogens. Some aerobic spoilage may be desirable, as a warning sign to consumers.

Effect of N_2

N_2 is used primarily as a filler gas, to keep the package from collapsing as CO_2 is dissolved into the product (Hotchkiss, 1989). For some poultry products which are packaged in flexible containers with a minimum of headspace, the addition of N_2 becomes important because a vacuum will be created in the container as the CO_2 dissolves in the meat. This vacuum can become strong enough to cause excessive water to be squeezed from the product.

Effect of Vacuum Packaging

Vacuum packaging (VP) is also considered a form of MAP in which the O_2 tension is decreased by the removal of air from the package without replacement by other gases. Although the atmosphere in the package is evacuated prior to sealing, CO_2 is generated as anaerobic organisms and the product itself respire (Brody, 1985).

Effect of Temperature and Microbial Age

The effectiveness of CO_2 as a bacteriostat is also dependent on its solubility in the aqueous portion of the food. As the storage temperature increases, solubility and thus the ability of the gas to inhibit growth decreases. For this reason, MAs containing elevated concentrations of CO_2 are more effective at low temperatures (Carr and Marchello, 1986).

Several researchers have demonstrated that the physiological state of the microorganisms also determines the effect of CO_2. Bacteria in the lag phase of growth are especially susceptible to CO_2, while those in the exponential phase are less so. The earlier the application of CO_2 to a food, the longer its potential shelf life (Silliker, 1981).

Effect of Package Barrier

The ability of the package to contain the MA also influences the effectiveness of MAP. Gases used in MAP permeate all common food packaging films by diffusion and so the shelf life is directly affected by the rate of diffusion. Common packaging plastics used for poultry and cooked meats vary in permeation rates by three orders of magnitude, but the cost of the films increases with increasing barrier (Robertson, 1993). Thus, a trade off in shelf life versus cost must be made. Higher barrier films contain the CO_2 longer than low barrier films, but at an increase in cost.

Effect of Product to Headspace Ratio

The total amount, rather than concentration of CO_2 relative to the mass of food in a package, the length of time the gas is in contact with the food, the condition of the product when the CO_2 is applied, and the storage temperature all contribute to the effectiveness of the MA. The amount of CO_2 required for maximum shelf life is, in part, a function of the ratio of headspace volume to mass of product. If this ratio is low, (i.e., <1) then a higher concentration of CO_2 may be required. If the ratio of headspace to product volume is large (i.e., >1), the concentration of CO_2 can be as low as 10–20% and still be effective.

Types of Packaging Equipment

The equipment used in MAP of poultry and cooked meats can be divided into two general types. The first is a "snorkel-type" machine. In this machine, product is placed in a flexible bag and the opening of the bag is placed in the jaws of an induction heat-sealing machine. A flat hollow tube, (i.e., snorkel) enters the bag through the jaws of the machine which are coated with a resilient silicone rubber material. The air in the bag is evacuated through the tube and the desired MA introduced into the bag through the same tube. The tube is then quickly withdrawn and the jaws quickly heated to seal the bag. If a vacuum is desired, the tube is withdrawn without reintroducing any gas and the bag sealed. These machines have the advantage of simplicity. However, the inside of the bag will be under vacuum while the outside will be at atmospheric pressure when the air is being withdrawn. For many products, particularly formed and cooked products, the collapse of the bag around the product as the air is removed from the package may cause the product to change shape or to stick together.

The second machine is a "chamber-type" process. In this machine one or more bags containing product are placed inside a large chamber which also contains several heat sealing jaws. The chamber is closed and a large pump evacuates the air from the entire chamber including the bags inside the chamber. The MA is introduced into the chamber (and bags), and when it reaches the desired pressure, the bags are heat-sealed inside the chamber before opening. The chamber is opened and the sealed bags containing the product and MA removed. These machines can be manual or highly automated, and have the advantage of not placing the product under external pressure during vacuuming because the pressure inside the bag and in the chamber is the same at all times. They can, however, be more expensive machines.

COMBINATION OF MAP AND OTHER TECHNOLOGIES

Recently published work on MAP and poultry has focused on combining MA and other potential microbial inhibitors. For example, Gray et al. (1984) inoculated fresh chicken thighs with nalidixic acid resistant *S. enteritidis* and then dipped them in a potassium sorbate solution (0, 1.0, 2.5, or 5.0%, pH 6.0). Treated samples were placed in barrier bags and flushed with air, 100% CO_2, 80% CO_2/20% air, 20% CO_2/80% air, or vacuum packaged, then stored at 10°C for 10 d. For all atmospheres, increasing concentrations of potassium sorbate decreased the final log counts of *Salmonella*. For the 0% sorbate treatment, log CFU for *Salmonella* were over 1 unit lower for the 100% CO_2 treatment than for both the air and vacuum treatments. The inhibition of *Salmonella* by the other CO_2 MA treatments was intermediate to this. The combination of CO_2 and sorbate was more effective at inhibiting microbial growth than either treatment alone. Similar experiments have demonstrated an extended shelf life of 3 d at 10°C (Elliott et al., 1985). Zeitoun (1991, 1992) combined a decontamination dip composed of a lactic acid/sodium lactate buffer with a high CO_2 MA to extend the shelf life of fresh chicken legs by 1 to 4 d. *Listeria monocytogenes* numbers initially were reduced by this combination of inhibitors. Growth occurred after 2 d, but in most cases, the *L. monocytogenes* only grew to numbers that were similar to the initial inoculum for up to 13 d.

Perhaps the most intriguing work combines MAP with low doses of ionizing irradiation. For example, workers at the U.S. Department of Agriculture's Eastern Regional Research Center have investigated the effect of 0 to 3.6 kGy irradiation on *Salmonella typhimurium* in chicken meat in the presence of air or under vacuum (Thayer and Boyd, 1991). Predictive equations were derived from the empirical data which indicated that irradiation was slightly more lethal at higher temperatures and in air than under vacuum. At −20°C, 1.5 kGy reduced the *S. typhimurium* counts by 2.53 and 2.12 logs in air and vacuum, respectively. However, this work only considered reduction and not potential for growth by the surviving organisms. MAP would likely reduce the potential for growth during storage. Grant and Patterson (1991) have also studied the microbiology of combined irradiated and MAP chicken and found that gram-positive predominated.

Workers at McGill University in Canada have extensively studied the combined effect of irradiation and MAP on the spoilage and pathogenic microorganisms of raw pork (Lambert et al., 1992a, 1992b). The longest shelf life was achieved by a combination of irradiation and anaerobic MA. High levels (45−75%) of CO_2 inhibited, and irradiation delayed toxin production, in samples inoculated with *Clostridium botulinum* (Lambert et al., 1991c).

SAFETY OF POULTRY AND COOKED MAP MEATS

As pointed out above, MAP can greatly reduce spoilage and thus increase shelf life for a variety of fresh and processed products including poultry. This extension of shelf life results from a reduction in the growth of spoilage organisms which would otherwise render the product organoleptically unacceptable. However, inhibition of organoleptic spoilage does not necessarily indicate that the time in which the product can be safely consumed has likewise been extended. Many pathogens do not render poultry or cooked meats organoleptically unacceptable and their growth could actually be promoted by the reduction in competitive spoilage organisms and/or the lengthened storage period under MAP.

This latter concern is pertinent given recent knowledge that some human pathogens continue to grow, albeit slowly, at refrigeration temperatures. Regulatory authorities (U.S. FDA, Kvenberg, 1990), food industry groups (National Food Processors Association, 1988), and others (Hintlian and Hotchkiss, 1986; Tompkin, 1986) have advised caution and called for further research into the safety of refrigerated extended shelf life products.

The major safety concern associated with MAP of poultry products is whether pathogenic microorganisms might increase to significant levels or produce toxin while organisms which would give organoleptic indications of spoilage are suppressed (Hotchkiss et al., 1994). The long storage period of MAP products could allow pathogen growth which would not occur at the shorter storage time of conventionally packaged product. The potential of pathogenic anaerobes such as *C. botulinum* as well as psychrotrophic facultative anaerobes such as *L. monocytogenes* to grow in MAP products has not been fully investigated.

Baker et al. (1986) compared the growth of *Pseudomonas fragi* as an indicator of spoilage to three pathogens or surrogate pathogens (*Clostridium sporogenes*, *S. aureus*, and *S. typhimurium*) in ground chicken held at 2, 7, and 13°C. High levels of CO_2 (%) reduced the growth rate when compared to air for all organisms except *C. sporogenes* which failed to grow under any tested condition. However, greater survival of *C. sporogenes* was seen in the high CO_2 MA compared to air, suggesting that the CO_2 MA had a protective effect on *C. sporogenes*. However, inhibition was greater for *P. fragi* than for the pathogens, raising the possibility that these pathogens could outgrow spoilage organisms if the chicken were held long enough.

The relationship between the growth of spoilage organisms and *L. monocytogenes* has been investigated. This relationship becomes of particular interest because of the ability of this pathogen to grow at refrigeration temperatures and its ability to compete well with other organisms. Wimpfheimer et al. (1990) found that a MA of 72.5% CO_2, 5%O_2, balance N_2 substantially inhibited the aerobic microbial populations in ground raw chicken compared to air, thus increasing shelf life. However, this MA had little effect on *L. monocytogenes*, which grew at the same rate in the MA as in air. Marshall et al. (1991, 1992) similarly compared the growth of *L. monocytogenes* and *P. fluorescens* on cooked chicken stored under aerobic and anaerobic composition MAs. *P. fluorescens* was inhibited to a greater extent than *L. monocytogenes*. *L. monocytogenes* grew in both the aerobic and anaerobic MA.

Additional work on pathogens in MAP poultry has focused on *L. monocytogenes*. Hart et al. (1991) compared aerobic and anaerobic MAs containing 30 or 100% CO_2 levels on the growth of *L. monocytogenes* on raw chicken. MAs containing CO_2 were inhibitory to the growth of *L. monocytogenes* as compared to growth in air, especially at 6°C. Line and Harrison (1992) investigated the thermal death of *L. monocytogenes* in poultry nuggets which were cooked to an internal temperature of 71°C after being sealed in gas-flushed (30% CO_2; 70% N_2) bags. This temperature was sufficient to destroy the pathogen.

There has been less work published on the effects of MAP on *Salmonella* spp., despite the apparent importance of poultry as a vector for this pathogen. Gray et al. (1984) found that 100% CO_2 inhibited the growth of *S. enteritidis* on raw chicken compared to air or vacuum while Baker et al. (1986) inoculated ground chicken with *Salmonella* spp. and stored samples under air and 80% CO_2/20% air at 2, 7, and 13°C. *Salmonella* counts increased in both cases but at a slower rate in the MA as compared to air. Aerobic counts (indicators of spoilage) were generally inhibited to a greater extent than the pathogen, particularly at the lower temperatures. Eklund and Jarmund (1983) found that both vacuum and CO_2 inhibited *Salmonella* growth as compared to growth in air. Langston et al. (1993) compared an aerobic MA containing 75% CO_2 and 5% O_2 to air at 13 and 27°C for both aerobic plate counts and growth of *S. enteritidis*. As in the previous work of Baker et al. (1986), the MA inhibited the aerobic spoilage bacteria but had less effect on *Salmonella* which grew in all atmospheres.

Hintlian and Hotchkiss (1987) studied the effect of MAs on the relative growth of spoilage and pathogenic organisms in cooked beef held at abusive temperatures. *S. aureus* was inhibited by the high CO_2 atmosphere while *S. typhimurium* was less effected. MA containing O_2 reduced the recovery of *Clostridium perfringens*. Eyles et al. (1993) have used a model system consisting of Brain-Heart Infusion agar to extensively investigate the growth of spoilage organisms in MAs. MAs containing CO_2 had little effect on lag phase time but decreased the numbers of organisms attained at the end of exponential growth.

SUMMARY

Our estimate based on information from the packaging materials suppliers is that 10 to 20% of the fresh poultry produced in the U.S. is shipped using some form of MAP which may make poultry the most commonly MAP product. In addition, several processed poultry products also make use of MAP to provide sufficient shelf life for distribution. This represents a very large amount of poultry. Considerable published scientific data as well as commercial experience indicates that MAP extends shelf life when compared to ice packing or other technologies. The degree of extension depends on several factors but extensions of two to four-fold are feasible under strict control. In most commercial applications, the shelf life of fresh poultry is extended from 10–12 d to 18–21 d. The most common technology is to pre-pack carcasses or parts in low-barrier consumer packages which are placed into large groups inside polyolefin bags. The large bags are evacuated and refilled with 100% CO_2. However, the resulting mixture of gases is less than 100% due to incomplete evacuation of the polyolefin bag.

To our knowledge, no food-borne disease outbreak has resulted from this commercial practice. However, not all aspects of safety are understood. Particularly of interest is the effect of spoilage organisms on the growth and/or survival of psychrotrophic human pathogens. It is possible that the extended storage period and the reduced competition from the spoilage microorganisms could allow for slow growth of psychrotrophic organisms.

REFERENCES

Anderson, K. L., Fung, D. Y. C., Cunningham, F. E. and Proctor, V. A. 1985. "Influence of modified atmosphere packaging on microbiology of broiler drumsticks," *Poultry Sci.*, 64(2):420–422.

Bailey, J. S., Reagan, J. O., Carpenter, J. A. and Schuler, G. A. 1979. "Microbiological condition of broilers as influenced by vacuum and carbon dioxide in bulk shipping packs," *J. Food Sci.*, 44(1):134–137.

Baker, R. C., Hotchkiss, J. H. and Qureshi, R. A. 1985. "Elevated carbon dioxide atmospheres for packaging poultry. I. Effects on ground chicken," *Poultry Sci.*, 64:328–332.

Baker, R. C., Qureshi, R. A. and Hotchkiss, J. H. 1986. "Effect of an elevated level of carbon dioxide containing atmosphere on the growth of spoilage and pathogenic bacteria at 2, 7, and 13 C," *Poultry Sci.*, 65:729–737.

Brody, A. L. 1985. "Controlled-atmosphere food packaging/Part 1," *Food and Drug Packaging*, 49(12):8, 10, 11, 43.

Brody, A. L., ed. 1989. *Controlled/Modified Atmosphere/Vacuum Packaging of Foods.* Food and Nutrition Press, Inc., Trumbull, CT, 179 pp.

Carr, T. P. and Marchello, J. A. 1986. "Microbial changes of precooked beef slices as affected by packaging procedure," *J. Food Prot.*, 49(7):534–536.

Coyne, F. P. 1933. "The effect of carbon dioxide on bacterial growth," *Proc. Royal Soc. London Series B*, 113:196–217.

Cunningham, F. E. 1987. Methods of preservation of poultry products, In *The Microbiology of Poultry Meat Products*, F. E. Cunningham and N. A. Cox, eds. Academic Press, Orlando, FL, pp. 275–292.

Dixon, N. M. and Kell, D. B. 1989. "The inhibition by CO_2 of the growth and metabolism of microorganisms," *J. Appl. Bacteriol.*, 67(2):109–136.

Eklund, T. and Jarmund, T. 1983. "Microculture model studies on the effect of various gas atmospheres on microbial growth at different temperatures," *J. Appl. Bacteriol.*, 55:119–125.

Elliott, P. H., Tomlins, R. I. and Gray, R. J. H. 1985. "Control of microbial spoilage on fresh poultry using a combination potassium sorbate-carbon dioxide packaging system," *J. Food Sci.*, 50(5):1360–1363.

Erichsen, I. and Molin, G. 1981. "Microbial flora of normal and high pH beef stored at 4 C in different gas environments," *J. Food Prot.*, 44(11):866–869.

Eyles, M. J., Moir, C. J. and Davey, J. A. 1993. "The effect of modified atmospheres on the growth of psychrotrophic pseudomonads on a surface in a model system," *Int. J. Food Microbiol.*, 20(2):97–107.

Farber, J. M. 1991. "Microbiological aspects of modified atmosphere packaging technology review," *J. Food Prot.*, 54(1):58–70.

Frazier, W. C. and Westhoff, D. C. 1988. *Food Microbiology*, Fourth Edition. McGraw-Hill, New York, 539 pp.

Gardner, F. A., Denton, J. H. and Hartley, S. E. 1977. "Effects of carbon dioxide environments on the shelf life of broiler carcasses," *Poultry Sci.*, 56:1715–1716.

Genigeorgis, C. A. 1985. "Microbial and safety implications of the use of modified atmospheres to extend the storage life of fresh meat and fish," *Int. J. Food Microbiol.*, 1:237–251.

Gill, C. O., Harrison, J. C. L. and Penney, N. 1990. "The storage life of chicken packaged under carbon dioxide," *Int. J. Food Microbiol.*, 11(2):151–158.

Grant, I. R. and Patterson, M. F. 1991. "A numerical taxonomic study of lactic acid bacteria isolated from irradiated pork and chicken packaged under various gas atmospheres," *J. Appl. Bacteriol.*, 70(4):302–307.

Gray, R. J. H., Elliott, P. H. and Tomlins, R. I. 1984. "Control of two major pathogens on fresh poultry using a combination potassium sorbate/carbon dioxide packaging treatment," *J. Food Sci.*, 49(1):142–145.

Guise, B. 1985. "Packaging techniques for fresh meat and poultry," *Converter*, 22(5):28, 30–32.

Haines, R. B. 1933. "The influence of carbon dioxide on the rate of multiplication of certain bacteria as judged by viable counts," *J. Soc. Chem. Ind.*, 52:13.

Hart, C. D., Mead, G. C., and Norris, A. P. 1991. "Effects of gaseous environment and temperature on the storage behavior of *Listeria monocytogenes* on chicken breast meat," *J. Appl. Bacteriol.*, 70(1):40–46.

Hintlian, C. B. and Hotchkiss, J. H. 1986. "The safety of modified atmosphere packaging: A Review," *Food Technol.*, 40(12):70–76.

Hintlian, C. B. and Hotchkiss, J. H. 1987. "Comparative growth of spoilage and pathogenic organisms on modified atmosphere-packaged cooked beef," *J. Food Prot.*, 50(3):218–223.

Hotchkiss, J. H. 1989. Modified atmosphere packaging of poultry and related products, In *Controlled/Modified Atmosphere/Vacuum Packaging of Foods*, A. L. Brody, ed., Food & Nutrition Press, Inc., Trumbull, CT, pp. 39–58.

Hotchkiss, J. H., Baker, R. C. and Qureshi, R. A. 1985. "Elevated CO_2 atmospheres for packaging poultry. II. Effects of chicken quarters and bulk packages," *Poultry Sci.*, 64:333–340.

Hotchkiss, J. H., Hendricks, M. T. and Chen, J. H. 1994. "Modeling the effects of carbon dioxide on spoilage and pathogenic bacteria," *Proceedings of the Food Preservation 2000 Conference*, Natick, MA, October 1993.

Jay, J. M. 1992. *Modern Food Microbiology*, Fourth Edition. Van Nostrand Reinhold, New York, 701 pp.

Jurdi, D., Mast, M. G. and MacNeil, J. H. 1980. "Effects of carbon dioxide and nitrogen atmospheres on the quality of mechanically deboned chicken meat during frozen and non-frozen storage," *J. Food Sci.*, 45:641–644, 666.

Kraft, A. A. 1984. "Modified atmosphere packaging (MAP) microbiology of meats and poultry," *J. Food Prot.*, 47:826.

Kvenberg, J. E. 1990. "Microbiological criteria and regulatory aspects of minimally processed refrigerated foods," *J. Food Prot.*, 53:910.

Labuza, T. P. 1982. *Shelf-Life Dating of Foods*. Food & Nutrition Press, Inc., Westport CT, 500 pp.

Lambert, A. D., Smith, J. P. and Dodds, K. L. 1991a. "Shelf-life extension and microbiological safety of fresh meat: A review," *Food Microbiol.*, 8(4):267–297.

Lambert, A. D., Smith, J. P. and Dodds, K. L. 1991b. "Effect of headspace CO_2 concentration on toxin production by *Clostridium botulinum* in MAP, irradiated fresh pork," *J. Food Prot.*, 54(8):588–592.

Lambert, A. D., Smith, J. P. and Dodds, K. L. 1991c. "Effect of initial O_2 and CO_2 and low-dose irradiation on toxin production by *Clostridium botulinum* in MAP fresh pork," *J. Food Prot.*, 54(12):939–944.

Lambert, A. D., Smith, J. P. and Dodds, K. L. 1992a. "Physical, chemical and sensory changes in irradiated fresh pork packaged in modified atmosphere," *J. Food Sci.*, 57(6):1294–1299.

Lambert, A. D., Smith, J. P., Dodds, K. L. and Charbonneau, R. 1992b. "Microbiological changes and shelf life of MAP, irradiated fresh pork," *Food Microbiol.*, 9(3):231–244.

Langston, S. W., Altman, N. S. and Hotchkiss, J. H. 1993. "Within and between sample comparisons of Gompertz parameters for *Salmonella enteritidis* and aerobic plate counts in chicken stored in air and modified atmosphere," *Int. J. Food Microbiol.*, 18:43–52.

Lawlis, T. and Fuller, S. L. 1990. "Modified atmosphere packaging incorporating an oxygen-barrier shrink film," *Food Technol.*, 44(6):124.

Line, J. E. and Harrison, M. A. 1992. "*Listeria monocytogenes* inactivation in turkey rolls and battered chicken nuggets subjected to simulated commercial cooking," *J. Food Sci.*, 57(2):787–788.

Marshall, D. L., Wiese-Lehigh, P. L., Wells, J. H. and Farr, A. J. 1991. "Comparative growth of *Listeria monocytogenes* and *Pseudomonas fluorescens* on precooked chicken nuggets stored under modified atmospheres," *J. Food Prot.*, 54(11):841–843, 851.

Marshall, D. L., Andrews, L. S. Wells, J. H. and Farr, A. J. 1992. "Influence of modified atmosphere packaging on the competitive growth of *Listeria monocytogenes* and *Pseudomonas fluorescens* on precooked chicken," *Food Microbiol.*, 9(4):303–309.

Mead, G. C., Griffiths, N. M., Grey, T. C. and Adams, B. W. 1986. "The keeping quality of chilled duck portions in modified atmosphere packs," *Lebensm-Wiss. Technol.*, 19(2):117–121.

National Food Processors Association. 1988. "Safety considerations for new generation refrigerated foods," *Dairy and Food Sanitat.*, 8(1):5–7.

Penney, N., Hagyard, C. J. and Bell, R. G. 1993. "Extension of shelf life of chilled sliced roast beef by carbon dioxide packaging," *Int. J. Food Sci. & Technol.*, 28(2):181–191.

Robertson, G. L., ed. 1993. *Food Packaging: Principles and Practice.* Marcel Dekker, Inc., New York, 688 pp.

Sander, E. H. and Soo, H.-M. 1978. "Increasing shelf life by carbon dioxide treatment and low temperature storage of bulk pack fresh chickens packaged in nylon/surlyn film," *J. Food Sci.*, 43:1519–1523, 1527.

Seideman, S. C. and Durland, P. R. 1984. "The utilization of modified gas atmosphere packaging for fresh meat: A Review," *J. Food Qual.*, 6:239–252.

Silliker, J. H. 1981. The influence of atmospheres containing elevated levels of CO_2 (carbon dioxide) on growth of psychrotrophic organisms in meat and poultry (*Pseudomonas* and closely related bacteria), In *Psychrotrophic Microorganisms in Spoilage and Pathogenicity*, T. A. Roberts et al., eds., Academic Press, London, pp. 369–375.

Studer, P., Schmitt, R. E., Gallo, L. and Schmidt-Lorenz, W. 1988. "Microbial spoilage of refrigerated fresh broilers. II. Effect of packaging on microbial association of poultry carcasses," *Lebensmittel-Wissenschaft Technol.*, 21(4):224–228.

Tan, K. H. and Gill, C. O. 1982. "Physiological basis of CO_2 (carbon dioxide) inhibition of a meat spoilage bacterium, *Pseudomonas fluorescens*," *Meat Sci.*, 7(1):9–17.

Thayer, D. W. and Boyd, G. 1991. "Effect of ionizing radiation dose, temperature, and atmosphere on the survival of *Salmonella typhimurium* in sterile, mechanically deboned chicken meat," *Poultry Sci.*, 70(2):381–388.

Tompkin, R. B. 1986. "Microbiological safety of processed meat: New products and processes–New problems and solutions," *Food Technol.*, 40(4):172–176.

Wabeck, C. J., Parmelee, C. E. and Stadelman, W. J. 1968. "Carbon dioxide preservation of fresh poultry," *Poultry Sci.*, 47:468–474.

Wimpfheimer, L., Altman, N. S. and Hotchkiss, J. H. 1990. "Growth of *Listeria monocytogenes* Scott A, serotype 4 and competitive spoilage organisms in raw chicken packaged under modified atmospheres and in air," *Int. J. Food Microbiol.*, 11:205–214.

Wolfe, S. K. 1980. "Use of CO- and CO_2-enriched atmospheres for meats, fish and produce," *Food Technol.*, 34(3):55–58, 63.

Zeitoun, A. A. M. 1991. "Inhibition, survival and growth of *Listeria monocytogenes* on poultry as influenced by buffered lactic acid treatment and modified atmosphere packaging," *Int. J. Food Microbiol.*, 14(2):161–169.

Zeitoun, A. A. M. 1992. "Decontamination with lactic acid/sodium lactate buffer in combination with modified atmosphere packaging effects on the shelf life of fresh poultry," *Int. J. Food Microbiol.*, 16(2):89–98.

Zeitoun, A. A. M. and Debevere, J. M. 1992. "Packaging of fresh poultry: Influence of modified atmosphere on the shelf life of fresh poultry," *Fleischwirtschaft*, 72(12):1686–1688.

CHAPTER 7

Fish and Shellfish Products in Sous Vide and Modified Atmosphere Packs

D. M. GIBSON*—*Ministry of Agriculture Fisheries and Food, U.K.*
H. K. DAVIS**—*Ministry of Agriculture Fisheries and Food, U.K.*

FISH COMPOSITION

PHYLOGENETICALLY, the trivial term *fish* covers a far wider range of genera and species than any other food group. Fish is taken to cover fin fish, elasmobranches such as sharks and rays, crustacean and molluscan shellfish, and tunicates such as squids. The environment from which fish are harvested varies tremendously as well. The temperature of the water these animals live in covers the range 0 to 30°C. Salinity varies from near zero in fresh waters, rising in estuarine waters to 35°C in marine environments. Some species migrate from marine to fresh waters and back again, others from temperate to tropical zones, or from surface waters to depths equivalent to many atmospheres within a day. A single species can be caught or harvested from diverse environmental conditions.

Within a species, even from the same location, the composition of fish varies. Generally, fish are composed of protein (approximately 20%), lipid, water, and, with the exception of shellfish, little carbohydrate. Low molecular weight material accounts for 1–5% of the edible parts of the specimen, higher values being found in shellfish. The composition of selected species is shown in Table 7.1. Variations in fat or lipid content are highest in migratory fish. Much of the lipid consists of polyunsaturated fatty acids, nutritionally regarded as desirable, but which may also be inhibitory to microbes if present in sufficient concentrations. Thus, the composition of a particular species is not constant and can influence their post-mortem bacterial decomposition, and may account for some of the variability in experimental results. As lipids have some considerable influence on flavour, they have a role in the determination of shelf life.

*Current address: BIODON, 43, Brighton Place, Aberdeen AB1 6RT, United Kingdom.
**Current address: Central Science Laboratory, Food Science Laboratory-Torry, P.O. Box 31, 135 Abbey Road, Aberdeen AB9 8DG, United Kingdom.

Table 7.1. Composition of some seafoods (after Murray and Burt, 1977).

Seafood	% Protein	% Water	% Fat
Cod, haddock	15–20	78–84	0.1– 0.9
Pollack	16–20	79	0.6– 0.8
Halibut	18	75–79	0.5– 9.6
Atlantic salmon	22	67–77	0.3–14.0
Pacific salmon	22	67–78	2.7–10.6
Trout	19	70–79	1.2–10.6
Skate	18–24	77–82	0.1– 1.6
King crab	7–14	81–91	0.4– 1.7
Mussels	9–12	80–84	0.8– 2.3
Shrimp	15–19	68–70	0.9

The supply of raw material, both in terms of quantity and of suitable quality, is becoming a problem. James (1992) has reported that world fish production has reached 100×10^6 t/y, of which 70% is used for human consumption. The demand could increase by 30×10^6 t/y in this decade. The main demand comes from the developed countries which have fully exploited their own resources, and continue to increase their imports from developing countries, whose own requirements are also increasing. James believes that the wealthy countries will take an increasing proportion of the world catch. It is mainly in such countries that there is a demand for value added products such as MAP and sous vide fish. There could be an innovative technology and market for sous vide products in developing countries because the processing partially stabilizes the product and only edible portions are transported to market.

Because of their environment, fish have unique osmoregulatory mechanisms to avoid dehydrating in marine environments and, conversely, waterlogging tissues in fresh waters. One major osmoregulant is trimethylamine oxide (TMAO), found at concentrations of up to 1.0% in teleosts, and at up to 1.5% in elasmobranches together with 2% urea (Love, 1980). It occurs in the feed of fish and is pH neutral and non-toxic. It is an N-oxide and also a simple betaine, compounds used osmotically by eucaryotes. As an N-oxide, it can be reduced by gram-negative bacteria to trimethylamine (TMA) in much the same way as nitrate. The important point is that such reductions are bioenergetic to the organism, in that they can obtain energy from them. While TMAO is nonodorous, TMA is a component in the odour of stale fish. It has been shown conclusively that bacteria isolated from fish can carry out this reduction and grow at its expense. The system is constitutive but repressed in the presence of O_2. However, at low O_2 tensions it is derepressed and allows the bacteria to maintain an aerobic lifestyle. The electron donors for TMAO reduction include formate, NADH, lactate, etc., all likely to be present in fish tissue either from autolytic (self digestion) mechanisms or from the activities of other microbes. The behaviour of the reductase system at high (>20%) O_2 levels has not been systematically examined (Barrett and Kwan, 1985). In MAP at O°C, the delay before TMAO is reduced is prolonged due to the effects of dissolved CO_2, as the activity of TMAO reductase decreases with lowered pH. O_2 also exerts an inhibitory effect, repressing the enzymes involved in the reduction of TMAO (Easter et al., 1982) to TMA and of demethylation to dimethylamine (Lundstrom et al., 1982).

As elasmobranch fish contain up to 2.5% urea, if any bacteria possessing the enzyme urease are present, the urea is broken down to ammonia and CO_2, the pH rises rapidly, and TMAO reductase is again inhibited.

In summary, seafoods have some unique properties when compared with other natural proteinaceous foods of similar composition: they have a built-in electron acceptor which is active when O_2 is restricted; a small temperature differential between the temperature of growth of the animal and that of its post-mortem storage; and a variability in composition characteristic of wild stocks.

NORMAL AEROBIC SPOILAGE PATTERN

In view of the species, ecological and environmental diversity, as well as the effects of processing into a multitude of products, the intrinsic bacterial populations of fish fall into but a few distinctive patterns. The first feature is the general division between cold/temperate water and warm water fish. In the former, gram-negative genera predominate, and in the latter, gram-positive. The first step in handling technology after capture or harvest is chilling, usually by the use of ice. Ice, despite its apparent lack of nutrients, harbours gram-negative bacteria and iced fish from temperate waters soon support a population in which TMAO reducers are prominent. Spoilage takes about 16 d. On the other hand, chilled fish from fresh water tropical regions, even in contact with locally produced ice, can take 28 d to spoil, thought to be due to the near absence of TMAO and the consequent lack of amine odours. This is disputed by some authors. For all types of fish, it is the gram-negative organisms which are the predominant cause of fish becoming unpleasant to eat.

The taxonomy and nomenclature of these gram-negative bacteria is comparatively poor because it is based on negative test results due to their lack of reactivity in conventional bacteriological/biochemical tests. They have been little studied by taxonomists using molecular biological techniques, but it is evident that many species have been wrongly named in the literature. As an example, the organism thought to be the principal spoiler of temperate water marine low fat fish was originally named as a *Pseudomonas* spp. It was then marginalized as a *Pseudomonas* spp. group iii/iv, became *Alteromonas putrefaciens*, and is now *Shewanella putrefaciens*, in honour of James M. Shewan, the noted fish microbiologist and taxonomist. These organisms have the ability to reduce TMAO to TMA, and to produce H_2S from the sulphur-containing amino acids. These two properties make them potent spoilers and their metabolic end products organoleptically offensive.

When these active spoilage organisms are eliminated by some kinds of processing, such as irradiation, then organisms formerly named "achromobacters" grow and predominate. They eventually cause spoilage, but the malodours and flavours are less pronounced because their metabolism centers on the tricarboxylic acid cycle with no organoleptically significant end products.

Fish taken from warm fresh waters have a mainly gram-positive bacterial population which can persist during storage. The taxonomy of these bacteria is also uncertain but unrelated to any pathogens. They are variously described as "coryneform/arthrobacter"-like. The gram-negative bacteria present may be *Aeromonas* and *Vibrio* spp., which are usually non-pathogenic but can be difficult to distinguish from some pathogenic members of the same genera (Liston, 1992).

The microorganisms found on fresh fish reflect their environment. Fish are not generally taken from obviously polluted waters and therefore do not harbour pathogens. Shellfish can be taken from waters polluted with sewage, a practice banned by legislation. Pathogens regarded as intrinsic and natural and found on aquatic products include *Clostridium botulinum*, especially non-proteolytic types E and B, *Vibrio parahaemolyticus* and more recently *V. vulnificus*, *Listeria monocytogenes*, and *Aeromonas hydrophila* (Gibson, 1992). Other emerging pathogens such as *Yersinia enterocolitica* have also been reported. Many of the common pathogens such as *Escherichia coli* and *Salmonella* spp.

do not survive well in the marine environment and seem to be outcompeted or eliminated in freshwater environments, despite their ability to gain energy from TMAO reduction.

The gram-negative spoilage bacteria of most fish are heat sensitive. It is well known that in the case of temperate fish, the aerobic bacterial counts obtained after incubation at 37°C are one-tenth those found at 20–25°C. Heat treatments chosen to eliminate pathogens, that is 70°C for 2 min or equivalent, are effective against non-spore formers. However, because fish are poikilothermal animals growing at relatively low temperatures, their proteins are less stable to heat than those of mammals, and fish taken from temperate zones appear to be "cooked" at temperatures as low as 35°C due to opacity caused by protein denaturation. Thus, relative to other proteinaceous foods, there is concern that fish and fish products may be undercooked and that pathogens could survive in apparently cooked products. However, because of the potential presence of *C. botulinum*, it is evident that if pasteurized products are stored at >4°C, there is the risk of toxicity without spoilage. Without sterilization treatments, the risk cannot be eliminated, but a precept of conventional handling and processing of fish and fish products is that the risk of botulism is not increased. Packaged fish and fish products have been extensively examined to ensure that this precept is obeyed and, despite the evident hazards, commercially packaged raw and pasteurized fish have not been associated or implicated in any food poisoning/toxicity incidents.

Another source of intoxication following fish consumption is referred to as histamine poisoning or scombrotoxicity, and there is no evidence that packaging or heat treatment has been a primary factor nor increases the toxicity. Although histamine, which is heat-stable, arises from bacterial decarboxylation of histidine found in significant concentrations in certain fish, it has been shown to be an improbable cause, alone, of scombrotoxicity. However the storage conditions which lead to production of histamine and of scombrotoxicity appear to be much the same. Hence, measurement of histamine content provides a useful measure of potential risk.

Shellfish toxins (paralytic, diarrhoeic, amnesiac, neurological, etc.) are being detected more frequently worldwide. These heat resistant toxins are found on harvesting, and there is controversy as to their true origins, in that there may be a bacterial as well as an algal factor. There are no reports of increases in toxin content during storage of products, whether packaged or not.

Microbial Spoilage in Vacuum-Packed and MAP Products

It is evident from the foregoing that vacuum packaging of raw products will have less inhibitory effects on fish spoilage than certain other commodities. Even with fresh water fish containing low concentrations of TMAO, packaging does not have such a dramatic effect with fish as with other foods, the shelf life extension usually being in the late rather than early stages of spoilage (Davis, 1993).

SENSORY PERCEPTION OF SPOILAGE AND DEFINITION OF SHELF LIFE

Odour, Flavour and Texture

Any discussion on the effects of packaging on shelf life depends on its definition. Perhaps more than with other foods, measurement of shelf life of fish products is difficult and debatable. The criterion adopted as the most reliable for describing and quantifying the state of spoilage of fish is its cooked flavour, which can be expressed in a numerical scale from, say 10 (fresh) to 0 (putrid) (Shewan et al., 1953). For fish stored at 0°C there is, for most of the practical part of the scale, a linear relationship between score and storage time. For processed fish and those not stored in ice, the relationship is reasonable, and still of value for comparing different treatments of the same batch or lot of fish, and for setting rejection standards. The spoilage of fish, shown in Figure 7.1, can be regarded as having four phases. Newly harvested fish have a fresh, often sweet, flavour, with a background flavour characteristic of the species, (which can persist, to some extent, throughout spoilage). Flavours are lost due to biochemical changes brought about by intrinsic enzymes, and by purely chemical reactions yielding a bland product of neutral flavour. Microbes are generally located on the surfaces of fish unless introduced into the flesh by, say, cutting. The process of autolysis also reduces the mechanical and biological barriers to microbial attack. This first phase of spoilage ends what can be termed "high quality shelf life." In the U.K., many supermarket retailers set their product quality rejection point at the end of this stage. In the second phase, as storage time and spoilage progress, bacterial numbers increase and more end products accumulate. Spoilage continues in the third phase when the products become regarded as stale and eventually putrid. Rejection by public health inspectors occurs at the start of the fourth phase as the

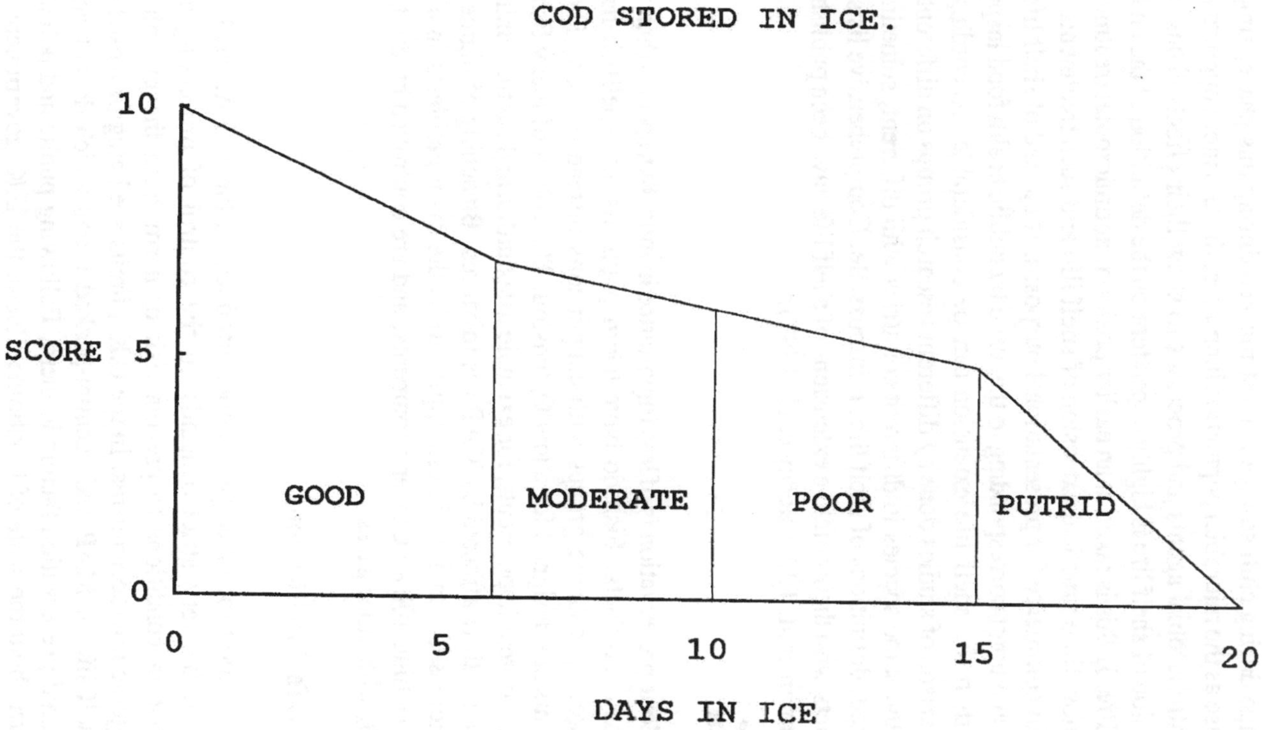

FIGURE 7.1 Sensory quality changes during fish spoilage.

product becomes putrid, but not unsafe. Any inhibition of bacterial spoilage during chill storage is most marked during this phase, simply because as the microbial population increases, there is more opportunity for antimicrobial agents and processes to show their effects. Thus, the extension of shelf life is highly dependent on the definition of the end of shelf life. If this is taken at an earlier position, antimicrobial treatments can show little practical extension of shelf life and such treatments are used in commerce for presentational purposes. If the end of shelf life is taken at a point corresponding to that used by public health food inspectors, then the shelf life extension can be considerable. Accordingly, comparison of studies done by different research groups on different or even the same species in different countries with different, sometimes unstated, definitions of shelf life, is impossible. Comprehensive lists of products with their putative extensions of shelf life have been published (Stammen et al. 1990; Reddy et al.,1992).

Colour

Colour preservation is of less importance in most fish species than for red meat products. Fish do have haem pigments and shellfish have analogous prosthetic groups with copper ions instead of iron. These pigments are designed for a low O_2 tension. The colour of many fish is due to carotenoid pigments, for example astaxanthin and canthaxanthin, which are little affected by MAP gas mixtures. Bleaching of pigments and increased opacity of flesh appear to be the main problems in high (approaching 100%) CO_2 environments, and are controlled by dilution of CO_2 with other gases.

Shelf Life Specifications

The market for packaged seafood products in the U.S.A. has been inhibited by very strict demands on the control of processing and distribution conditions to prevent risk of harm from the growth of pathogenic microorganisms. In the U.K., limits are being proposed on the shelf life of MAP and vacuum-packed foods which do not have additional preservation factors in them. Following public and political concern about the scale of foodborne illness, the U.K. government set up a committee to report on the microbiological safety of foods. It produced two excellent reports (Richmond, 1990; 1991). Subsequently it has published a report on the safety of vacuum packed chilled foods

(Anon., 1992). The conclusions are that the main public health concern from vacuum-packed, MAP and sous vide foods is the possible growth and toxin formation by non-proteolytic strains of *C. botulinum*. Because smoked fish have featured prominently in the small number of cases of botulism, it is recommended that prepared chilled foods, including hot and cold smoked fish, be given a maximum shelf life of 10 d at a maximum temperature of 10°C unless any, or combinations, of the following conditions are met throughout the food. 1) The product has been given a heat treatment of 90°C for 10 min or equivalent (based on a $D_{90°C}$ value of 1.1 min with a z-value of 9). 2) A pH value of 5 or less. 3) A minimum salt level of 3.5% NaCl in the aqueous phase. 4) An a_w of 0.97 or less. 5) A combination of heat and preservative factors or components which can be shown consistently to prevent growth and toxin production by psychrotrophic botulinum strains at temperatures up to 10°C.

The rationale for these recommendations is that although it is known that longer shelf lives are safe at lower temperatures, the group believed that with current industrial and domestic technologies, such temperatures cannot be guaranteed and that, accordingly, temperature alone is not a control for the prevention of botulism. Also, cooking is not regarded as a preventative measure as the time-temperature profiles are not sufficient to destroy preformed toxin (80°C for 10 min or equivalent). Stammen et al. (1990) and Reddy et al. (1992) list the time taken for a variety of fish products stored at set temperatures to become toxic and spoiled. Table 7.2 has been constructed from their sources but only for storage temperatures at or below 10°C. The shelf lives quoted at higher temperatures are generally very short and based on few sensory assessments. It can be seen that there are many instances when the proposed 10 d/10°C shelf life would not ensure safety if products contained 50–100 spores/g, the inoculum used in many studies.

The time-temperature profiles are based on a 6D (a reduction in numbers by a factor of 10^6) process. Hauschild (1989) has tabulated the quantitative incidence of botulinum spores in fish and fish products. Fresh fish contain 3–70 spores/kg and smoked fish 1–24/kg. As many fish products cannot withstand the time-temperature regimes proposed without deleterious effects on their appearance, then this option for extending the shelf life beyond 10 d is not possible. Again, without changing the nature of the product, lowering the pH to below 5 is not practical. For smoked products, salt levels of >3.5% are common, but there is a trend towards lower salt levels both for health reasons and

Table 7.2. Production of Clostridium botulinum *toxin and spoilage of some packaged fish products stored at or below 10° C.*

Product (reference)	Storage Temperature (°C)	Gas Atmosphere	Time to Toxicity (days)	Sensory Quality* When Toxic or Sensory Shelf Life (days)
Rockfish **	7.2	100% CO_2	>29	S*
	4.4	100% CO_2	>29	S
	1.7	100% CO_2	>29	S
Salmon†	4.4	60% CO_2 25% O_2 15% N_2	>57	S
	4.4	Air	>57	6
Salmon‡	10.0	60% CO_2 25% O_2 15% N_2	10	>10
	10.0	90% CO_2 10% air		>10
	10.0	Air	10	4
	5.0	60% CO_2	>21	>10
	5.0	90% CO_2	>21	>10
Salmon§	8.0	Vacuum	6	**
	8.0	100% CO_2	9	**
	8.0	70% CO_2 30% air	12	**
	4.0	Vacuum	15	**
	4.0	100% CO_2	>60	—
	4.0	70% CO_2 30% air	>60	—
Salmon§§	8.0	Vacuum	12	>21
	8.0	100% CO_2	9	12
	8.0	70% CO_2 30% air	9	15
	4.0	Vacuum	>60	48
	4.0	100% CO_2	>60	48
	4.0	70% CO_2 30% air	>60	24
Herring¶	10.0	40% CO_2 30% O_2 30% N_2	7–9	2
	10.0	Vacuum	6–7	2
	10.0	60% CO_2 40% N_2	6–8	2

Table 7.2. (continued).

Product (reference)	Storage Temperature (°C)	Gas Atmosphere	Time to Toxicity (days)	Sensory Quality* When Toxic or Sensory Shelf Life (days)
Smoked mackerel¶	10.0	40% CO_2 30% O_2 30% N_2	>14	6
	10.0	Vacuum	12–14	12
	10.0	60% CO_2 40% N_2	>14	14
Cod¶	10.0	40% CO_2 30% O_2 30% N_2	10–11	3
	10.0	Vacuum	8	3
Cod¶¶	8.0	Air	>10	6
	8.0	Vacuum	20	16
	8.0	100% N_2	17	17
	8.0	100% CO_2	19	23
	8.0	90% CO_2 8% N_2 2% O_2	8	16
	8.0	65% CO_2 31% N_2 4% O_2	9	16
	4.0	100% CO_2	18–21	40
Whiting¶¶	8.0	100% CO_2	20	15
	8.0	100% N_2	17	10
	8.0	Air	>12	4
	8.0	Vacuum	17	10
	8.0	90% CO_2 8% O_2 2% N_2	8	13
	8.0	65% CO_2 31% O_2 4% N_2	5	7
	4.0	100% CO_2	27	15
Flounder #	10.0	100% CO_2	6	A
	8.0	Air	>12	5
	8.0	Vacuum	>21	7
	8.0	100% N_2	>21	4

(continued)

Table 7.2. (continued).

Product (reference)	Storage Temperature (°C)	Gas Atmosphere	Time to Toxicity (days)	Sensory Quality* When Toxic or Sensory Shelf Life (days)
	8.0	100% CO_2	23	10
	4.4	100% CO_2	>21	S
	4.4	70% CO_2 30% air	>21	S
Rockfish**	8.0	Vacuum	12	15
	8.0	100% CO_2	12	21
	8.0	70% CO_2 30% air	9	19
	4.0	Vacuum	21	15
	4.0	100% CO_2	21	21
	4.0	70% CO_2 30%	21	19
Rockfish##	7.2	0–100% CO_2	>29	S

*A, acceptable; S, spoiled.
**Ikawa and Genigeorgis (1986).
†Stier et al. (1981).
‡Eklund (1982).
§Garcia and Genigeorgis (1987).
§§Garcia et al. (1987).
¶Cann et al. (1983).
¶¶Post et al. (1985).
#Lobrera (1990).
##Lindsay (1982).

sensory acceptability, as well as lower weight loss during processing. For fresh fish, such salt levels are impractical. Dried fish products have sufficiently low water activities. It is worth noting that there have been cases of botulism following the consumption of kapchunka, a fermented salted, air-dried uneviscerated white fish product where the water activity/salt/pH combination failed (Slater et al., 1989). The product was not packaged but had probably been stored at >20°C for over 24 h.

If MAP fish packs are stored at temperatures of <10°C the only other pathogen of concern is *L. monocytogenes*. It is readily killed by heat, but there are some cold smoked fish products such as smoked salmon which are processed at 25°C and are not cooked before consumption. It can grow in up to 10% NaCl and so the salt levels found in such products, 4–6%, are not inhibitory. Some countries have set a zero tolerance, i.e., in effect requiring their absence, in ready-to-eat foods and so their survival even without growth is of importance. Kvenberg (1988) has reported that *L. monocytogenes* is found in about 5% of all fish products and natural incidences of 10% are not uncommon (Dillon and Patel, 1992). MAP does not affect survival of the organism.

PROCESSING PARAMETERS

Effect of Gas Atmosphere

Different gas atmospheres have been used for fish products, as is evident from Table 7.2. CO_2 dissolves readily in fish tissue and, while apparently effective in increasing the shelf life, is not used on its own commercially because the amount of gas dissolving from the headspace can cause packs to become concave or even collapse, surfaces of fish products become bleached, and effervescent taste sensations may be detected. Other atmospheric gases are much less soluble in aqueous systems and are used as diluents in combination with CO_2. The most popular mixture for lean fish products is CO_2, O_2 and N_2, often in the ratio 40:30:30. Equal (50%: 50%) ratios of CO_2 and O_2 are also used, and there seems to be little difference in the resulting quality. For fatty fish, O_2 is omitted to reduce the possibility of chemical oxidation of the lipids. For species such as herring, mixtures of CO_2 and N_2 (80%/20%, 60%/40%, respectively) are used. There is no pack collapse and, with a film of low O_2 permeability, little lipid oxidation.

Besides the sensory assessment systems mentioned previously, there are some chemical indices of quality which have been correlated to eating quality (Davis, 1990). TMA values are often used. In unpackaged fish, when due to bacterial metabolism the pH of the fish is >7.0, the TMA is volatile and escapes. In packaged fish, the TMA is trapped and its concentration can be much higher than the values used to mark the end of shelf life of unpackaged fish. For example, the TMA content of MAP fish well within their sensory shelf life can be 100 mg TMA-N/100 g, which is well above the value of 15 mg TMA-N/100 g used by some public health inspectors as corresponding to the end of phase 3 spoilage. Conversely, the inhibitory effects of CO_2 on TMAO can lead to a relative inhibition of amine production which becomes evident when products are stored at 0°C (where CO_2 inhibition is maximal). It may well be that if there is a significant rate of TMAO reduction when fish are packed it continues, but if the system is still repressed, it remains so. Non-volatile indices such as hypoxanthine (Hx) have also been shown to be affected by MAP, giving a different relationship between sensory score and nucleotide degradation products in packaged fish stored at 0°C. The cause appears to be the result of the inhibitory effect of lowered pH on the breakdown of inosine monophosphate (IMP) to inosine and then Hx (Lindsay et al., 1987; Davis, 1990). These observations are particularly important because of the synergistic effects of IMP on flavour. Provided that fish are very fresh when packed, CO_2-MAP affords the opportunity of some extension of shelf life in the early as well as the later microbial stage of decay.

Effect of Storage Temperature

Fish spoilage rates are temperature dependent. Gibson and Ogden (1987) and Davis (1990) have analysed the sensory and bacteriological data from an extensive series of experiments on MAP and vacuum-packed fish products, using the square root relationship of Ratkowsky et al. (1983) as a model. This relationship has been shown to be a better base for modelling microbial growth rates than the Arrhenius equation over the temperature range of interest. The data modelled well, indicating that the effect of storage temperature on spoilage rate was the same as that for bacteria in culture. Gibson (1984) also used a rapid method for bacteriological assay, conductance measurements, and as the results were available in a matter of hours, the models could be used to predict shelf life. The best results were obtained with media containing TMAO (Gibson, 1984). However, Davis (1990) showed that the relative rates for production of TMA, inosine and Hx were all higher for MAP packed (40% CO_2, 30% O_2, 30% N_2), when compared with vacuum-packed fillets. The differences were due to the greater inhibitory effects on the rates of increase in tissue concentrations which can be achieved at 0°C, reflecting the greater solubility of CO_2 at this temperature.

Baker and Genigeorgis (1990) stored MAP fish inoculated with *C. botulinum* spores at different temperatures. They found that, of all the factors analysed, the storage temperature had the greatest effect on lag time before toxigenesis, accounting for 74.6% of the total variability. Their model could also confirm the lag time from experimental data in the literature. Thus, storage temperatures above 4°C, the limit for the growth of *C. botulinum*, are critical in limiting the safe shelf life for MAP fish.

Packaging Film

In trials of packaged fish in the 1960s, it was sometimes found that the shelf life was shorter than that of conventionally stored unpackaged products. These findings were attributed to the high O_2 permeability rates of the thin polyethylene films used. Probably, the spoilage bacteria were able to utilise the O_2 diffusing through the film and the TMAO present, as electron acceptors. Labuza et al. (1992) have determined the relationship between the O_2 permeability of plastic films, the O_2 partial pressures within the pack, and the rate of consumption of O_2 within the pack. When very permeable films are used, the supply of O_2 can exceed the demand of the microbes and so fuel spoilage.

Product:Gas Ratio

There are complex, but undetermined inter-relationships between the composition of the gas mixtures used and the gas-product ratios. A volume ratio of about 2:1 was seen to be the minimum for cod packed under a 40/30/30 $CO_2/O_2/N_2$ mixture (Davis, H.K., unpublished observations). Although MAP can have significant inhibitory effects on spoilage of many fish products, it is more effective as a form of presentation of fish than as a method of preservation of high quality shelf life. Thus, the ratio of product to headspace is chosen more for the appearance and marketing considerations than for technological reasons. During storage, fish lose moisture and the juices or drip accumulate in pack. Some packers place an absorbent pad under the product to absorb the drip and others treat the product, after all cutting is complete, with polyphosphate to inhibit drip formation. When the headspace is too large, the product can move in the pack during handling and reveal the pads. In the case of smoked fish which have a dry surface, the portions can easily move. Some packaging systems incorporate an extra plastic film to hold the products in position.

Packaging Equipment

The range of packaging techniques and equipment is much the same for fish as for meat and many other products. Vacuum skin packaging (VSP) (Day, 1992) offers no advantage over conventional VP other than, perhaps, the higher gloss appearance which is not always regarded as beneficial. VSP of fish has been criticized for presenting potential harm to many consumers who were likely, mistakenly, to perceive from a presentational style which is frequently used for stable non-food products, a greater storage stability and security. The presentational aspect of MAP has led to semi-rigid, thermoformed packs, produced on a variety of continuous forming/packing/sealing/trimming machines, dominating the market for MAP fish. Final products can be stacked neatly and, given relatively large spaces between items of labelling information, the product can be viewed through a smooth top film. A tendency for the top film to be pulled down towards the product as CO_2 is absorbed is often overcome by adjusting the machinery to give a slight excess of gas mixture when packs are sealed. The convex top film then causes some temporary instability when packs are stacked.

There are difficulties in automating placement of fish products into the preformed trays and, as many products are wet and/or rich in lipids, care is needed to avoid smearing the heat sealing edges. The pack floor is usually formed with large dimples which permit greater access of gases around products.

Integrity of the seals is often checked by simply applying pressure manually but is uncertain and can be made more reliable by the use of a water immersion chamber or "exicator" where bubbles can be seen as the pressure falls inside a transparent chamber containing the test pack immersed in water (Hastings, 1993). Other methods are given by Alli (1993).

MAP gases are usually blended on-line and regular checks are needed on the mixtures actually applied to packs. This can be done via random checks on product-filled packs, in which case it needs to be remembered that delays between packing and testing will result in inaccurate readings due to the absorption of CO_2 by the product. If some compartments are left free of product in order to provide dummy packs for testing, preliminary checks are needed to ensure that the gas flushing sequence is adequate to cope with the greater volume being flushed. Test instruments need to be able to measure the proportions of all the component gases and several devices, with varying degrees of portability, are available. They are sufficiently accurate but their operating principles need to be understood by users, often production staff, who may be unaware of problems when they arise. As well as ensuring that gas mixtures are correct, precise testing can be used to detect the early stages of developing mechanical faults in packaging equipment (Davis, H. K., unpublished observations).

Good refrigeration is essential as the added expense of MAP is soon wasted if low product temperatures are not maintained. Ideally, products should be held at 0°C but the practicalities impose limitations, particularly at retail display with the need for illumination and frequent access. Modern display units should be able to maintain air temperatures at between −1°C (at which fish will remain unfrozen) and +3°C. As processing inevitably results in some warming, post-packaging chilling needs cold air to be blown directly over packs which are supported in open-sided racks and which are separated as much as possible.

SHELF LIFE EXTENSION WITH SOUS VIDE PROCESSING

As fish cooks so easily, only short cooking or pasteurization times are used. The fish have to be cut and all bones removed before packaging to avoid punctures. Typical cooking times are 70°C for 40 min, but much shorter times, a few minutes, are known anecdotally. Products are cooled in forced air and/or iced water and refrigerated. Most sous vide products are consumed within a week, but Beauchemin (1990) reports shelf lives of 21–42 d.

The heat treatment will kill all the vegetative microorganisms present except for thermophiles and sporeformers. Mulak (1990) found that *Pseudomonas paucimobilis* survived some time-temperature treatments. Sous vide products do not show typical fish spoilage odours or flavours; the amines and sulphydryl compounds are absent. Instead, the rather bland products develop a mild mustiness, perhaps by chemical rather than bacteriological mechanisms, and it can be difficult to define the end of shelf life or acceptability on sensory grounds. Extensions of shelf life relative to raw fish of >200% are typical. However, the proper limit to shelf life is the possibility of botulinum toxicity. The heat treatments are not sufficient to ensure safety. Experiments in progress (Taylor and Gibson, unpublished) are showing that the botulinum hazard relates to the storage temperature of the products. It is possible for products to be toxic before they are overtly spoiled. The data are being modelled and the predicted safe shelf lives will probably be similar to those calculated from the Baker and Genigeorgis equations. Such cooked foods are likely to support growth and toxigenesis similar to bacteriological media. Thus, because of the risk of fish containing spores of *C. botulinum*, the apparent extension of shelf life as judged organoleptically is unsafe and criteria such as proposed by the U.K. working group described earlier should be adopted.

IMPLICATIONS OF MAP AND SOUS VIDE PROCESSING ON PRODUCT SAFETY

Microbiologists have adopted the experimental approach to determining product safety and have recently taken to the technique of mathematically modelling their own and other researchers' data to predict safety (Gould, 1990). There are now many research groups worldwide who are producing probabilistic and kinetic models to predict the safe shelf life of foods with respect to many pathogens under defined storage conditions. The safety and predicted storage lives of MAP fish have been reported by Baker and Genigeorgis (1990) who carried out 927 experiments yielding 18700 data points on the growth of non-proteolytic *C. botulinum* types B, E and F. The products were packed in 100% CO_2, or 70% CO_2, 30% N_2 and stored for up to 60 d at 4–30°C. They reported the time to toxin detection (lag time). The variables used were storage temperature, the number of spores inoculated and their type, the microbial competition, the gas atmosphere, and the species of fish. They found that the most significant factor effecting lag time was the storage temperature, which accounted for 74.6% of the total variability. They were able to predict the lag time (*LT*) before toxigenesis from the following equation

$$\log(LT) = 0.974 - 0.042(\text{temp}) + (2.741/\text{temp}) - 0.091(\log \text{inoculum}) + 0.035(\log \text{initial APC})$$

Models under development (various, unpublished) also include factors such as % NaCl, pH, presence of antimicrobials, etc. Data from such safety models override sensory shelf life predictions especially for products consumed without further cooking. Simple reheating is not sufficient to ensure toxin destruction. It is expected that in the future most shelf lives will be determined from such models.

CURRENT USE OF MAP AND SOUS VIDE

In the U.K., supermarkets have replaced the traditional high street fishmonger, or specialty store, as the retailers of the largest market share of chilled fish. While some of these outlets have a wet fish counter, most of the sales are of packaged fish, either MAP or vacuum packed. The packs are prepared at processing factories or at specialist packing factories. In contrast, in France, most fish is sold and then packaged for carriage to the home. The use of MAP for fish and fish products is variable but low in many other European countries.

The use of sous vide for fish is difficult to assess. A report by Schellekens and Martens (1993) for the Commission of the European Communities describes the results of a survey of 118 companies throughout Europe. Only 8 companies, regarded as small or medium sized, provided full data. Extrapolating from their responses, the sous vide market is worth \$2.67 × 10^8 per annum, and is forecast to grow by 24% per annum. The products of fifteen companies were listed: 24% of these contain fish or shellfish products, many being complete dishes with other foods, e.g., sauces, vegetables, etc. The relative product volumes were not stated.

OTHER PRESERVATION TECHNOLOGIES

Smoked Fish

Depending mostly on the species, fish are smoked at a low temperature (cold) or a high temperature (hot). Most cold smoked fish are cooked before consumption but some such as smoked salmon are eaten without further heating. During hot smoking, the products are cooked and can be consumed without reheating. The normal spoilage microorganisms are usually destroyed during smoking or cannot grow at the salt concentration of the product. Shelf life is extended if the salt concentration in the water phase is >3.5% NaCl or the a_w <0.97, because these conditions increase the lag phase before botulinum toxigenesis. The time-temperatures achieved during hot smoking should be sufficient to destroy any toxin in the starting material. Thus, in smoked fish, the microorganisms normally present are inhibited and the safety margin is increased and so the shelf life is extended.

Marinades

Because fish have a high buffering capacity and lack fermentable carbohydrate, marinades are made by the addition of acids (and sometimes salts and sugars) to the fish. The pH is generally below pH 5.0. Both spoilage and botulinum toxigenesis are inhibited. Great care is needed as the example of botulism caused by the consumption of kapchunka described earlier has shown. Authorities in the U.S. have banned the production of this product because there have been too many production failures.

Additives

Nitrites are used in meat products to inhibit botulinum toxigenesis. They cannot be used in fish because they react with dimethylamine (DMA), a product of TMAO demethylase, to form dimethylanitrosamines which are regarded as carcinogenic. Nisin is reported to inhibit botulinum toxin production. It is most effective at pH values below that of fish, and is presently of limited applicability. Sorbates are active over a similar pH range. Thus, there are currently no additives of practical or legal use (in the U.K.) for extending the shelf life of MAP fish.

Irradiation

Irradiated fish must be labelled as such and need protection from contamination, so packaging has been a prerequisite for such products. Only low levels of irradiation, either from Co^{60} or electromagnetic field sources, can be tolerated in fish because otherwise off-flavours are noticeable. Irradiation also increases the amount of drip accumulating during storage. In effect practical levels of irradiation, of < 100 kGy, can reduce the total bacterial population by a factor of 1000 and have no effect on botulinum spores. Thus any shelf life extension is dependent for safety on chill temperature storage.

REFERENCES

Alli, I. 1993. Quality control of MAP products, In *Principles and Applications of Modified Atmosphere Packaging of Food*, R. T. Parry, ed., Blackie, London, pp. 101–112.

Anon. 1992. *Report on Vacuum Packaging and Associated Processes*. HMSO, London.

Baker, D. A. and Genigeorgis, C. 1990. "Predicting the safe storage of fresh fish under modified atmospheres with respect to *Clostridium botulinum* toxigenisis by modelling length of the lag phase," *J. Food Prot.*, 53:131–140.

Barrett, E. L. and Kwan, H. S. 1985. "Bacterial reduction of trimethylamine oxide," *Ann. Rev. Microbiol.*, 39:131–149.

Beachemin, M. 1990. Sous vide technology, In *Advances in Fisheries Technology and Biotechnology for Increased Profitability*, M. N. Voight, ed., Technomic, Lancaster, PA, pp. 169–188.

Cann, D. C., Smith, G. L. and Houston, N. C. 1983. "Further studies on marine fish stored under modified atmosphere packaging," MAFF, Aberdeen.

Davis, H. K. 1990. "Some effects of modified atmosphere packaging gases on fish and chemical tests for spoilage," International Institute of Refrigeration, C2, pp. 201–207.

Davis, H. K. 1993. Modified atmosphere packaging of fish, In *Principles and Applications of Modified Atmosphere Packaging of Food*, R. T. Parry, ed., Blackie, London, pp. 189–228.

Day, B. P. F. 1992. Chilled food packaging, In *Chilled Foods: A Comprehensive Guide*. C. Dennis and M. Stringer, eds., Chicester: Ellis Horwood, pp. 147–163.

Dillon, R. M. and Patel, T. R. 1992. "*Listeria* in seafoods: A review," *J. Food Prot.*, 55:1009–1015.

Easter, M. C., Gibson, D. M. and Ward, F. B. 1982. "Induction and location of trimethylamine-*N*-oxide reductase in *Alteromonas* sp. NCMB 400," *J. Gen. Microbiol.*, 129:3689–3696.

Eklund, M. W. 1982. "Significance of *Clostridium botulinum* in fishery products preserved short of sterilization," *Food Technol.*, 36:107–112.

Garcia, G. W. and Genigeorgis, C. 1987. "Quantitative evaluation of *Clostridium botulinum* non-proteolytic types B, E and F growth risk in fresh salmon homogenates stored under modified atmospheres," *J. Food Prot.*, 50:390–397.

Garcia, G. W., Genigeorgis, C. and Lindroth, S. 1987. "Risk of growth and toxin production by *Clostridium botulinum* non-proteolytic types B, E and F in salmon fillets stored under modified atmospheres at low and abuse temperatures," *J. Food Protect.*, 50:330–336.

Gibson, D. M. 1984. "Predicting the shelf life of packaged fish from conductance measurements," *J. Appl. Bacteriol.*, 58:465–470.

Gibson, D. M. 1992. Pathogenic microorganisms of importance in seafood, In *Quality Assurance in the Fish Industry*, H. H. Huss, M. Jakobsen, and J. Liston, eds., Elsevier, Amsterdam, pp. 197–209.

Gibson, D. M. and Ogden, I. D. 1987. Estimating the shelf life of packaged fish, In *Seafood Quality Determination*, D. E. Kramer, and J. Liston, eds., Elsevier, Amsterdam, pp. 437–451.

Gould, G. W. 1990. Modelling for shelf life and safety, In *Processing and Quality of Foods*, Vol. 3, P. Zeuthen, ed., Elsevier, London, pp. 41–60.

Hastings, M. J. 1993. Packaging machinery, In Principles and Applications of Modified Atmosphere Packaging of Food, R. T. Parry, ed. Blackie, London, pp. 41–62.

Hauschild, A. H. W. 1989. *Clostridium botulinum*, In *Foodborne Bacterial Pathogens*, M. P. Doyle, ed. Marcel Dekker, New York, pp. 111–189.

Ikawa, J. K. and Genigeorgis, C. 1986. "Probability of growth and toxin production by non-proteolytic *Clostridium botulinum* in rockfish stored under modified atmospheres," *Int. J. Food Microbiol.*, 4:167–181.

James, D. 1992. Seafood technology in the 90s: The needs of developing countries, In *Seafood Science and Technology*, E. G. Bligh, ed., Fishing News Books, Oxford, pp. 12–23.

Kvenberg, J. E. 1988. "Outbreaks of listeriosis/*Listeria*-contaminated foods." *Microbiol. Sci.*, 5:355–358.

Labuza. T. P., Fu, B. and Taoukis, P. S. 1992. "Prediction for shelf life and safety of minimally processed CAP/MAP chilled foods: A review," *J. Food Prot.*, 55:741–750.

Lindsay, R. C., Josephson, D. B. and Olafsdottir, G. 1987. Chemical and biochemical indices for assessing the quality of fish packaged in controlled atmospheres, In *Seafood Quality Determination*, D. E. Kramer, and J. Liston, eds., Elsevier, Amsterdam, pp. 221–234.

Liston, J. 1992. Bacterial spoilage of seafood, In *Quality Assurance in the Fish Industry*, H. H. Huss, M. Jacobsen, and J. Liston, eds., Elsevier, Amsterdam, pp. 93–105.

Lobrera, A. T. Quoted by Stammen et al., 1990.

Love, R. M. 1980. "*The Chemical Biology of Fishes*" Academic Press, London.

Lundstrom, R. C., Correia, F. F. and Wilhelm, K. A. 1982. "Dimethylamine production in fresh hake (*Urophycis chuss*): The effect of packaging material, oxygen permeability and cellular damage," *J. Food Biochem.*, 6:229–241.

Mulak, V. M. 1990. La cuisson sous vide de preparations a base de produits de la mer; aspects microbiologiques. Ph. D. thesis, University of Lille, France.

Murray, J. and Burt, J. R. 1977. *Composition of Fish.* Torry Research Station Advisory Note, HMSO London.

Post, L. S., Lee, D. A., Solberg, M., Furgang, D., Specchio, J. and Graham, C. 1985. "Development of botulinum toxin and sensory deterioration during storage of vacuum and modified atmosphere packaged fish fillets." *J. Food Sci.*, 50:990–996.

Ratkowsky, D. A., Lowry, R. K., McMeekin, T. A., Stokes, A. N. and Chandler, R. E. 1983. "Model for bacterial culture growth rate throughout the entire biokinetic temperature range," *J. Bacteriol.*, 154:1222–1226.

Reddy, N. R., Armstrong, D. J., Rhodehamel, E. J. and Kautter, D. A. 1992. "Shelf life extensions and safety considerations about fishery products packaged under modified atmospheres: A review," *J. Food Safety*, 12:87–118.

Richmond, M. 1990. "*The Microbiological Safety of Food: Part 1.* HMSO, London.

Richmond, M. 1991. *The Microbiological Safety of Food. Part 2.* HMSO, London.

Schellekens, W. and Martens, T. 1993. "Sous-vide" cooking Part 2, Feedback from practice. Publication No. EUR 15018 EN. Commission of the European Communities, Luxembourg.

Shewan, J. M., Macintosh, R. G., Tucker, C. G. and Ehrenberg, A. S. C. 1953. "The development of a numerical scoring system for the sensory assessment of the spoilage of wet fish stored in ice," *J. Sci. Fd. Agric.*, 4:283–298.

Slater, P. E., Addiss, D. G., Cohen, A., Leventhal, A., Chassis, G., Zehavi, H, Bashari, A. and Costin, C. 1989. "Foodborne botulism: an international outbreak," *Int. J. Epidemiol.*, 18:693–696.

Stammen, K., Gerdes, D. and F. Caporaso. 1990. "Modified atmosphere packaging of seafood," *Crit. Rev. Food Sci. Nutr.*, 29:301–331.

Stier, R. F., Bell, L., Ito, K. A., Shafer, B. D., Brown, L. A., Seeger, M. L., Allen, B. H., Porcuna, M. N. and Lerke, P. A. 1981. "Effect of modified atmosphere storage on *C. botulinum* toxigenisis and the spoilage microflora of salmon fillets," *J. Food Sci.*, 46:1639–1642.

CHAPTER 8

Principles and Practice of Modified Atmosphere Packaging of Horticultural Commodities

DEVON ZAGORY—*Postharvest Technology Consultants, U.S.A.*

CHANGING EXPECTATIONS AND CHANGING MARKETS

CHANGES in the technology and practice of packaging fresh horticultural commodities has been driven by a rapidly evolving confluence of changes in consumer expectations, globalization of produce markets, and technological advances that are resulting in the rapid development of ever more sophisticated ways to maintain produce in a "fresh" condition for extended periods of time. As the market for fresh fruits and vegetables has become global, consumers have come to expect and to demand safe, clean, high quality fresh produce every day of the year. Surveys of consumers have repeatedly shown that "freshness" of produce is one of the most important factors in selection of where people shop and of the produce that they purchase (GPMC, 1988; Zind, 1989). This requires that products be kept "fresh" throughout often lengthy shipping and distribution cycles. At the same time, many of the chemicals that have been used to reduce defects, suppress disease and extend shelf life have come under suspicion by the public and increased scrutiny by various regulatory agencies. These changed expectations have required innovative approaches to packaging. In order to meet the challenges of these changing conditions, the produce industry is looking increasingly to sophisticated plastic film packaging to ensure consistent high quality with ever increasing shelf life.

Fresh produce is not a single commodity. There are hundreds of different fruits and vegetables, each with its own particular requirements for package performance. The produce business has traditionally been a high volume, low margin, commodity business. Therefore, new packages have had to be inexpensive or produce managers would not buy them. However, this is changing and value-added packaging is gaining ground. Minimally processed, precut produce and other value-added presentations of fresh fruits and vegetables are in increasing demand by individuals and, especially, by restaurants and other foodservice operations. Packages that can increase shelf life, maintain quality, protect from injury, ensure cleanliness, reduce disease, carry a recognized brand label, and attractively present fresh fruits and vegetables can help satisfy these newly evolving demands of the marketplace. For these reasons, many produce packers and processors have renewed their interest in modified atmosphere packaging (MAP) to help them anticipate the demands of the market.

Modified atmosphere packaging is a technology appropriate to the needs of the produce industry, particularly the precut or minimally processed industry which is the fastest growing sector. *The Packer*, a weekly trade magazine of the produce industry, reported on December 28, 1991 that the fresh produce processing industry had doubled in size in 1991, accounting for sales of $4 billion. Most of this product is packaged in plastic films. How fresh produce benefits from MAP, how MAP works, and how MAP is currently being used in the North American fresh produce industry are the subjects that will occupy the balance of this chapter. Mention of companies and products is not meant as an endorsement of those products. Neither is this review meant to be a comprehensive list of products available, but rather a sample of some of the major products and players at this time.

It is important to keep in mind that MAP seeks to delay deterioration of foods that are not sterile, and whose enzymatic systems are still operative. Thus, we are not discussing food preservation here, but merely methods to maintain foods in a "natural" condition while slowing specific deteriorative processes. For this reason, rigorous temperature control is absolutely essential for the technologies to work properly. The capabilities of the distribution system will define the degree to which extended shelf life packaging can be used. These technologies have fit very well into the vertically integrated, geographically restricted distribution systems in Western Europe. They are often more difficult to implement in the more fragmented and sprawling distribution systems of North America.

QUALITY AND THE MAINTENANCE OF QUALITY

The quality of fruits and vegetables is comprised of several different elements. Flavour, nutrition, texture, aroma, appearance, and safety are all important elements of produce quality and when any one of these parameters falls below a certain desirable level, the quality of the product is compromised. When the quality falls significantly, the end of product shelf life results. This point will be different for different kinds of produce and will be different for produce bound for different markets. For example, size and appearance requirements are often different for fruit bound for the fresh market and fruit bound for processing.

Components of Quality in Fresh Fruits and Vegetables

There are many different components to produce quality (Table 8.1). The relative importance of each component will depend upon the particular commodity and its intended use. An important aspect of produce quality is the "freshness" of the product. Freshness is due to a combination of properties that is difficult to define precisely but is typified by the quality of a fruit or vegetable when it is freshly harvested. "Fresh" fruits and vegetables are expected to be crisp, not tough, sweet (where appropriate), juicy, nutritious, and free from defects. The challenge of produce marketing is to maintain these properties of freshness during long transportation and marketing periods. The maintenance of crispness and juiciness will depend on the prevention of moisture loss. The maintenance of sweetness will depend on reducing the rate of sugar metabolism. The maintenance of nutrition is complex but often involves the prevention of oxidation and water loss. The prevention of defects will involve protection from mechanical damage and avoidance of conditions that encourage the development of decay.

Packaging Cannot Hide Poor Quality

In most cases, the highest quality is attained at the moment that the fruit or vegetable is harvested. When the fruit or vegetable is separated from the plant (or the soil), it loses its source of water, of sugar and of nutrition. After harvest, quality will decline. MAP can reduce the rate of quality loss, but it cannot reverse the process. Because MAP creates a value-added product, only the highest quality produce should be packed in MAP. It makes no sense to invest in expensive packaging for poor quality product. You cannot add substantial value through packaging to a product that lacks inherent value. Start with high quality and packaging can help maintain that quality. Start with low quality, and you will be selling low quality.

Table 8.1. Quality components of fresh fruits and vegetables. *

Main Factors	Components
Appearance (visual)	Size: dimensions, weight, volume
	Shape and form: diameter/depth ratio, smoothness, compactness, uniformity
	Colour: uniformity, intensity
	Gloss: nature of surface wax
	Defects: external, internal
	Morphological
	Physical and mechanical
	Physiological
	Pathological
	Entomological
Texture (feel)	Firmness, hardness, softness
	Crispness
	Succulence, juiciness
	Mealiness, grittiness
	Toughness, fibrousness
Flavour (taste and smell)	Sweetness
	Sourness (acidity)
	Astringency
	Bitterness
	Aroma (volatile components)
	Off-flavours and off-odours
Nutritive Value	Carbohydrates (including dietary fiber)
	Proteins
	Lipids
	Vitamins
	Minerals
Safety	Naturally occurring toxicants
	Contaminants (chemical residues, heavy metals)
	Mycotoxins
	Microbial contamination

*From Kader, 1992, with permission.

Effects of MAP on Produce Quality Components

There is extensive literature documenting the effects of MAP and of controlled atmospheres on various aspects of produce quality. This literature has been recently reviewed (Zagory and Kader, 1989).

Appropriate atmospheres have been shown to delay loss of chlorophyll and, in some cases, the loss of orange carotenoid pigments in some fruits and vegetables. Elevated CO_2 and/or reduced O_2 reduces chlorophyll loss in many fruits and vegetables (Weichmann, 1986; Yang and Henze, 1988). Several hypotheses have been advanced to account for this reduction in chlorophyll breakdown including increased cell pH in elevated CO_2 environments leading to reduced conversion of chlorophyll to colourless pheophytin, and reduced production of chlorophyllase as a result of reduced ethylene synthesis (Martens and Baardseth, 1987). Ethylene can directly accelerate chlorophyll breakdown and the reduced sensitivity to ethylene of plant tissues in elevated CO_2 may also play a role in the maintenance of chlorophyll by modified atmospheres.

Carotenoids are fat-soluble pigments comprised of isoprene units and, in plants, generally associated with membranes. The carotenoids most important in imparting colour to fruits and vegetables are derivatives of α- and β-carotenes and lycopene. Carotenes are important to nutrition, flavour and appearance as precursors of vitamin A, precursors of flavour volatiles and as pigments (Watada, 1987). Carotenoids vary in their stability but, due to their unsaturated nature, they are generally susceptible to oxidation. Low O_2 generally delays or inhibits the synthesis of lycopene, β-carotene, and xanthophylls in tomato fruit (Salunkhe and Wu, 1973; Goodenough and Thomas, 1980). In sweet pepper, high CO_2 delayed development of red colour equally whether combined with 21% or 3% O_2 (Wang, 1977). Ethylene is known to accelerate the biosynthesis of carotenoids. Elevated CO_2, which reduces ethylene sensitivity, may antagonize this effect. Lipoxygenase appears to catalyze the direct oxidation of fatty acids with the concurrent bleaching of carotenoids (Eskin, 1979). Carotenoids are also sensitive to nonenzymatic oxidation with concurrent loss of colour. Carotenoids can lose their colour through bleaching after loss of moisture in the presence of O_2 (Mackinney et al., 1958; Chou and Breene, 1972). The low O_2 high moisture conditions common in MAP may alleviate some of these changes.

Modified atmospheres can reduce oxidative browning and discolouration and can slow softening of many fruits. Phenolic compounds are important contributors to colour and flavour (Van Buren, 1970). Phenolic compounds, particularly flavonoids and derivatives of chlorogenic acid, play a role in the development of a number of postharvest disorders through their oxidation to brown compounds that discolour many fruits and vegetables and substantially reduce their quality. The activities of these compounds and the severity of the disorders they are involved in are affected by modified atmospheres. CO_2 can influence phenolic metabolism in a number of ways (Siriphanich, 1984). Low, but not high, concentrations of CO_2 competitively inhibited polyphenol oxidase (PPO) and browning of mushrooms (Murr and Morris, 1974). Ethylene-induced browning in pea seedlings was inhibited by CO_2 concentrations above 5% (Hyodo and Yang, 1971). Twenty per cent CO_2 inhibited browning of mechanically damaged snap-beans and reduced activity of PPO, while 15% CO_2 caused brown stain to develop in lettuce, but symptoms developed only after the removal of the CO_2 (Siriphanich and Kader, 1985). Fifteen percent CO_2 was associated with an increase in phenylalanine ammonia lyase (PAL) activity in lettuce, but this may have been a stress reaction and not the primary cause of browning (Siriphanich and Kader, 1985). CO_2 was found to suppress the production of phenolic compounds, but CO_2 injury scores did not correlate well with total phenolic content (Siriphanich and Kader, 1985).

Because MAP is usually created within a plastic package that is relatively impermeable to moisture, MAP generally reduces moisture loss and thus maintains crispness and juiciness. MAP can slow metabolic processes (Kader et al., 1989) and slow loss of sweetness and flavour (Zagory and Kader, 1989). Elevated CO_2 and reduced O_2 can prevent the growth of microorganisms and reduce decay. Finally, MAP has been shown, in some cases, to maintain higher levels of vitamin C (ascorbic acid) (Zagory and Kader, 1989).

Temperature Management

The loss of produce quality after harvest is generally due to the progression of normal metabolic processes that result in the consumption of carbohydrate reserves, loss of moisture, production of heat, oxidation of vitamins and destruction of cell membranes. The rate of cell metabolism, and thus the rate of all of these processes, is strongly dependent on temperature. For this reason, the single most important factor in maintaining product quality is the rigorous control of product temperature. Generally, it is desirable to maintain temperature as close as possible to 0°C (32°F) throughout distribution. Though this is rarely possible in practice, it is the ideal. Exceptions to this would be those fruits and vegetables of tropical or subtropical origin that may be chilling sensitive. Such commodities as bananas, mangoes, tomatoes, eggplant, cucumbers and others will be damaged by exposure to temperatures below about 10° – 12°C (50° – 55°F) for extended periods of time. The symptoms of chilling injury can include failure to ripen properly, sunken epidermal lesions, and increased susceptibility to decay organisms. The severity of chilling injury is dependent on time, temperature, maturity and inherent sensitivity to chilling temperatures. However, for all fruits and vegetables, the best way to maintain quality is to keep them at the lowest possible temperature without causing low temperature damage. After temperature control has been optimized, it may then be worthwhile to use MAP to further retard quality deterioration and extend shelf life.

RESPIRATION

Fruits and vegetables differ from most of the commodities discussed in this book because they are still alive up to the time of consumption. Because they are alive, they continue to breathe, taking in O_2 and producing CO_2, water vapour and heat. The CO_2, water vapour and heat come from the oxidation of carbohydrate (usually sugars) or other respiratory substrates such as organic acids or lipids. There are several important consequences of respiration that must be considered when handling fresh fruits and vegetables (and ornamental plants, which will not be treated further here). Firstly, living plant tissues must have access to adequate O_2 to satisfy their respiratory requirements. If insufficient O_2 is available, anaerobic metabolism will result with the consequent production of off flavours, off odours, and metabolic damage to the tissues. If O_2 is withheld for long periods of time, the plant tissues will die. Secondly, respiration uses up the storage products of the tissues, usually sugars or starch. This results in a progressive reduction in sweetness and often undesirable changes in texture. Thirdly, respiration produces water vapour which, if it accumulates as free water, provides favourable conditions for the growth of decay-causing microorganisms. Finally, respiration produces heat as a byproduct. The production of respiratory heat can defeat our best efforts to maintain proper temperature control during distribution. It is worth reiterating that the rate of respiration is strongly dependent on temperature. Respiration rate increases by two to three times (or more) for each temperature increase of 10°C. Deterioration of the product occurs much faster at higher temperatures. Respiration rate provides a good reflection of the metabolic activity of the tissues. Thus, respiration rate is a reasonably good gauge of the potential storage life of the commodity (Haard, 1985; Ryall and Lipton, 1972). The higher the respiration rate, the shorter the life. Because respiration releases the energy necessary for the metabolic processes that support life, some respiration is necessary. The challenge in handling fresh produce is to reduce respiration to a minimum without harming the plant tissues.

SHELF LIFE EXTENSION BY MAP

MAP is able to slow quality loss and extend shelf life of fresh fruits and vegetables through suppression of a number of the processes that would otherwise deplete the tissues of their components or hasten ripening and subsequent senescence. MAP is able to have such a diversity of effects on plant tissues due to the differential effects of low O_2 and elevated CO_2, suppression of the effects of the ripening hormone ethylene (C_2H_4), and reduction of moisture loss due to the moisture barrier properties of the plastic film.

The Importance of O_2

Because O_2 acts as the terminal electron acceptor in many metabolic reactions, the rates of some essential metabolic processes are sensitive to O_2 concentration. Reducing O_2 concentrations below about 10% around many fresh fruits and vegetables slows their respiration rate and indirectly slows the rates at which they ripen, age and decay. Reducing the O_2 concentration can, in some cases, reduce oxidative browning reactions which can be of particular concern in precut leafy vegetables. Reduced O_2 can delay compositional changes such as fruit softening, pigment development, toughening of some vegetables (such as asparagus and broccoli), and development of flavour (Kader, 1986). Finally, there is a great deal of interest in the use of low O_2 as a quarantine treatment to disinfest fresh produce of insects and insect larvae. Proper combinations of low O_2, low temperature and time may be effective against some of the most troublesome insect pests of concern in international commerce (Ke and Kader, 1992).

However, O_2 is required for normal metabolism to proceed. O_2 concentrations below about 1–2% can lead to anaerobic metabolism and associated production of ethanol and acetaldehyde resulting in off flavours, off odours and loss of quality. Of even greater concern is the potential growth of anaerobic bacteria, some of which are pathogenic to humans, under low O_2 conditions. The proper O_2 concentration will depend upon the fruit or vegetable and its tolerance to low O_2, the temperature (which will affect the product's tolerance to low O_2), and the time that the product will be exposed to low O_2.

The Importance of CO_2

CO_2 occurs in small amounts in air (~0.03%) but, at elevated levels, has important metabolic effects on both produce and microorganisms.

At concentrations above 1–2%, CO_2 reduces the sensitivity of plant tissues to the ripening hormone ethylene. Ethylene can cause premature ripening, fruit softening, yellowing of leafy vegetables, increased respiration rate and senescence of many fruits and vegetables.

Elevated CO_2 can, like reduced O_2, slow respiratory processes thereby extending shelf life. Although the effects of elevated CO_2 on respiration are not as dramatic as those of low O_2, high CO_2 and low O_2 together can, in some cases, reduce respiration more than either gas alone (Kader et al., 1989).

CO_2 at relatively high concentration (>10%) has been shown to suppress the growth of a number of decay-causing fungi and bacteria. For example, 15–20% CO_2 is routinely applied around strawberries during shipment primarily to suppress growth of the mold *Botrytis cinerea*, which would otherwise greatly reduce the postharvest life of strawberries. However, these levels of CO_2 do not suppress some human pathogenic bacteria of potential concern on fresh produce. For example, *Clostridium botulinum* and *Listeria monocytogenes* are relatively resistant to the effects of CO_2 (Farber, 1991). There is some concern that elevated CO_2 could suppress spoilage microorganisms that would otherwise signal microbial growth and product spoilage, while allowing potentially hazardous pathogens to continue to grow. For this reason, MAP should always work in conjunction with an excellent program of sanitation and quality assurance.

The Importance of Ethylene

Ethylene is a colourless, odourless, tasteless gas that has many effects on plant physiology and is active in such small amounts (parts per million) that it is considered a plant hormone. Ethylene has many effects on plant tissues, with some of the most important effects including induction of rapid ripening, and overripening in many fruits, and premature yellowing in many vegetables. The prevention of these ethylene effects is important in the maintenance of quality attributed to MAP. With some fruits, such as bananas and tomatoes, ethylene is routinely applied as a postharvest treatment to ensure rapid and uniform ripening. With many other fruits, vegetables and ornamental plants, it is important to prevent exposure to ethylene and its subsequent deleterious effects.

Ethylene is normally produced by many kinds of ripening fruit. In addition, ethylene is produced by any aerobic combustion such as fires, auto exhaust or diesel and propane forklifts. Such sources should be eliminated from areas where produce is stored or handled. Elevated CO_2 ($> \sim 2\%$) can help reduce the damaging effects of ethylene by rendering plant tissues insensitive to ethylene (Herner, 1987; Kader et al., 1989). This may be one of the primary benefits of modified atmospheres for many commodities.

Microbiological Concerns

There are three important reasons for concern about microorganisms related to the use of MAP. Firstly, high populations of bacteria and/or fungi result in unacceptable deterioration of quality and signal the end of shelf life. Secondly, the longer shelf lives possible with MAP may provide sufficient time for dangerous human pathogens to reproduce that would not have been able to reach dangerous population levels in shorter time periods. Thirdly, the altered atmospheres within MAP may differentially suppress spoilage microorganisms that would organoleptically signal the end of shelf life, while favouring some pathogens that do not alter the organoleptic properties of the product and thus are not likely to be detected by the consumer. Shelf life concerns have driven efforts to improve sanitation of raw product and within processing facilities. It is axiomatic that product with a low initial microbial load will last longer and be of better quality than equivalent product with a higher initial microbial load. However, human health concerns are now driving the fresh cut industry to rigorously reevaluate their sanitation programs.

The ability of elevated CO_2 and reduced O_2 to retard the microbiological spoilage of fresh cut produce is one of the most compelling reasons for the use of MAP. However, the desired suppression of spoilage microorganism, mostly bacteria in the genus *Pseudomonas* (Hotchkiss and Banco, 1992), may create opportunities for slower growing, potentially dangerous microorganisms. Although, in general, produce has not previously been considered a major vector for foodborne disease, there is concern with MAP as to whether or not spoilage (inedibility) occurs before or after foodborne pathogens reach potentially hazardous levels (Hotchkiss and Banco, 1992). Fresh, whole produce is not normally associated with foodborne disease. Microorganisms capable of causing human disease are animal pathogens, not plant pathogens. As such, they normally lack the ability to penetrate plant defenses, such as the plant cell wall. However, when fruits and vegetables are minimally processed by peeling, cutting, shredding, etc., the nutrients required by microorganisms are released, providing conditions conducive to microbial growth.

There is particular concern about *L. monocytogenes* because of the following: 1) it is found in agricultural soils and has been found on several types of fresh produce including lettuce and cabbage; 2) it can survive for very long periods of time both in the environment and in foods; 3) it can grow at low temperatures; 4) it can cause illness and death in infants, pregnant women and immunocompromised individuals; and 5) it does not cause organoleptic changes in the produce that would indicate its presence to the consumer. In air storage, *L. monocytogenes* grows more slowly on foods at refrigeration temperatures than spoilage bacteria and has not been a cause of concern because other microorganisms were likely to cause spoilage before significant growth of *L. monocytogenes* had occurred. In MAP, if other microorganisms were suppressed, *L. monocytogenes* could grow and potentially cause disease. There is evidence that MAP atmospheres usually used for produce do not suppress the growth of *L. monocytogenes* (Berrang et al., 1989). There is also evidence that *L. monocytogenes* is more prevalent in prepared shredded lettuce and cabbage salads than in other types of fruits and vegetables (Hotchkiss and Banco, 1992). These are precisely the vegetables most likely to be packaged in MAP. Furthermore, *L. monocytogenes* is capable of growing to high levels in lettuce, cabbage and other vegetables, given sufficient time (Berrang et al., 1989; Beuchat and Brackett, 1990a; Steinbruegge et al., 1988). Carrots appear to be inhibitory to *L. monocytogenes* (Beuchat and Brackett, 1990b). In inoculation studies by Beuchat and Brackett (1991), whole and chopped tomatoes were inoculated with *L. monocytogenes* and sampled over time. The bacteria grew slowly in whole tomatoes and died off slowly in chopped tomatoes stored at 21°C. Although tomatoes would not normally be expected to support growth of this pathogen, growth might occur given sufficient time (Beuchat and Brackett, 1990a). The extended shelf life conferred by MAP could provide the necessary time.

Also of concern are other pathogenic bacteria such as *Salmonella* spp., *Shigella* spp., and some strains of enteric bacteria as well as enteric viruses (Brackett, 1992). All have been linked to outbreaks of foodborne illness associated with consumption of fresh produce [*Salmonella* on melons (CDC, 1979, 1991; Gayler et al., 1955; Ries et al., 1990), *Shigella sonnei* on shredded lettuce (Davis et al., 1988; Martin et al., 1986), *Klebsiella* spp. on various plant products (Madden, 1992), and viruses on vegetables (Konowalchuk and Spiers, 1975)]. These pathogens are not commonly found in agricultural environments but can survive on produce subsequent to contact with infected workers. Because of the increased handling associated with processing fresh produce, the opportunities for contamination are increased (Brackett, 1992).

Of particular concern in MAP products are toxigenic, obligate anaerobic pathogens such as *C. botulinum*, the causative agent of botulism. These bacteria are commonly found in agricultural soils and on the surfaces of fruits and vegetables. Normally, these bacteria do not show significant growth nor become toxigenic in aerobic environments. In MAP products that have been kept at inappropriately high temperatures, the package could become depleted of O_2 allowing the naturally-occurring spores of *C. botulinum* to outgrow and produce toxin. It is also notable that elevated CO_2 may not inhibit growth of *C. botulinum* and in some cases may even stimulate it (Farber, 1991). A recent incidence of botulism was traced to shredded cabbage that had been packaged anaerobically and left at room temperature. Investigators determined that toxigenesis, under these conditions, preceded organoleptic spoilage (Soloman et al., 1990). Fresh mushrooms and fresh tomatoes have also been shown to contain spores of *Clostridium* spp. (Hotchkiss et al., 1992; Kautter et al., 1978; Malizio and Johnson, 1991).

Because of the unique environments created with MAP and the ability of dangerous pathogens to survive and compete in these environments, it has become necessary for the fresh cut produce industry to initiate comprehensive sanitation programs to ensure the microbiological integrity of their products. Such programs may include tracking product from source to destination. Such sources of contamination as the use of organic animal manures as fertilizers in the field, worker sanitation practices, and handling practices in the processing plant must be monitored. Packages must be designed in such a way that anaerobic conditions do not evolve and allow *C. botulinum* to grow and produce toxin. Temperature control must be maintained and monitored throughout the distribution cycle to ensure that high temperatures don't stimulate growth of microorganisms. The many requirements of such comprehensive sanitation systems may seriously tax the resources of smaller processors. Many processors are establishing HACCP systems within their facilities. A commitment to sanitation and microbiological safety ought to be an integral part of any MAP program.

SELECTING THE PROPER PACKAGING MATERIALS

Modified atmospheres can be implemented in a number of different ways. Changing the atmosphere surrounding apples and pears in storage rooms has been done for many years and is now so rigorously controlled that the technique is referred to as controlled atmosphere storage. Atmospheres may be modified inside refrigerated trailers or marine containers to facilitate quality maintenance during transportation. Because, in most cases, there is some form of monitoring of the atmosphere and adjustment of gas concentrations based on that monitoring, these would rightfully be called controlled atmospheres and will not be discussed further here. This chapter will concern itself with the establishment and maintenance of modified atmospheres inside packages. Such MAP can be established and maintained in a number of ways. The three principal techniques employed can be described as passive MAP, active MAP and vacuum packaging.

Passive MAP relies on the respiration of the commodity to consume the O_2 in a sealed bag and replace it with CO_2, a byproduct of normal aerobic respiration. The bag itself restricts the movement of gases in and out of the sealed package due to its selective permeabilities to O_2 and to CO_2. Over time, the system achieves an equilibrium modified atmosphere with the O_2 lower than that found in air (20.9%) and the CO_2 concentration higher than that in air (0.03%). Active MAP introduces a desired gas mixture into the bag prior to sealing, thereby accelerating the process of achieving an equilibrium atmosphere. Vacuum packaging draws a slight vacuum prior to sealing the bag, thereby reducing the headspace in the bag and accelerating the process of achieving an equilibrium atmosphere. In all three cases the result is the same. It is only the time to reach an equilibrium atmosphere that is different.

Packaging Films

The films used in MAP include various kinds of plastic polymers that provide protection, strength, sealability, clarity, and a printable surface.

However, their unique function is to restrict the movement of O_2 and CO_2 through the bag and allow the establishment of a modified atmosphere. They maintain a gradient between the gas concentrations in air and those inside the bag. It is the interaction of the respiration of the product and the gas gradient formed by the bag, that results in the formation of the modified atmosphere. The gradient that results is not dependent on the initial gas concentrations inside the bag, but rather on the respiration rate of the product and the gas permeabilities of the bag. It is important to recognize that by adding gas mixtures to the bag (active MAP) or by drawing a vacuum before sealing (vacuum packaging), the equilibrium atmosphere is not affected. These measures may allow the desired atmosphere to evolve more rapidly and this may be desirable in some cases. But the atmosphere that will occur inside a MAP is a function of the film and the product. For this reason it is essential to know the respiratory requirements of the product (how much O_2 the product will consume under specified conditions), and the permeability properties of the plastic bag that will be used.

The most commonly used films for MAP of fresh produce are made of low density polyethylene (LDPE) or of polypropylene (PP). Both plastics can be formulated into films of adequate strength with the requisite gas permeabilities. LDPE typically has higher gas permeability values while PP has greater clarity. In addition, both LDPE and PP are relatively inexpensive. Some of the important properties of different films are compared in Table 8.2.

Oxygen and Carbon Dioxide Permeabilities of the Film

A modified atmosphere package is formed by the ability of the plastic film to maintain a gas gradient between the ambient air and the inside of the package. The product respiration depletes the O_2 and enriches the CO_2 inside the package. The film permeability ensures that sufficient O_2 enters the package to prevent anaerobic conditions and that sufficient CO_2 exits the package to prevent CO_2 injury to the product. The actual amount of O_2 that needs to enter and the CO_2 that needs to exit will be product-specific (Table 8.3).

Because both product respiration and film permeability are sensitive to temperature, it is essential to determine O_2 and CO_2 permeabilities at the low temperatures at which the packages will be kept. Too often film manufacturers supply permeability values determined at 21°C (73°F). Such information is of limited value for films that are going to be used

*Table 8.2. Properties of major packaging resins based on one mil film thickness.**

	HDPE	LDPE	LLDP	EVA	Ionomer	OPP**	PET**	PVC	PVDC	OPS**	EVOH
Density (g/cc)	0.945–0.967	0.91–0.925	0.918–0.923	0.93	0.94–0.96	0.90	1.4	1.22–1.36	1.6–1.7	1.05	1.14–1.19
Yield (m^2/kg; 1mil)	41.2	42.6	42.5	41.9	42.0	44	28.4	28	24	38	32.7–34.7
Tensile strength (kpsi)	2.5–6	1.5–5	3–8	2–3	3.5–5	20–30	25–33	4–8	8–16	8–12	1.2–1.7
Tensile modulus 1% secant (kpsi)	125	20–40	25	8–20	10–50	350	700	350–600	50–150	400–475	300–385
Elongation at break (%)	200–600	200–600	400–800	500–800	300–600	50–275	70–130	100–400	50–100	2–30	120–280
Tear strength (Graves) (lb/in)	—	100–500	—	100–500	—	1000–1500	1000–2000	100–300	2	300–1000	—
Tear strength (Elmendorf) (gm/ml)	200–350	100–200	150–900	40–200	20–40	340	20–100	400–700	10–90	2–15	400–600
WVTR (g/100 in^2-day) at 100F & 90% RH	0.4	1–2	1–2	2–3	1.5–2	0.4	1–1.5	2–30	0.05–0.3	7–10	3–6

(continued)

Table 8.2. (continued).

	HDPE	LDPE	LLDP	EVA	Ionomer	OPP**	PET**	PVC	PVDC	OPS**	EVOH
O_2 permeability (cc/100 in^2-day-atm) at 77°F & 0% RH	100–200	500	450–600	700–900	300–450	100–160	3–6	30–600	0.1–1	200–350	0.01–0.02
Haze (%)	3	5–10	6–13	2–10	1–15	3	2	1–2	1–5	1	1–2
Light transmission (%)	—	65	—	55–75	85	80	88	90	90	92	90
Heat seal temperature range (F)	275–310	250–350	220–340	150–350	225–300	200–300†	275–350**	280–340	250–300	250–350	350–400
Service temperature (F)	−40 to 250	−70 to 180	−20 to 220	−100 to 150	−150 to 150	40 to 250	−100 to 300	−20 to 150	0 to 275	−80 to 175	0 to 300

HDPE = high density polyethylene; LDPE = low density polyethylene; LLDP = linear low density polyethylene; EVA = ethylene vinyl acetate; OPP = oriented polypropylene; PET = polyethylene terepthalate; PVC = polyvinyl chloride; PVDC = polyvinylidene chloride; OPS = oriented polystyrene; EVOH = ethylene vinyl alcohol.

*From Jenkins and Harrington, 1991, with permission.

**Biaxially oriented.

†Using polyolefin or PVDC as sealant layer.

*Table 8.3. Classification of fruits and vegetables according to their tolerance to low O_2 concentrations.**

Gas Sensitivity	Commodities
Minimum % O_2 Tolerated	
0.5	Tree nuts, dried fruits and vegetables
1.0	Some cultivars of apples and pears, broccoli, mushrooms, garlic, onion, most cut or sliced fruits and vegetables
2.0	Most cultivars of apples and pears, kiwifruit, apricot, cherry, nectarine, peach, plum, strawberry, papaya, pineapple, olive, cantaloupe, sweet corn, green bean, celery, lettuce, cabbage, cauliflower, brussels sprouts
3.0	Avocado, persimmon, tomato, pepper, cucumber, artichoke
5.0	Citrus fruits, green pea, asparagus, potato, sweet potato
Maximum % CO_2 Tolerated	
2	Apple (Golden Delicious), Asian pear, European pear, apricot, grape, olive, tomato, sweet pepper, lettuce, endive, Chinese cabbage, celery, artichoke, sweet potato
5	Apple (most cultivars), peach, nectarine, plum, orange, avocado, banana, mango, papaya, kiwifruit, cranberry, pea, chili pepper, eggplant, cauliflower, cabbage, brussels sprouts, radish, carrot
10	Grapefruit, lemon, lime, persimmon, pineapple, cucumber, summer squash, snap bean, okra, asparagus, broccoli, parsley, leek, green onion, dry onion, garlic, potato
15	Strawberry, raspberry, blackberry, blueberry, cherry, fig, cantaloupe, sweet corn, mushroom, spinach, kale, Swiss chard

*From Kader et al., 1989, with permission.

at 1–2°C (33°–36°F). Film permeabilities should be measured and reported at low temperatures to allow evaluation of films for use in MAP of fresh produce. If the expected respiration rate of the product at the desired temperature is known, then it is possible to calculate the desired film permeabilities necessary to accommodate the product respiration [see Equation (1)].

CO_2/O_2 Permeability Ratio

Most plant tissues, under normal aerobic conditions, produce about one unit of CO_2 for every unit of O_2 consumed. If plastic films were equally permeable to O_2 and to CO_2, to reduce the O_2 concentration inside the package from 21% (the amount in air) to, say, 3% (a change of 18%), the package would accumulate 18% CO_2. Some produce commodities are benefitted by elevated CO_2 and others are harmed by it. Similarly, some are benefitted by low O_2 while others can be harmed by it. For this reason, the relative permeabilities to O_2 and CO_2 of a MAP film are as important as the absolute permeabilities. In many cases, it is desirable to reduce O_2 concentration without allowing significant CO_2 to accumulate inside the package. The determinant of the relative proportions of CO_2 and O_2 in the package is the ratio of film permeabilities to CO_2 and O_2. This ratio is referred to as β (P_{CO_2}/P_{O_2}), and is one of the most useful descriptive parameters of a plastic film. Films with a high β value will allow CO_2 to escape the package relatively easily, resulting in an atmosphere with low CO_2. Films with lower β values will allow greater buildup of CO_2 in the package (Figure 8.1). For most low density polyethylene films (the most common in use for MAP in the produce industry) $\beta = 2-4$. It will be different for different kinds of film

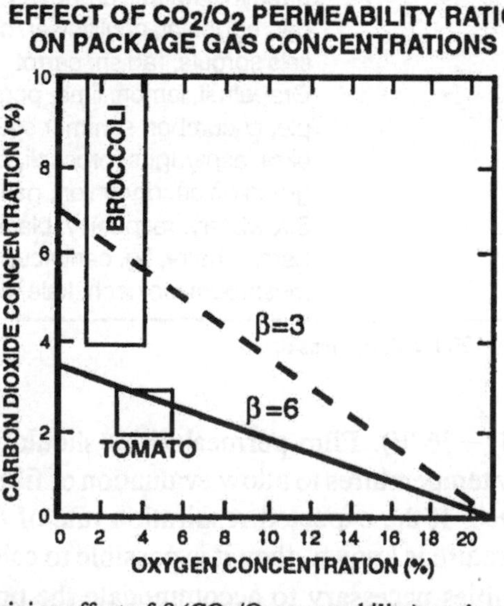

FIGURE 8.1 Determining effect of β (CO_2/O_2 permeability) on the possible concentrations of gases in a modified atmosphere package.

polymers. The permeability ratio (β) determines the possible combinations of O_2 and CO_2 concentrations inside the package. Gas flushing, vacuum packing, changing the size of the bag, or changing the amount of the product in the bag will not affect these possible gas concentrations. The β value of a film is the determinant of the possible combinations of gas concentrations within a package. Since fruits and vegetables vary in their tolerance to CO_2 and in their ability to benefit from high CO_2, the β value of a film is very important as a predictor of the relative amounts of O_2 and CO_2 that will accumulate in the package. The absolute amounts of O_2 and CO_2 will be determined by the absolute permeability values of the film. Figure 8.1 shows the possible gas concentrations in a MAP made from films with β values of 3 or of 6. The boxes labeled broccoli and tomato represent the zone of desirable gas concentrations for those two commodities. The $\beta = 3$ and $\beta = 6$ lines represent the possible combinations of O_2 and CO_2 concentration that could evolve within sealed packages made from films with those β values. A film with a $\beta = 3$ would be suitable for broccoli, but not for tomatoes. The opposite would be true for a film with $\beta = 6$. Broccoli is benefitted by high CO_2 and low O_2 concentrations. Tomato is less tolerant of high CO_2 but will benefit from low O_2. Tomatoes, therefore, should be packaged in a film that will allow the escape of CO_2 (high β), while broccoli should be packaged in a film that allows accumulation of CO_2 (low β).

The Importance of Film Thickness and Film Surface Area

Besides lending strength to the film, thickness limits gas movement across the film. Thicker film restricts gas movement more than thinner film and thus can maintain a greater gas gradient between the air and the atmosphere inside the package. Doubling the thickness of a film will roughly halve the amount of gas that can move across the film in a given amount of time. Making the film thinner will, of course, increase gas movement across the film. Once the β of the film is known, the thickness of the film will help determine where on the β line (Figure 8.1) the actual gas concentrations in the package will lie. For example, if we have a film with a β value of 3 and we wish to use it for a broccoli package, we know from Figure 8.1 that the film *can* maintain an appropriate atmosphere. But where on that $\beta = 3$ line will our actual equilibrium gas concentrations lie? If we make a package with the film and put a given weight of broccoli in it, we may find that the equilibrium atmosphere has 9% O_2 and 4% CO_2. This atmosphere will not be optimal for broccoli. We can move the gas equilibrium concentrations up the β line by increasing the thickness of the film, or by decreasing the surface area of the film. Either will have the effect of reducing gas movement across the film. Thus, less O_2 will enter the package and less CO_2 will exit the package. Increasing the weight of product inside the package will have a similar effect because more broccoli will consume more O_2 and produce more CO_2. In this way, several factors interact to determine the equilibrium gas concentrations inside the package. The quantitative relationship of these factors, at a given temperature, can be expressed as Equation (1):

$$P_{O_2} = RR_{O_2} * t * W/A * (O_{2atm} - O_{2pkg}) \quad (1)$$

where

P_{O_2} = Oxygen permeability of the film (ml-mil/m²-d-atm)
RR_{O_2} = Respiration rate as product consumption of O_2 (ml/kg-h)
t = Film thickness (mils)
W = Product weight (kg)
A = Film surface area (m²)
$(O_{2atm} - O_{2pkg})$ = Desired O_2 gradient between the air and the inside of the package (%)

Using Equation (1), it is possible to calculate the desired permeability of a plastic film to supply sufficient O_2 to a product in order to prevent anaerobic conditions. The use of Equation (1) presupposes knowledge of the product respiration rate at the O_2 concentration desired within the MAP. Information about product respiration rates in modified atmospheres is available from published reports in the scientific literature or from consultants, trade groups or extension agents active in the area of MAP of produce.

Market Requirements

Most of this discussion has been concerned with the technical requirements of packages to create a desired atmosphere within a MAP. Knowledge and application of this information is essential in the design of MAP. However, films must meet many other requirements for a MAP product to be successful in the competitive market environment of the modern produce industry. Films should have the following characteristics. 1) Adequate strength to withstand packaging and distribution without tearing or splitting (e.g., oriented polypropylene or oriented polystyrene); 2) Sufficient slip to work well with bagging machines (conferred by acrylic coatings or stearate additives); 3) Flex resistance so that the film can withstand repeated flexing without pinholing or delaminating (oriented polypropylene); 4) Puncture resistance (low density polyethylene or linear low density polyethylene); 5) Tensile strength (oriented polypropylene or polyethylene terephthalate); 6) Ability to form a complete and durable seal, usually through heat impulse sealing (EVA copolymers or ionomers); 7) Sparkling clarity for retail markets (oriented polypropylene, polyvinyl chloride or polyethylene terephthalate); 8) Printability (several kinds of coatings); 9) Resealability for many markets through either ziploc or sticky seals (ethylene vinylacetate additives); and 10) Easy opening for retail markets.

Because no single polymer can deliver all, or even most of these properties to a film, modern films for MAP are often composite films and/or laminates of several very thin polymers. In this way desirable properties can be added to the film in very thin layers, any one of which would lack the strength to perform on its own. Because each layer is very thin, the barrier properties of even relatively impermeable films are acceptable and the entire laminate can breathe adequately to accommodate the respiration of the product. The actual commercial formulations are, of course, trade secrets and the polymers mentioned above are just possible satisfactory examples.

PACKAGING EQUIPMENT

MAP packaging of fresh produce is generally done in one of three ways: vacuum packaging, gas flush packaging, or passive modified atmosphere packaging. In all three cases fresh product is put in a permeable bag and the bag is sealed to maintain the modified atmosphere. Vacuum packaging draws a vacuum to remove the headspace air and seals the vacuum inside the bag to instantly reduce the O_2 content of the package atmosphere. Gas flush packaging also draws a vacuum but rather than sealing the vacuum in the package, a desired gas mixture is injected into the bag to replace the withdrawn air, thereby rapidly establishing a desirable atmosphere. Passive modified atmosphere packaging relies solely on the product respiration and the package film permeability to alter the atmosphere in the package. All three rely on similar types of packaging machines that have been in use for many years. Incremental improvements in speed, ability to handle new films, and reliability of the seal and gas delivery, have maintained competitive innovation in the packaging machinery business.

Most packaging equipment systems are variations on horizontal form/fill/seal, vertical form/fill/seal, tray/stretch overwrap or pre-made bag filling systems. In the fresh cut produce industry, which is currently leading the industry in importance and innovation, most machinery relies on pre-made bags and vertical fill/gas flush systems. A few suppliers of packaging equipment dominate the produce market and some of their products will be discussed below.

CVP Systems (Downers Grove, IL) developed a machine to facilitate vacuum packaging in order to obviate the need for metal clips on the bags. This machine, the A-300, is a snorkel-type machine that draws a vacuum prior to sealing. It is thus suitable for vacuum packs, vacuum/gas flush, and forming a gas-tight seal. The machine is designed to be used on handpacked product. Bags are filled with product at another station and then brought to the modular bagging machine. A single operator can control up to three A-300 machines at a time allowing a throughput of as many as sixteen 5-pound bags per minute for that operator. The A-300 can also handle 10-pound bags. There is a newer horizontal fill-and-seal machine, the C-1200, that can do 5- and 10-pound bags of vacuum or gas-flushed produce. The C-1200 uses special large tubes for drawing the vacuum and thus is gentler on fluffy and delicate products such as cut lettuce. It is also gentle enough to accommodate thin and delicate films such as 0.5 mil polyethylene. It can run up to 5 mil polyethylene or nylon so it has great flexibility to run different kinds of films. Pre-made bags are fed from a roll attached to the side of the machine. The top lip of the bag is sandwiched between belts so that it rides smoothly. The bag is filled, a slight vacuum is drawn and the bag is sealed, all at different stations so that the machine can attain speeds of 25–30 bags per minute. The C-1200 is very exact in the vacuum it pulls and the amount of gas that it injects, and is designed to handle bulk quantities. It can handle diverse film types but the films must be made properly to work with the machine. Therefore, CVP Systems recommends and sells CVP T-grade films for use with the C-1200. The T-grade films come in a number of thicknesses and the permeabilities can be varied to suit an application. The C-1200 can run zip loc bags down to 1-pound and can change from 1-pound to 5-pound bags in five minutes.

M-Tek Inc. (Elgin, IL) makes the Corr-Vac MAP system. This

machine can do vacuum and gas flushing and sealing and is gentle enough to use for lettuce and other sensitive commodities. The company generally recommends N_2 gas flush because the cost of other gas mixtures is uneconomical. M-Tek does not supply bags but their machinery is versatile enough to work with many kinds of bags. The Corr-Vac system can seal a wide variety of bags including low density polyethylene, as well as bags containing EVA or other additives. Sealing is done with an impulse sealer that is vacuum cooled between sealing cycles to reduce seal variability, even when sealing through wrinkles and moisture contamination. The seal bar area is made from heavy stainless steel components to give a strong pressure profile to prevent the plastic from moving when it gets hot, and also to drive moisture and other contaminants out of the seal. The Mark-1 model can do sixteen 5-pound bags of lettuce per minute with one operator working two machines. M-Tek has already sold more than 20 Corr-Vac MAP systems to the produce industry. They are planning to market a system with a 54-inch seal bar to seal full cartons for foodservice customers.

A number of companies offer various bagger/sealer systems and some supply plastic bags appropriate to produce packaging. Among these are such companies as Automated Packaging Systems (Twinsburg, OH), Ilapak Inc. (Newton, PA), Bemis Machinery Co. (Green Bay, WI), Multivac Inc. (Kansas City, MO), Fresh Cut Solutions Inc. (London, Ontario), General Packaging Equipment Company (Houston, TX), Rovema Packaging Machines (Lawrenceville, GA), Hayssen Manufacturing Company (Sheboygan, WI), Packaging Concepts Corp. (Hagerstown, MD), and Packaging Aids Corporation (San Rafael, CA), to name a few.

COMMERCIAL USE OF MAP FOR FRESH PRODUCE

The use of MAP in the produce industry is rapidly growing. This is, in large part, due to the explosive growth of the "fresh cut" industry that now supplies a large part of the fresh produce used in the foodservice sector. Because the produce has been cut, sliced, diced or shredded, its shelf life is likely to be reduced due to the physical and physiological injury inherent in cutting living tissues and the fact the nutrients are released, permitting the growth of spoilage microorganisms. Added shelf life and maintenance of quality can be attained through the proper use of MAP and the "fresh cut" industry now uses MAP as a primary tool to achieve this end. A great diversity of package films are for sale, but the choice of package types used by the industry is driven by both technological and market considerations.

Tom-Ah-Toes

One of the earliest commercial applications of MAP of produce was introduced by Natural Pak Produce (Closter, NJ) with the Tom-Ah-Toe package. This long, narrow box typically holds four tomatoes and is overwrapped with a gas permeable film. Inside the package is a sachet containing a predetermined amount of calcium chloride and activated lime. The calcium chloride maintains the relative humidity at a high level but below saturation to avoid formation of condensate and ensuing decay. Activated lime absorbs CO_2 and keeps the level low enough to prevent injury to the tomatoes which are sensitive to CO_2. Natural Pak Produce is now also marketing avocadoes and mangos in similar packages and has plans to introduce melons in the future.

FreshHold

The FreshHold packaging system was developed by Hercules Chemical Company but has been licensed to Fresh Western Marketing (Salinas, CA). The system relies on a polypropylene label that has calcium carbonate embedded in it. The film is stretched in a proprietary way to develop channels within the film that facilitate gas movement. The channels can confer much greater permeabilities than the film would otherwise have. The polypropylene label can be customized to accommodate the respiratory requirements of virtually any product. The patch is then applied over holes in a 2−3 mil polyethylene or polypropylene/polyethylene laminated bag or, in some cases, an impermeable rigid polystyrene tray. The label then mediates the gas exchange through the package. The label can be printed with brand and commodity information allowing the use of a standard bag or tray for a variety of products. Fresh Western is currently using the FreshHold system to package broccoli, asparagus, cauliflower and cherries and they may package roses in the future. The bags run on a form/fill and seal machine and are sealed with an impulse sealer.

Cryovac

Cryovac Division of W.R. Grace & Co. (Duncan, SC) is one of the largest suppliers of breathable bags to the produce industry. They have been very active in developing new packaging materials for cut lettuce, broccoli and cauliflower products. Of these, cut lettuce makes up the bulk of the market. They believe that the film for a lettuce bag should be about 1.5 mils thick to maintain good flavoured lettuce, but most films this thin don't ship or seal well. Consequently, most bags on the market are 2–2.5 mils thick but bags this thick can cause anaerobic conditions. The Cryovac bag, called PD-961, is 1.25 mils thick but is very strong and seals well. It is a multilayer coextruded bag made of several layers of polyethylene related polymers. Crosslinking gives it great strength and also makes the bag tear in only two directions. The bag will tear either down or across for easy opening. There is a 90 degree curve and a notch on one corner of the bag to allow a zipper-type opening mechanism. The oxygen transmission rate of the film (OTR) is 6000–8000 cc/m^2/day (normally about 6,500 cc/m^2) and the CO_2 transmission rate is 19,000–22,000 cc/m^2/day (at 21°C and 1 atm). Cryovac also developed PD-941 film specifically for broccoli and cauliflower but this bag has also been used for other uses such as spinach. This is also a coextruded, multilayer polyolefin film. The PD-941 bag is 0.75 mils thick with a very high OTR of about 17,000 cc/m^2/day (at 21°C, 1 atm). The CO_2 transmission rate is about 3–4 times the OTR. These high transmission rates are necessary because the respiration rates of broccoli and cauliflower are much higher than that of lettuce. The PD-900 has a lower OTR (3000 cc/m^2/day at 21C, 1 atm). It is a bag that was developed specifically for vacuum packaging a variety of fresh produce for foodservice markets. The PD-900 is designed for chopped carrots, celery and other processed vegetables and includes a built-in aroma barrier that makes it suitable for fresh cut onions. The bag is a very strong multilayered polyolefin with superior seal strength. It lacks the sparkling clarity of some other bags but this is not an important consideration in the foodservice market. Cryovac also offers a barrier bag for peeled potatoes (The B-900, 2.5 mils thick, OTR is 30 cc/m^2) without sulfites. This bag is used with antioxidants and can keep cut potatoes white for two weeks. Finally, RD-106 is a multilayered, cross-linked polyolefin film with antifog properties applicable for shrink-packaging fresh fruits and vegetables. The film has a reasonably high OTR (8,500–11,500 cc/m²/day, 21°C, 1 atm) and CO_2 transmission rate (22–26,000 cc/m²/day, 21°C, 1 atm), and is very resistant to punctures and tears.

ICI

ICI (Imperial Chemical Industries PLC) markets a series of polypropylene-based films under the brand Propafilm. Propafilm CR and CK were developed for packaging fresh cut lettuce and are used mostly in France for salads, but could be used for other vegetables. These films are made with a core of coextruded polypropylene that is biaxially oriented and PVDC coated to give sparkling clarity, a wide heat sealing range and the ability to be readily printed and processed at high speeds. Propafilm CR film has antifog properties and is available in 1 or 1.25 mil thickness. Propafilm CK is also antifog and available in 1 and 1.25 mil thicknesses as well. The gas transmission properties of these films are poor and so they are likely to be suitable for most produce only when kept at low temperature. These films are not likely to compensate well for higher temperatures. The manufacturer suggests that, when used for lettuce, packages always be kept under 4°C. Propafilm CK has good antistatic properties so it machines very well on Vertical Form Fill and Seal machines. It also has slightly better gas transmission properties than Propafilm CR.

Courtaulds Packaging

Courtaulds Packaging markets a system for equilibrium modified atmosphere packaging using their P-Plus films. These films are spark perforated resulting in non-uniform perforations throughout the film to facilitate gas exchange. Microperforated films can achieve high permeabilities by allowing gases to move across the film via mass flow which is much faster than the usual permeation processes. However, mass flow does not provide the differential permeabilities to O_2 and CO_2 that non-perforated films offer. With resulting β values close to 1, these films will not provide low CO_2 atmospheres. The depletion of O_2 through respiratory processes will, of necessity, result in accumulation of high levels of CO_2 in these packages. For those commodities that are benefitted by high CO_2 and low O_2, these films may be useful. P-Plus film packages are used in England for brussels sprouts, lettuce, broccoli, and fresh mushrooms. In Japan they have been used for bean sprouts. P-Plus films offer O_2 permeability of 3000–4000 cc/m²/24 h.

CVP Systems

CVP Systems is primarily a manufacturer and vendor of food packaging equipment, but they also manufacture and supply packaging films designed to be used on their machinery. The CVP films are sold under the name T-Grade. The OTRs of the T-Grade plastic vary from 6,400–11,200 cc/m²/day with CO_2 transmission about four times higher. The T-Grade films are coextruded bilayer films that offer excellent strength and sealing properties. They are available in thicknesses of 1.0, 1.25, 1.5 and 1.75 mils.

DuPont

DuPont markets a series of films under the Clysar brand name that, although they were designed for meat packaging, are appropriate for produce packaging applications. These include Clysar EHC, EH, ECL, and LLP films. These films are biaxially oriented, heat shrinkable polyethylene or polyolefin films that offer good toughness, sealing and clarity properties combined with reasonably high OTR (4,150–5,200 cc/m²/day).

Laminated Boxes

A number of manufacturers of corrugated cardboard cartons are beginning to offer cartons with films laminated within the cardboard or coated on the inside of the cardboard liner. These films reduce moisture loss and can mediate gas exchange through the carton and create a modified atmosphere inside. These boxes are being promoted for packaging of strawberries and broccoli among other things. Georgia Pacific (Lamilux), Weyerhaeuser and Tamfresh Ltd. (the packaging division of Tampella) are currently marketing boxes of this type.

Film Convertors

In addition, there are a number of convertors making plastic packages for produce. These companies typically buy resin or film from a plastics manufacturer and extrude and/or laminate, irradiate, heat treat or coat plastics into films with desirable properties. A large proportion of the bags used for modified atmosphere packaging of fresh produce are, in fact, made and sold by convertors who can sometimes be more sensitive and responsive to the needs of regional produce processors than the large film manufacturers. This category includes such companies as Cypress Packaging (Rochester, NY), Ultratech Plastics Inc. (Mansfield, OH), Durapak Company Inc. (Baltimore, MD), Fredman Bag Company (Milwaukee, WI), Golden Eagle Extrusions Inc. (Loveland, OH), Packaging Concepts Corp. (Hagerstown, MD), and Plastics Plus Inc. (Cincinnati, OH).

Activated Earth Films

In the past few years several produce bags have been sold in Japan with the claim that they have the property of maintaining produce freshness. These bags are typically polyethylene bags with powdered clay material incorporated into the film matrix which lends a cloudy appearance to the film. The clay materials are a variety of finely powdered aluminum silicates. There has been speculation that the putative salutary effects of these produce bags is due to the ethylene adsorbing properties of the clay materials incorporated into the bags. However, research has not confirmed any ethylene absorbing ability for these bags (Joyce, 1988; Zagory et al., 1989). The incorporation of clay particles in the film matrix opens pores within the film that facilitate gas flux across the film. It may be that, because of these pores, ethylene is able to more rapidly escape the bag and O_2 enters more rapidly, thus maintaining favorable conditions for fresh produce. The incorporation of clay particles is known to alter the permeabilities of plastic films, generally increasing gas transmission properties (Ahrens, 1992). This may be beneficial for some products. Activated earth films are marketed in the USA by Sealed Fresh Incorporated (Denver, CO) and Double B Produce (Pinedale, CA).

Temperature Responsive Films

Respiration rates of fruits and vegetables increase dramatically as the temperature increases. Permeabilities of plastic films also increase as the temperature increases. However, respiration and permeability do not increase at similar rates. Typically, product respiration rates go up much faster than film permeabilities as temperature increases. This results in an important limitation in the use of films for produce packaging. A film package that maintains an appropriate atmosphere at low temperature may generate an inappropriate, or even dangerous anaerobic atmosphere at higher temperatures. This is a particular problem for retail products where temperature control is generally poor. Temperature responsive films, under development by Landec Labs (Menlo Park, CA), can increase their gas permeabilities in response to temperature increases in tandem with increases in product respiration rates. Thus, the makeup of the modified atmosphere within the package remains essentially unchanged at different temperatures. Through their patented technology they are able to engineer changes in crystallinity of polymer side chains that reversibly change their structure in response to temperature changes. The temperature sensitivity of the reaction can be designed into the polymer to accommodate the product to be packaged. Such innovations in plastic film technology are likely to open a new world of possibilities for modified atmosphere packages at the retail level.

FUTURE MARKET POTENTIAL

With the twin concerns of freshness and freedom from chemicals impacting the produce market, the future use of MAP is bound to increase. The advantages of extended shelf life, convenience and freshness that can be delivered free of chemicals will continue to be leveraged in today's international marketplace. Although many consumers believe that pesticide residues are a significant threat on fresh produce, the dangers of pathogenic microorganisms are likely to play a more important role in the development of products and procedures among processors and packagers. Vigorous sanitation programs to ensure microbiological safety will play a more important role in the future. The adoption of HACCP programs is likely to become standard in the fresh cut produce industry.

Improved temperature control during distribution and marketing of MAP produce will be crucial to the further growth of the market. Distribution channels to foodservice are generally considered to maintain adequate temperature control. However, temperature management within retail channels is inconsistent at best and abusive at worst. Because of the increasing importance of fresh cut packaged produce in retail markets, produce managers will devote more attention to the special requirements of MAP produce and their devotion to temperature management will no doubt improve. At the same time, technical advances in film formulation will provide packages that better compensate for changes in temperature and thus prevent the development of anaerobic atmospheres inside packages even in instances of poor temperature management.

The market for MAP salads and cut vegetables is already large, though it is still growing rapidly. However, explosive growth is expected in MAP fruit products. The current barriers to commercialization of fresh cut fruit include enzymatic browning, loss of texture and expression of undesirable amounts of juice within packages. These issues will be addressed by a combination of more precise development of modified atmospheres and treatment with antioxidants to minimize browning, better temperature control to maintain texture, and better handling practices to minimize loss of juice. If such advances are realized, the market for cut and sliced fruit, especially tropical fruit, will be an important area of growth in the next ten years.

REFERENCES

Ahrens, J. 1992. Personal communication.

Berrang, M. E., Brackett, R. E. and Beuchat, L. R. 1989. "Growth of *Listeria monocytogenes* on fresh vegetables stored under a controlled atmosphere," *J. Food Prot.*, 52:702–705.

Beuchat, L. R. and Brackett, R. E. 1990a. "Survival and growth of *Listeria monocytogenes* on lettuce as influenced by shredding, chlorine treatment, modified atmosphere packaging and temperature," *J. Food Sci.*, 55:755–758, 870.

Beuchat, L. R. and Brackett, R. E. 1990b. "Inhibitory effects of raw carrots on *Listeria monocytogenes*," *Appl. Environ Microbiol.* 56:1734–1742.

Beuchat, L. R. and Brackett, R. E. 1991. "Behavior of *Listeria monocytogenes* inoculated into raw tomatoes and processed tomato products," *Appl. Environ. Microbiol.*, 57:1367–1371.

Brackett, R. E. 1992. "Shelf-stability and safety of fresh produce as influenced by sanitation and disinfection," *J. Food Prot.*, 55(10):808–814.

Centers for Disease Control. 1979. "*Salmonella oranienburg* gastroenteritis associated with consumption of precut watermelons," *Morbid. Mortal. Weekly Rep.*, 28:522–523.

Centers for Disease Control. 1991. "Multistate outbreak of *Salmonella poona* infections—United States and Canada, 1991," *Morbid. Mortal. Weekly Rep.*, 41:549–552.

Chou, H-E. and Breene, W. M. 1972. "Oxidative decoloration of β-carotene in low moisture model systems," *J. Food Sci.*, 37:66–68.

Davis, H., Taylor, J. P., Perdue, J. N., Stelma, G. N. Jr., Humphreys, J. M. Jr., Rowntree, R. III, and Greene, K. D. 1988. "A shigellosis outbreak traced to commercially distributed shredded lettuce," *Am. J. Epidemiol.*, 128:1312–1321.

Eskin, N. A. M. 1979. *Plant Pigments, Flavors and Textures.* Academic Press, New York, p. 219.

Farber, J. M. 1991. "Microbiological aspects of modified-atmosphere packaging technology—A review," *J. Food Prot.*, 54(1):58–70.

Gayler, G. E., MacCready, R. A., Reardon, J. P. and McKernan, B. F. 1955. "An outbreak of salmonellosis traced to watermelon," *Public Health Rep.*, 70:311–313.

Goodenough, P. W. and Thomas, T. H. 1980. "Comparative physiology of field-grown tomatoes during ripening on the plant or retarded ripening in controlled atmosphere," *Ann. Appl. Biol.*, 94:445–451.

GPMC. 1988. *Grocery Attitudes of Canadians*, 1988. Grocery Products Manufacturers of Canada, Don Mills, Ontario.

Haard, N. F. 1985. Characteristics of edible plant tissues, In *Food Chemistry*, Second Ed., O. R. Fennema, ed., Marcel Dekker, Inc., New York, pp. 857–911.

Herner, R. C. 1987. High CO_2 effects on plant organs, In *Postharvest Physiology of Vegetables*, J. Weichmann, ed., Marcel Dekker, New York, p. 239.

Hotchkiss, J. H. and Banco, M. J. 1992. "Influence of new packaging technologies on the growth of microorganisms in produce," *J. Food Prot.*, 55(10):815–820.

Hotchkiss, J. H., Banco, M. J., Busta, F. F., Genigeorgis, C. A., Kociba, R., Rheaume, L., Smoot, L.A., Schuman, J. D. and Sugiyama, H. 1992. "The relationship between botulinal toxin and spoilage of fresh tomatoes held at 13 and 23°C under passively modified and controlled atmospheres and air," *J. Food Prot.*, 55(7): 522–527.

Hyodo, H. and Yang, S. F. 1971. "Ethylene-enhanced synthesis of phenylalanine ammonia lyase in pea seedlings," *Plant Physiol.*, 47:765–770.

Joyce, D. C. 1988. "Evaluation of a ceramic-impregnated plastic film as a postharvest wrap," *Hort Sci.* 23(6):1088.

Jenkins W. A. and Harrington, J. P. 1991. *Packaging Foods with Plastics*. Technomic Publishing Company, Inc., Lancaster, PA., 326 pp.

Kader, A. A. 1986. "Biochemical and physiological basis for effects of controlled and modified atmospheres on fruits and vegetables," *Food Technol.*, 40(5):99–100, 102–104.

Kader, A. A. 1992. Quality and safety factors: Definition and evaluation for fresh horticultural crops, In *Postharvest Technology of Horticultural Crops*, Second Edition, A. A. Kader, ed., Publication 3311. University of California, Davis, CA, pp. 185–189.

Kader, A. A., Zagory, D. and Kerbel, E.L. 1989. "Modified atmosphere packaging of fruits and vegetables," *CRC Crit. Rev. Food Sci. Nutr.*, 28(1):1–30.

Kautter, D. A., Lilly, T. Jr. and Lynt, R. 1978. "Evaluation of the botulism risk in fresh mushrooms wrapped in commercial polyvinylchloride film," *J. Food Prot.*, 41:120–121.

Ke, D. and Kader, A. A. 1992. "Potential of controlled atmospheres for postharvest insect disinfestation of fruits and vegetables," *Postharvest News & Info.*, 3(2):31N–37N.

Konowalchuk, J. and Spiers, J. I. 1975. "Survival of enteric viruses on fresh vegetables," *J. Milk Food Technol.*, 38:469–472.

Mackinney, G., Lukton, A. and Greenbaum, L. 1958. "Carotenoid stability in stored dehydrated carrots," *Food Technol.*, 12:164–166.

Madden, J. M. 1992. "Microbial pathogens in fresh produce—The regulatory perspective," *J. Food Prot.*, 55(10):821–823.

Malizio, C. J. and Johnson, E. A. 1991. "Evaluation of the botulism hazard from vacuum-packaged Enoki mushrooms *(Flammulina velutipes)*," *J. Food Prot.*, 54:20–21.

Martens, M. and Baardseth, P. 1987. Sensory quality, In *Postharvest Physiology of Vegetables*, J. Weichmann, ed., Marcel Dekker, Inc., New York, pp. 427–454.

Martin, D. L., Gustafson, T. L. Pelosi, J. W., Suarez, L. and Pierce, G. V. 1986. "Contaminated produce–A common source for two outbreaks of *Shigella* gastroenteritis," *Am. J. Epidemiol.*, 124:299–305.

Murr, D. and Morris, L. 1974. "Influence of O_2 and CO_2 on *o*-diphenol oxidase activity in mushrooms," *J. Amer. Soc. Hort. Sci.*, 99:155–158.

Ries, A. A., Zara, S., Langkop, C., Tauxe, R. V. and Blake, P. A. 1990. "A multistate outbreak of *Salmonella chester* linked to imported cantaloupe," Abstr., *1990 Interscience Conference on Antimicrobial Agents and Chemotherapy*, p. 238.

Ryall, A. L. and Lipton, W. J. 1972. *Handling, Transportation and Storage of Fruits and Vegetables. Vol. 1. Vegetables and Melons.* AVI Publ. Co., Westport, CT.

Salunke, D. K. and Wu, M. T. 1973. "Effects of low oxygen atmosphere storage on ripening and associated biochemical changes of tomato fruits," *J. Amer. Soc. Hort. Sci.*, 98:12–14.

Siriphanich, J. 1984. Carbon dioxide effects on phenolic metabolism and intracellular pH of crisphead lettuce (*Latuca sativa* L). Ph.D. Thesis, Dept. of Pomology, University of California, Davis, CA.

Siriphanich, J. and Kader, A. A. 1985. "Effects of CO_2 on total phenolics, phenylalanine ammonia lyase, and polyphenol oxidase in lettuce tissue," *J. Amer. Soc. Hort. Sci.*, 110(2):249–253.

Solomon, H. M., Kautter, D. A., Lilly, T. and Rhodehamel, E. J. 1990. "Outgrowth of *Clostridium botulinum* in shredded cabbage at room temperature under modified atmosphere," *J. Food Prot.*, 53:831–833, 845.

Steinbruegge, E. G., Maxey, R. B. and Liewen, M. B. 1988. "Fate of *Listeria monocytogenes* on ready to serve lettuce," *J. Food Prot.*, 51:596–599.

Van Buren, J. 1970. Fruit Phenolics, In *The Biochemistry of Fruits and Their Products. Vol. 1*, A. C. Hulme, ed., Academic Press, London and New York. pp. 269–304.

Wang, C. Y. 1977. "Effect of CO_2 treatment on storage and shelf life of sweet peppers," *J. Amer. Soc. Hort. Sci.*, 102:808–812.

Watada, A. 1987. Vitamins, In *Postharvest Physiology of Vegetables*, J. Weichmann, ed., Marcel Dekker, Inc., New York, pp. 455–462.

Weichmann, J. 1986. "The effect of controlled atmosphere storage on the sensory and nutritional quality of fruits and vegetables," *Hort. Rev.* 8:101–127.

Yang, Y. J. and Henze, J. 1988. "Influence of CA-storage on external and internal quality features in broccoli (*Brassica oleracea* var. Italica). II. Changes in chlorophyll and carotenoid contents," *Gartenbauwissenschaft*, 53:41–44 (In German).

Zagory, D. and Kader, A. A. 1989. Quality maintenance in fresh fruits and vegetables by controlled atmospheres, In *Quality Factors of Fruits and Vegetables*. J. J. Jen, ed., ACS Symp. Series 405, American Chemical Society, Washington, DC, pp. 174–188.

Zagory, D., Brecht, B. and Kader, A. A. 1989. Unpublished results. Dept. of Pomology. University of California, Davis, CA.

Zind, T. 1989. *Fresh Trends '90–A Profile of Fresh Produce Consumers*. Packer Focus, Vance Publ. Corp., Lincolnshire, IL.

CHAPTER 9

Modified Atmosphere Packaging of Bakery and Pasta Products

J. P. SMITH–*McGill University, Quebec, Canada*
B. K. SIMPSON–*McGill University, Quebec, Canada*

INTRODUCTION

BAKERY and pasta products are an important part of a balanced diet and account for ~20% of the daily dietary protein of the average Canadian consumer (Personal Communication, Agriculture Canada). According to Hunt and Robbins (1989), the average weekly expenditure on bakery and pasta products was ~\$8.30 per person or about 13% of the total grocery dollar. Spending on bakery products accounted for approximately 9% of total food expenditures with ~1% being spent on pasta products. For every dollar spent on bakery and cereal products, bread was by far the most important product, accounting for 27 cents, followed by cookies, while pasta products accounted for only about 5 cents (Hunt and Robbins, 1989).

In 1987, the Canadian bakery industry sold \$1.33 billion worth of products, or approximately 4% of total food sales, with bread accounting for approximately 50% of the total value of bakery products sold (Ooraikul, 1991). For the same year, the bakery industry in the U.S. sold about \$22 billion worth of bakery products, i.e., 7% of the total value of food products sold during 1987. According to the U.S. Department of Commerce, sales of all baked products were expected to rise by ~1.5% annually between 1987 and 1992 (Ooraikul, 1991).

The per capita consumption of dry and packaged pasta in the U.S. is also increasing and is projected to double in value from \$1.5 billion in 1986 to just over \$3.0 billion in 1995 (Donnelly, 1991). Contributing to the increase in pasta consumption is the phenomenal growth of fresh pasta. Projections are that fresh pasta sales will increase an average 17% annually over the same ten year period to total \$720 million in 1998 or 10% of the total pasta market (Donnelly, 1991).

Classification of Bakery and Pasta Products

Bakery products consist of bread, unsweetened rolls and buns, doughnuts, meat pies, dessert pies, pizza pies and crusts, crackers, cookies and other products. Several methods can be used to classify bakery products. Classification can be based on product type, e.g., bread or sweet goods, the method of leavening, e.g., biological, chemical or unleavened, or on the basis of moisture content and water activity (a_w) (Seiler, 1988; Doerry, 1990). Examples of bakery products within each of three categories—low, intermediate and high moisture products, are shown in Table 9.1.

Pasta products can also be classified according to moisture content and a_w into dry, fresh and pre-cooked pasta as shown in Table 9.2 (Castelvetri, 1988). Dry pasta products can be further sub-divided into two categories, macaroni and noodles (Labuza, 1982). Macaroni includes such products as spaghetti, macaroni and vermicelli. This class of product is made from semolina, water, added vitamins and minerals and contains ~1.5% fat. The second category of dry pasta products,

Table 9.1. Classification of bakery products based on moisture content and water activity.

Product Type	Water Activity (a_W)
Low Moisture Products	
Cookies	0.20–0.30
Crackers	0.20–0.30
Intermediate Moisture Products	
Cake doughnuts	0.85–0.87
Chocolate coated doughnuts	0.82–0.83
Danish pastries	0.82–0.83
Cream-filled snack cakes	0.78–0.81
Pound cake	0.84–0.86
Banana cake	0.84–0.86
Soft cookies	0.50–0.78
High Moisture Products	
Bread	0.96–0.98
Yeast raised doughnuts	0.96–0.98
Fruit pies	0.95–0.98
Cheese cake	0.91–0.95
Carrot cake	0.94–0.96

Adapted from Doerry (1990) and Smith (1992).

Table 9.2. Classification of pasta products based on moisture content and water activity.

Product Type	Moisture Content (%w/w)	Water Activity (a_w)
Dry Pasta Products	<12	<0.6
Macaroni		
Spaghetti		
Vermicelli		
Noodles		
Egg spaghetti		
Egg vermicelli		
Fresh Pasta Products	26–34	0.92–0.95
Ravioli		
Tortellini		
Cannelloni		
Pre-cooked Products	40–50	<0.96
Lasagna		
Lasagna bolognese		
Macaroni and cheese		
Cannelloni with meat		
Cannelloni with vegetables		
Chicken fettucine		

Adapted from Castelvetri (1988) and Notermans et al. (1990).

i.e., noodles, is made from similar ingredients as macaroni products. However, they contain egg solids (~5.5%) to enhance colour, and therefore the lipid content of noodle products (~4.6%) is considerably higher than macaroni products. Examples of noodle products are noodles or egg noodles, egg macaroni, egg spaghetti and egg vermicelli. Fresh, filled pasta products include ravioli, tortellini and cannelloni while pre-cooked pasta products include lasagna, macaroni and cheese and chicken fettucine (Table 9.2).

SPOILAGE OF BAKERY AND PASTA PRODUCTS

Spoilage of bakery and pasta products can be sub-divided into (1) physical spoilage, e.g., moisture loss, staling, (2) chemical spoilage, e.g., rancidity and (3) microbiological spoilage, e.g., yeast, mold and bacterial growth. The pre-dominant spoilage problem is influenced by inter-related factors, specifically, storage temperature, level of preservatives, relative humidity, the gaseous environment surrounding the product and, most importantly, by the moisture content and a_w (Table 9.3). The common spoilage problems of both bakery and pasta products will be briefly reviewed.

Physical Spoilage

Moisture loss or pick-up is a serious problem in many bakery products irrespective of moisture level. Both moisture loss and moisture pick-up can be prevented by packaging products with proper sealing in moisture impermeable packaging material such as low density polyethylene (LDPE). However, the use of such films may result in conditions conducive to mold growth, particularly in high moisture bakery products.

The most common cause of physical spoilage of dry pasta products is moisture loss or gain and staling. The optimum moisture content for dry pasta is ~10–11%, i.e., equivalent to an a_w of 0.44 (Labuza, 1982). If pasta loses moisture to <6%, it becomes too fragile and unacceptable in quality. Conversely, if pasta gains moisture, to ~14–16%, mold growth and starch re-crystallization (retrogradation) occur. This latter defect results in an unacceptable tough product on cooking (Labuza, 1982). Moisture loss or gain can be prevented by packaging dry pasta products in plastic films such as cellophane, oriented polypropylene (OPP) or LDPE.

Table 9.3. Major causes of spoilage in bakery and pasta products based on water activity.

Major Cause of Spoilage	a_w Range of Product
Bacteria	0.91–0.95
Yeasts	0.87–0.91
Molds	0.80–0.87
Halophilic bacteria	0.75–0.80
Xerophilic molds	0.65–0.75
Osmophilic yeasts	0.60–0.65
Non-enzymatic browning	0.60–0.80
Enzymes (amylases)	0.95–1.00
Lipases	0.1
Oxidation of fats	0.01–0.50

Adapted from Doerry (1990).

Staling is a serious problem in bread and other fermented products but it is less important for other bakery products such as cakes and cookies. Staling has been defined as "all the physico-chemical changes that occur after baking and does not include changes that occur as a result of microbial spoilage" (Pomeranz and Shellenberger, 1971). The mechanism of staling has been the subject of many investigations. Several studies have suggested that staling is due to the moisture migration from the crumb to the crust and more specifically from swollen starch to gluten. Products with a higher moisture content, e.g., bread and cakes, stale faster than intermediate or low moisture products such as cookies and crackers. Staling, however, is not simply due to moisture loss or migration (Kulp, 1979). It has been shown that the degree and rate of crystallization (association) of starch components, specifically the non-linear amylopectin fraction, is mainly responsible for staling. Complex formation between starch polymers, lipids and flour proteins are thought to inhibit the aggregation of amylose and amylopectin (Kulp, 1979). Thus, the content of these components can influence the rate of staling. Cookies and biscuits for example, have a higher lipid content than bread and tend to stale more slowly. However, these products are more susceptible to lipid oxidation and the development of rancid flavors.

Staling is important economically to the bakery industry. It is estimated that ~1 billion dollars in bread sales are lost annually due to staling (Kulp, 1979). Measures to delay staling include reformulation, and the addition of an anti-staling enzyme, Multifresh™, developed by Enzyme Bio-Systems Ltd., Beloit, Wisconsin. This enzyme is active at temperatures above starch gelatinization and works by hydrolyzing the amylopectin fraction, thereby preventing re-crystallization and hence staling (Boyle and Hebeda, 1990).

Chemical Spoilage

Lipid degradation results in off-flavours and off-odours, commonly known as rancidity and these changes render products unpalatable and decrease shelf life. Two types of rancidity problems can occur—oxidative rancidity and hydrolytic rancidity.

Oxidative rancidity involves the breakdown of unsaturated fatty acids by O_2 through an autolytic free radical mechanism. This results in the formation of odourous aldehydes, ketones and short chain fatty acids. The free radicals and peroxides formed during lipid oxidation can also (1) bleach pigments, e.g., tomato paste in pizza, (2) destroy vitamins A and E, (3) breakdown protein, and (4) cause darkening of fat (Smith,

1992). Oxidative rancidity proceeds at its fastest rate at low a_w values, i.e., <0.3. (Table 9.3).

Hydrolytic rancidity, unlike oxidative rancidity, occurs in the absence of O_2. It results in the hydrolysis of triglycerides and the release of glycerol and malodourous short chain fatty acids. This type of rancidity, which is enhanced by the presence of moisture and endogenous enzymes such as lipases and lipoxygenases, has been shown to occur in frozen pizza. Lipases are present in the crust, in cheese, tomato sauce and vegetables while lipoxygenases are present in spices, sausage, wheat flour and vegetables. These enzymes catalyze the oxidation of unsaturated fats producing peroxides as well as volatile compounds which are heat stable and can survive the pre-cooking and baking process (Smith, 1992).

Pasta products such as noodles are more susceptible to lipid degradation due to their higher lipid content (~4.6%). Lipoxidase enzymes which are naturally present in semolina flour can oxidize lipids in the presence of light. The formed peroxides can also attack the carotene pigments resulting in colour loss (Labuza, 1982).

Microbiological Spoilage

While physical and chemical spoilage problems occur in many products, microbiological spoilage is often the major factor limiting the shelf life of both bakery and fresh pasta products. Microbiological spoilage is also a major cause of economic loss to the bakery industry. It has been estimated that in the U.S. alone, losses due to microbial spoilage are ~1 to 3%, or over 90 million kg of product each year (Ooraikul, 1991). This loss is based on in-plant and in-store spoilage only and if losses at the consumer level were also taken into account, the total loss could reach a staggering proportion (Ooraikul, 1991).

The most important factor influencing the microbial spoilage of bakery and pasta products is a_w (Table 9.3). For low moisture baked and dry pasta products (a_w < 0.6), microbiological spoilage is not a problem. In intermediate moisture products (a_w of 0.6–0.85), osmophilic yeasts and molds are the predominant spoilage microorganisms. In high moisture products (a_w of 0.94–0.99), almost all bacteria, yeasts and molds are capable of growth (Smith, 1992).

Bacterial Spoilage

Since most bacteria require a high a_w for growth, bacterial problems are limited to bakery products with a high moisture content. The major bacterial problem in bread is "rope" caused by *Bacillus mesentericus*, an aerobic spore forming bacteria. This microorganism, which is usually present in raw ingredients, e.g., flour, sugar, and yeast, survives the baking process and germinates upon cooling, growing under both aerobic and anaerobic packaging conditions. "Ropey" bread has a characteristic flavour not unlike ripe cantaloupe. The bread crumb becomes discoloured and sticky due to protein and starch degradation during growth of the bacteria (Smith, 1992). Rope problems can usually be overcome by the use of preservatives, such as propionates.

Pastries undergo deterioration similar to that in bread. However, when the pastries are filled they are subject to other types of spoilage. Most fillings support the growth of food spoilage microorganisms, especially if they contain egg or dairy products (Seiler, 1978). Custard-filled products are a potential health hazard due to the growth of *Bacillus cereus* and *Staphylococcus aureus* (Seiler, 1978). This latter pathogen has also been implicated in food poisoning outbreaks from cream-filled bakery products. Other bakery ingredients, such as chocolate, desiccated coconut and cocoa powder, have been implicated in outbreaks of *Salmonella* food poisoning from bakery products (Seiler, 1978). However, bacterial food poisoning outbreaks due to filled bakery products in both Canada and in the U.S. are rare. According to Todd et al. (1983), outbreaks of food poisoning involving cream-type pies accounted for ~0.6% of all foodborne outbreaks in Canada between 1973 and 1977 and 0.2% of outbreaks in the U.S. during the same time period. With the introduction of synthetic ingredients, e.g., imitation cream and processed ingredients such as pasteurized frozen or dried egg and skim milk, and with improved standards of both plant and personal hygiene, the opportunity for microbial contamination should be further reduced.

Bacterial problems in pasta products are mostly confined to fresh and pre-cooked pastas with moisture contents of 26–34% and 40–50%, respectively. The microbiological quality of fresh pasta depends on the quality of the raw ingredients as well as good manufacturing practices and temperature control during production and distribution.

Dry pasta has been reported to be contaminated with *Salmonella* species, *S. aureus*, *Clostridium perfringens*, molds and yeasts (Castelvetri, 1988). Therefore, these microorganisms are most likely to be found in fresh and pre-cooked pasta products. For fresh pasta products (moisture content 26–34%), bacteria of concern include most cocci, lactobacilli and aerobic sporeformers while gram-negative rods and anaerobic spore forming bacteria are the major bacteria of concern in pre-cooked pasta products with a moisture content of between 40 to 50%

(Castelvetri, 1988). The major spoilage bacteria isolated from these products are summarized in Table 9.4.

Yeast Spoilage

Yeast problems occur mainly in bakery products. According to Legan and Voysey (1991), yeast problems of bakery products can be subdivided into two broad types: 1) visible growth of yeasts on the surface of the products (white or pink patches); and, 2) fermentative spoilage of a wide range of products or ingredients manifested by alcoholic, ester or other odours and/or visible evidence of gas production, such as gas bubbles in jams and fondants or expansion of flexible packaging.

Visible yeast growth is generally associated with products of high a_w and short shelf life, while fermentative spoilage is usually associated with low a_w and long shelf life products, e.g., fruit cakes and Christmas puddings (Legan and Voysey, 1991). Yeasts, which cause surface spoilage of bread, are mainly *Pichia burtonii* ("chalk mold") and to a lesser extent *Candida guilliermondii*, *Hansenula anomala* and *Debaromyces hansenii* (Legan and Voysey, 1991). The most common osmotolerant yeast which causes spoilage of high sugar coatings and fillings such as jam, marzipan and mincemeat is *Zygosaccharomyces rouxii*. Contamination of products by osmophilic yeasts normally results from unclean utensils and equipment. Therefore, maintaining good manufacturing practices will minimize contamination by osmophilic

Table 9.4. Microbiological problems in pasta products.

Product Type	Moisture Content (%w/w)	Spoilage Microorganism(s)
Dry pasta products	<12 (a_w < 0.6)	Shelf-stable
Fresh pasta products	26–34 (a_w of 0.92–0.95)	*Staphylococcus* sp. *Streptococcus* sp. *Lactobacillus* sp. *Bacillus* sp. Molds
Pre-cooked products	40–50 (a_w < 0.96)	Gram-negative rods *Clostridium* sp. Yeasts

Adapted from Castelvetri (1988).

yeasts. Preservatives such as sorbates, benzoates and parabens are also effective in inhibiting the growth of yeasts.

Mold Spoilage

The most important microbial problem limiting the shelf life of high and intermediate bakery products is mold growth. Many molds are capable of growing at a_w values of >0.8 while a few xerophilic molds are capable of growth at a_w values as low as 0.65. Losses due to mold spoilage vary between 1 and 5% of products depending on season, type of product and method of processing (Malkki and Rauha, 1978). Mold spoilage of bakery products is therefore of serious economic concern to the industry. Although fresh bread and other baked products are free of viable vegetative molds and mold spores, products soon become contaminated as a result of post-baking contamination by mold spores from the air, bakery surfaces and equipment, food handlers and raw ingredients such as glazes, nuts, spices and sugars (Jarvis, 1972; Seiler, 1978; Seiler, 1988). Mold problems are most troublesome during the summer months due to airborne contamination and the warmer, more humid storage conditions. Furthermore, products may be wrapped prior to being completely cooled. This results in moisture condensing inside the package and on a product's surface, conditions which are conducive to mold growth.

The most common mold contaminants found on cake in the United Kingdom are *Wallemia sebi*, *Penicillium* species, *Cladosporium* species, *Eurotium (Aspergillus) glaucus* group and other aspergilli (Seiler, 1976). *Penicillium* species, in particular, *P. notatum*, *P. expansum* and *P. viridicatum*, were the predominant spoilage molds in products with a high Equilibrium Relative Humidity (ERH), i.e., products with an ERH > 86%. However, for products with ERH values below this level, *Eurotium glaucus* species, and in particular, *Eurotium amstelodami*, predominate (Seiler, 1976). During a one year study with bread obtained from forty-six bakeries and stored in plastic bags at 22°C for 5–6 d, *Penicillium* species were present in nearly all of the loaves while *Aspergillus* species and *Cladosporium* species occurred on approximately half the loaves (Legan and Voysey, 1991). Chalk molds (*P. burtonii*) were isolated from 5 to 30% of all loaves but peaked at 50% in the month of September. In another study with bread stored at 25°C and 70% ERH, visible yeasts caused less than 7% spoilage while *Penicillium roquefortii* caused 85% of the observed spoilage and other molds accounted for the remainder (Spicher, 1984). *P. roquefortii* is a rare contaminant in North

American and British bread but it is a common contaminant in German bread due to the sourdough breadmaking process (Spicher, 1984).

Methods for Shelf Life Extension of Bakery and Pasta Products

Since mold spoilage is often the main form of microbiological spoilage limiting the shelf life of the majority of high and intermediate moisture bakery products, methods to control mold spoilage are of significant economic importance to the bakery industry. Three basic approaches can be used to extend the mold free shelf life of bakery products. These are:

(1) Prevention of post baking contamination, e.g., packaging before baking or packaging after baking under aseptic conditions.
(2) Destruction of mold spores on the surfaces of products after wrapping, e.g., U.V. light, infra-red irradiation or microwave heating.
(3) Controlling mold growth in contaminated products, e.g., reformulation to reduce product a_w, the use of preservatives, e.g., sorbates or propionates incorporated directly into the product or applied to the surface as a spray (Seiler, 1988, 1989).

However, bacterial spoilage is the major problem limiting the shelf life of fresh and pre-cooked pasta products. While freezing can be used to extend the shelf life of such products, there is increasing consumer demand for refrigerated products with extended shelf life yet retaining "fresh like" characteristics. Furthermore, growing consumer concerns about preservatives and increasing energy costs associated with freezing has forced the food industry to look for alternative methods of food preservation. One such method is Modified Atmosphere Packaging (MAP), a technology used extensively to extend the shelf life of food in Europe and now gaining acceptance as a food preservation technique in the U.S. and Canada.

MODIFIED ATMOSPHERE PACKAGING

In recent years, Modified Atmosphere Packaging (MAP) has become increasingly popular as a preservation method to extend the shelf life of food products. MAP can be defined as "the enclosure of food products in a high gas barrier film in which the gaseous environment has been changed or modified to slow respiration rates, reduce microbiological growth and retard enzymatic spoilage with the intent of extending shelf life" (Young et al., 1988).

MAP technology is fast emerging as the packaging technology of the future. Currently, the United Kingdom leads the way in MAP technology followed by France and Germany. It is estimated that between 300–500 European companies use MAP technology for shelf life extension and distribution of food. With respect to the North American market, MAP technology is still in its infancy. However, it is estimated that the demand for MAP foods in North America could reach 11 billion packages by the year 2000 (Ooraikul, 1991; Smith, 1992).

Growth of MAP Technology

The anticipated growth of MAP technology, for both medium- and long-term preservation of food products, is due to a number of interrelated factors as follows.

Developments in New Polymeric Barrier Packaging Materials

The success of any packaging technology as a means of extending the shelf life of food is dependent on the permeability characteristics of packaging materials surrounding a product. Developments in polymer chemistry have resulted in production of packaging films such as polyvinylidene chloride (PVDC) and ethylene vinyl alcohol (EVOH). Both of these films have excellent water vapour and gas barrier characteristics and can be laminated to other polymers to give films with the desired strength, heat sealability and permeability characteristics for shelf life extension of products. In addition, developments in high speed continuous and thermoforming packaging equipment, compatible with the machinability characteristics of these films, have also promoted the growth of new packaging technologies.

Extended Market Areas

In the past, many food products were distributed on a regional basis. However, many larger companies are now distributing products on a national basis. Since high moisture bakery/pasta products have a limited shelf life due to microbiological spoilage, MAP in conjunction with proper storage conditions ensures that consumers nationwide have a constant supply of a variety of products which are shelf stable and nutritious. One Canadian success story was Forecrest Foods of Calgary, Canada, which produces crumpets. Prior to gas packaging, this product had a mold free shelf life of ~4 d. However, the mold free shelf life was extended to 28–35 d through reformulation and gas packaging. As a result of this extended shelf life, crumpets were distributed throughout Canada and into the U.S. market, and sales of crumpets have quadrupled.

Consumer Concerns about Preservatives

Consumers are becoming more aware and concerned about food additives and food preservatives. In a recent poll conducted by Agriculture Canada (personal communication), 75% of consumers were concerned about the levels of preservatives in food products while only 19% were concerned about food irradiation. The response by the food industry has been to use MAP as an alternative method of food preservation as the gases used in gas packaging are natural and do not need to be declared on the label.

Increasing Energy Costs

Increasing energy costs associated with traditional methods of food preservation/storage, such as freezing, have resulted in the growth of less energy intensive and more economical methods of short- and long-term preservation, e.g., MAP. It has been estimated that MAP is 18–20% less energy intensive compared to freezing for shelf life extension of bakery products (Aboagye, 1986).

Consumer Perceptions of MAP

Another recent consumer survey conducted by Agriculture Canada (personal communication) on consumers' perceptions of MAP, indicated that a significant proportion of the Canadian population favored MAP technology as a means of food preservation and would pay a premium for MAP food products. Most consumers agreed with the idea of foods preserved by natural gases such as CO_2. They also believed that MAP would save money by reducing waste and that the MAP process represented a significant improvement over current packaging technologies.

As a result of these inter-related factors, MAP is emerging as the preservation technology of the future for many food products, including bakery products and fresh and pre-cooked pasta products.

Several methods can be used to modify the atmosphere within the packaged products. These include vacuum packaging, gas packaging, the use of O_2 absorbents/gas generators and ethanol vapour generators.

The earliest form of MAP was Vacuum Packaging (VP). In VP, products are placed in high gas barrier flexible bags or rigid thermoformed trays, air is evacuated and the package sealed. Under conditions of a good vacuum, headspace O_2 is reduced to $<1\%$. This low level of headspace O_2 helps extend the shelf life of products by inhibiting oxidative rancidity and the growth of aerobic spoilage microorganisms. VP is used extensively to extend the shelf life of fresh and pre-cooked pasta products. However, the shelf life of these products depends on a number of inter-related factors, specifically the microbiological quality of the raw ingredients, packaging material permeability, packaging integrity, a_w, levels of preservatives and storage temperature. VP is not used to extend the shelf life of bakery products due to its "crushing" effect on softer products.

The injection of gas mixtures, more commonly known as gas packaging, is widely used by companies in both the United Kingdom and in Europe to extend the shelf life of food products. In Japan, however, and other Asian countries, O_2 absorbents/CO_2 generators and ethanol vapour generators are widely used to actively modify the headspace gas composition of the packaged bakery product.

Since gas packaging is now commonly used to extend the shelf life of bakery and pasta products, it will be discussed in more detail in this chapter. Novel methods of atmosphere modification involving O_2 absorbents/CO_2 generators and ethanol vapour generators will be reviewed in detail in a subsequent chapter.

Gas Packaging

There are essentially five elements in any MAP system involving gas packaging. These are: type of packaging equipment, gas composition, protective film, product quality and storage temperature. Each of these components will be briefly reviewed.

Gas Packaging Equipment

Gas packaging is simply an extension of vacuum packaging technology. The technique involves removing air from the pack and replacing it with a mixture of gases, the pressure of gas inside the package usually reaching about one atmosphere, i.e., equal to the external pressure. This is achieved by either continuous forming or thermoforming gas packaging equipment.

In the continuous forming or horizontal form, fill and seal equipment (Figure 9.1), the machine creates a tube of film which encloses the product. Gas is introduced in a continuous flow into the package to dilute the air present, the ends of the package are sealed and the packages are cut from each other (Figure 9.1). These machines are usually adapted to give a dwell time and are fitted with a gas analyzer to ensure that the correct gas mixture is flushed into the packs. The advantages of the form, fill and seal system are: (1) the style of pack used is similar for many bakery products; (2) the machines are versatile and can be used for products of various sizes and configurations using packaging materials which can be printed on all surfaces; and (3) a high production rate is achieved with as many as 120 packages per min being gas packaged. However, the disadvantages of the gas flushing technique are: (1) difficulty in complete removal of all traces of residual O_2; and (2) difficulty in obtaining gas tight seals, particularly where several layers of film are joined together at the end.

In the thermoforming technique (Figure 9.2), a compensated vacuum method is used to introduce the gas mixture. In this method, product is placed into a thermoformed tray and a vacuum is drawn to remove most of the air. The vacuum is broken by the appropriate gas mixture and the package heat sealed with a top web of film. The advantage of the thermoforming method of gas packaging is the high efficiency of removing O_2 to residual levels of $<1\%$, as well as the formation of a good gas-tight seal between the top web and the thermoformed tray. However,

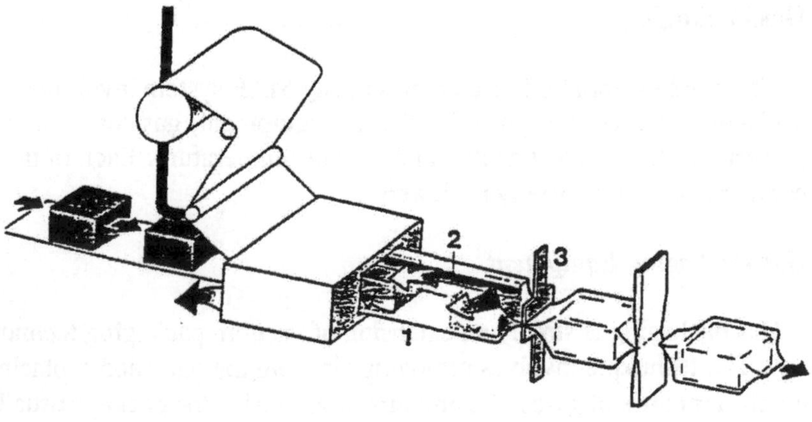

FIGURE 9.1 Continuous forming gas packaging equipment.

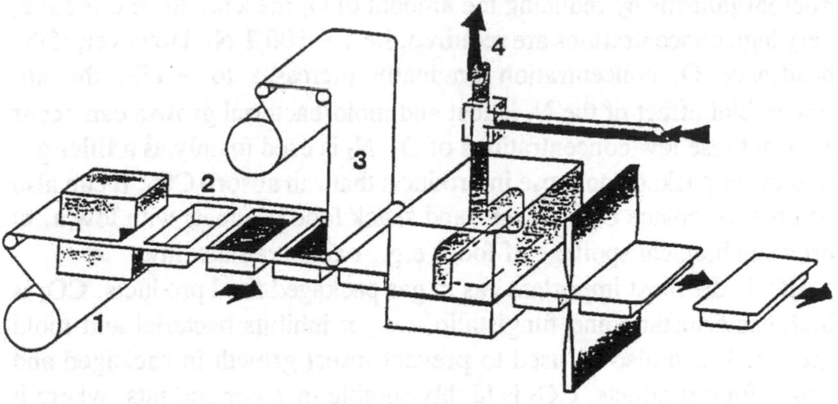

FIGURE 9.2 Thermoforming gas packaging equipment.

leakage problems may occur if the thermoformed tray is too thin or is damaged at the corners.

These two methods are commonly used for individual packaging of bakery/fresh pasta products where a long shelf life is desired. Bulk packaging of products is also possible by placing individual products packaged in O_2 permeable materials (LDPE) inside a masterpak consisting of a cardboard box lined with an impermeable liner. The individual packs are placed in the masterpak which is then evacuated and flushed with the appropriate gas concentration prior to sealing on a film lid. The disadvantage of this method is that protection only lasts until the products are put on display.

Gases Used in Gas Packaging

The gas packaging technique involves packaging of product in an impermeable film with the appropriate gases and heat sealing of packages. The gases commonly used for gas packaging of bakery products are N_2 and CO_2. These gases are neither toxic, nor dangerous, nor are they regarded as food additives. Other gases which have antimicrobial properties include carbon monoxide, sulphur dioxide, ethylene oxide, ozone, nitrous and nitric oxide. These are not included in MAP systems for a variety of reasons such as the stability of the gas, limited approval for use in foods, and the formation of toxic residues.

N_2 is an inert gas which has no effect on the food and has no antimicrobial effect per se. However, it may inhibit the growth of aerobic microorganisms by reducing the amount of O_2 present. To be effective, very high concentrations are required, i.e., ~100% N_2. However, if the headspace O_2 concentration gradually increases to ~1%, the antimicrobial effect of the N_2 is lost and mold/bacterial growth can occur even at these low concentrations of O_2. N_2 is used mainly as a filler gas to prevent package collapse in products that can absorb CO_2. It can also be used to replace O_2 in bakery and snack food products with low a_w to prevent chemical spoilage of food, e.g., oxidative rancidity.

CO_2 is the most important gas in gas packaged food products. CO_2 is both bacteriostatic and fungistatic, i.e., it inhibits bacterial and mold growth. It can also be used to prevent insect growth in packaged and stored food products. CO_2 is highly soluble in water and fats, where it forms carbonic acid. Its high solubility may lower the pH of the food product resulting in slight flavour changes. Absorption by the product may also cause package collapse.

Antimicrobial Effect of CO_2

Although the preservative action of CO_2 in foodstuffs has been known for many years, its mechanism of antimicrobial action has not been fully determined. However, several theories have been postulated. One theory suggested that displacement of O_2 was the main reason for the antimicrobial properties of CO_2. This theory was refuted by Coyne (1933) who showed that aerobic spoilage organisms of fish grew well in 100% N_2 but not in 100% CO_2, indicating that displacement of O_2 was not the only reason for the antimicrobial effect. Valley and Rettger (1927) suggested that CO_2 acted by lowering extracellular pH as a result of the dissolution of CO_2 in the aqueous phase of the product. However, when the pH was lowered by HCl to values equivalent to those achieved under CO_2 atmospheres, bacterial growth was less inhibited (Valley and Rettger, 1927). Furthermore, CO_2 will inhibit microbial growth in buffered media and in naturally buffered foods such as meat. In response to these observations, Wolfe (1980) suggested that the inhibitory effect of CO_2 may be due to intracellular, rather than extracellular pH changes, which could interfere with enzymatic activities associated with cell metabolism. A further theory is that CO_2 inhibits enzymes associated with aerobic metabolism such as oxaloacetate decarboxylase, succinate dehydrogenase and cytochrome oxidase activity. However, King and Nagel (1975) demonstrated that a 50% CO_2 atmosphere did not inhibit extracts of these enzymes. They observed that CO_2 specifically inhibited malic and isocitric dehydrogenase activity in vitro and concluded that the inhibitory effect of CO_2 may be due to its mass action effect on decarboxylases within the cell. Another theory suggests that CO_2 acts on the cell membrane, affecting its permeability characteristics and its external environment by redistribution of lipids at the surface. This has been demonstrated using a model system by Sears and Eisenberg (1961), and has been proposed as the mechanism by which CO_2 inhibits spore germination (Enfors and Molin, 1978).

In conclusion, while many studies have been done on the effect of CO_2 on microorganisms, there is little conclusive evidence on its mechanism of action. In an excellent review on the effects of CO_2 on microbial growth and food quality, the following appear to be the salient points of all previous investigations (Daniels et al., 1985).

(1) The exclusion of O_2 by replacement with CO_2 may contribute slightly to the overall antimicrobial effect by slowing the growth of aerobic spoilage microorganisms.
(2) The CO_2/HCO_3^- ion has an observed effect on the permeability of cell membranes.
(3) CO_2 is able to produce a rapid acidification of the internal pH of the microbial cell with possible ramifications relating to metabolic activities.
(4) CO_2 appears to exert an effect on certain enzyme systems.

Whatever the reason for its antimicrobial effect, CO_2 is effective for extending the shelf life of food products by retarding the growth of aerobic spoilage microorganisms. The overall effect of CO_2 is to increase both the lag phase and generation time of spoilage and pathogenic microorganisms.

Factors Influencing the Antimicrobial Effect of CO_2

Factors that influence the antimicrobial effect of CO_2 include types and numbers of microorganisms, gas concentration, temperature, and packaging film permeability.

Types of Microorganisms

The numbers and types of organisms present in a food product influence the antimicrobial effect of CO_2. Microorganisms differ considerably in their sensitivity to CO_2 and this sensitivity is related to their requirements for O_2. It has been shown that CO_2 is most effective against aerobic spoilage microorganisms, such as aerobic bacteria and molds, with concentrations of CO_2 as low as 5−10% being used to suppress growth of these spoilage microorganisms (Ooraikul, 1991; Smith, 1992).

Generally, gram-negative bacteria are more sensitive to CO_2-enriched atmospheres than gram-positive types. However, while certain gram-positive bacteria, e.g., *Brochothrix thermosphacta* and lactic acid bacteria are very resistant to CO_2 and can grow in atmospheres containing 75−100% CO_2, other bacteria, e.g., *Micrococcus* and *Bacillus* species are very sensitive to CO_2. Anaerobic bacteria, such as food poisoning strains of *Clostridium botulinum*, are not inhibited by elevated levels of CO_2 in the gas atmosphere and indeed their growth may actually be stimulated by CO_2. There is concern that these microorganisms may represent a public health hazard in minimally processed gas packaged fresh and pre-cooked pasta products, particularly at temperature abuse storage conditions (Smith, 1992).

Mold species also vary in their sensitivity to the inhibitory effects of CO_2. The *Eurotium (Aspergillus) glaucus* group of molds was observed to be more susceptible to the effect of CO_2 than *Penicillium* species, the predominant spoilage molds in bakery products with an ERH of 86% and above (Seiler, 1989). Furthermore, certain *Penicillium* species, in particular *P. roquefortii*, a common contaminant in German rye bread, was more CO_2 resistant than others. Studies with products of differing ERH inoculated with the same *Penicillium* species failed to show that the activity of CO_2 increases as the ERH decreases (Seiler, 1989). It was concluded that the type of mold present is a more important factor affecting the antimycotic effect of CO_2 than the ERH of the product (Seiler, 1989).

When baked products of high ERH are gas packaged, the predominant spoilage microorganisms may be yeasts or lactic acid bacteria due to the microaerophilic, fermentative nature of these spoilage microorganisms and hence their resistance to high levels of CO_2. Surface growth of the filamentous yeast *P. burtonii* limited the shelf life of gas packaged rye bread, while the shelf life of gas packaged apple turnovers was limited due to gas production and swelling of packages by CO_2 resistant yeasts (Smith et al., 1987). Lactic acid bacteria, in particular *Leuconostoc mesenteroides*, were responsible for the spoilage and termination of shelf life of gas packaged crumpets (Smith et al., 1983). These secondary spoilage problems of gas packaged bakery products are discussed in more detail in this chapter.

The numbers and age of the microbial population also influence the inhibitory effect of CO_2. Gas packaged baked products contaminated with a high spore load will have a reduced shelf life compared to similarly packaged products contaminated with a low spore load. Gas packaging should not be regarded as a panacea for hygiene. The shelf life of gas packaged bakery products will only be as effective as the manufacturing conditions under which they are produced, cooled and packaged. Furthermore, as microorganisms move from the lag phase to the log phase, the inhibitory effects of CO_2 are reduced (Ooraikul, 1991). Thus, the antimicrobial effect of CO_2 will be enhanced the earlier a product is gas packaged.

Concentration of CO_2

Early experiments clearly demonstrated that the success in controlling the growth of aerobic spoilage microorganisms in food was not simply due to the elimination of O_2; rather there was a definite need for CO_2 in the gas atmosphere (Valley and Rettger, 1927; Coyne, 1933). Furthermore, the growth of aerobic bacteria and molds can be inhibited by low concentrations of CO_2. For example, the growth of *Aspergillus niger*, *Rhizopus nigricans* and *Penicillium expansum* could be controlled for ~2 to 3 d in bread stored in 10 and 17% CO_2 at 28°C (Skovholt and Bailey, 1933). However, a concentration of 50% CO_2 was needed for complete inhibition of mold growth on bread. Upon transfer from the CO_2 enriched atmosphere to air, mold growth occurred as if no treatment had been applied, confirming the fact that CO_2 exerts a fungistatic effect and not a fungicidal one (Skovholt and Bailey, 1933). Similar observations have been reported in gas packaging studies of crumpets (Ooraikul, 1982). Mold growth was evident after 7 d in products packaged under 25% CO_2 and stored at 25°C. However, higher concentrations of CO_2 in the package headspace (50–100%) resulted in complete inhibition of mold growth throughout 14 d storage at 25°C (Ooraikul, 1982).

It is evident from these studies that the concentration of CO_2 in the gas mix is critical if the desired extension in shelf life of a product is to be achieved. For most products, a minimum of 20–30% CO_2 (v/v) is required to inhibit the majority of aerobic spoilage microorganisms while for longer shelf life extensions, a concentration of ~60% should be used. This is based on studies which have shown that the inhibitory effect of CO_2 increases linearly with increasing concentration, with little or no increased effect being shown at concentrations above 50–60% (Ooraikul, 1982). However, while there is little or no increased antimicrobial effect or extension in product shelf life at concentrations of CO_2 above 50−60%, slightly higher concentrations (70−80%) are sometimes used to compensate for losses of CO_2 through packaging films or its absorption by products. However, higher concentrations of CO_2 (100%) can result in the formation of a partial vacuum within the product and package collapse (Ooraikul, 1982). While this problem can be overcome by using mixtures of CO_2:N_2, Seiler (1989) believes there is little justification in using gas mixtures for the preservation of bakery products, apart from lower cost. Seiler (1989) recommends the use of 100% CO_2 in conjunction with packaging films which are less permeable to CO_2, to prevent/reduce the vacuum package effect.

Temperature

Storage temperature also affects the antimicrobial activity of CO_2. It has been shown that CO_2 is a very effective antimicrobial agent at low storage temperatures, but less effective at higher temperatures. This increased inhibitory effect at lower storage temperatures has been attributed to the greater dissolution of CO_2 in the aqueous phase of products and resultant changes in the intra-cellular pH and enzymatic activities of microorganisms. Therefore, the decrease in inhibition at higher storage temperatures results from the lower solubility of CO_2 in the aqueous phase of a product. MAP should not be regarded as a substitute for proper storage temperature. While MAP slows down the deterioration of a food product, it never totally arrests these deteriorative reactions.

The effects of temperature abuse on gas packaged products are twofold. Firstly, microorganisms will overcome the inhibitory effect of the CO_2-enriched atmosphere and spoilage may occur within the "best buy" date of the product. Secondly, higher temperatures may change the permeability characteristics of a high barrier film resulting in an increase in headspace O_2. This will enhance both microbial spoilage and chemical spoilage (oxidative rancidity) problems in gas packaged products.

Temperature control is also essential if gas packaging is used with any meat- or cream-filled baked products. Several studies have shown that *Salmonella* and *Staphylococcus* species can grow under anaerobic conditions at 10 to 12.5°C, i.e., conditions of mild temperature abuse (Palumbo, 1986). Strict temperature control is also critical to ensure the safety of minimally processed fresh and pre-cooked pasta products packaged under a modified atmosphere.

Film Type

One of the most important factors influencing the antimicrobial effect of CO_2 is packaging film permeability. The success or failure of MAP foods depends on both the O_2 and CO_2 impermeability of packaging materials necessary to maintain the correct gas mixture in the package headspace. In addition, films used in gas packaging should also have low water vapour transmission rates to prevent moisture loss or moisture gain. Polymers commonly used for gas packaging of food include polyester (nylon), polypropylene (PP), PVDC, EVOH, and LDPE (Smith et al., 1990).

If only a short shelf life is desired, e.g., 2–3 d for a gas packaged bakery product such as bread, LDPE bags are suitable. However, if a longer shelf life is desired, individual polymers are laminated to one another since all the desired characteristics of a packaging film for MAP applications, i.e., strength, impermeability and heat sealability, are seldom found in one polymer. Examples of laminated structures for gas packaging of non-respiring products include nylon/PE, nylon/PVDC/PE or nylon/EVOH/PE. These composite structures have all the desired characteristics of a packaging film for gas packaging applications, specifically, strength, provided by the outermost layer of nylon, gas and moisture vapour impermeability, provided by EVOH or PVDC and heat sealability, provided by LDPE or ethylene vinyl acetate (EVA) or an ionomer (Surlyn). The important attributes of laminated films for MAP foods are: 1) high lamination bond strength; 2) consistent and uniform thickness; 3) consistent seal strength; and 4) consistent barrier to O_2 and moisture vapour (Smith et al., 1990). This latter attribute is very critical since the O_2 level in the package headspace may soon reach concentrations of 1% or more, even using high barrier films. Several studies have shown that molds can tolerate and grow in such low concentrations of headspace O_2, even in the presence of elevated levels of CO_2 (Smith et al., 1986, 1987; Ellis et al., 1993).

Laminated films commonly used for gas packaging of bakery products are nylon/LDPE (medium barrier film) and PVDC-coated polypropylene laminated with polyethylene or ionomer (high barrier film). Surlyn is preferable to polyethylene as the heat sealant layer since the ionomer has been shown to give consistent seals when crumbs and other debris are lodged in the heat sealing layers (Seiler, 1989).

The actual packaging materials used for fresh pasta products depend on whether or not the product is pasteurized (in which case the package must be able to withstand the pasteurization process without deforming). Furthermore, if the product is to be re-heated in its package in a microwave by the consumer, then the packaging material must be able to withstand domestic microwave temperatures (Robertson, 1993).

For air packaged products which are not pasteurized nor intended to be heated in their package, a rigid tray of PVC/LDPE onto which is sealed a nylon/LDPE lid is common (Robertson, 1993). Fresh pasta products are gas packaged in a thermoformed polyvinyl chloride (PVC) rigid-base tray (Ernst, 1989). The trays are then sealed after flushing with the appropriate CO_2:N_2 mixture with ethylene vinyl alcohol (EVOH), a high gas barrier lidding material. As a further control measure, an O_2 absorber can be added to each package to completely remove any residual headspace O_2.

However, if the MAP pasta products are pasteurized and re-heated using microwave energy, the rigid tray is usually constructed from co-extruded polyester (CPET) and sealed with a high barrier top web of a polyvinyl/polyvinylidene (PVC/PVDC) copolymer coated polyester or polypropylene (PET or PP) (Robertson, 1993).

Use of Gas Packaging for Shelf Life Extension of Bakery Products

The use of CO_2-enriched atmospheres for shelf-life extension of food is not a new concept in food preservation. As early as the 19th century, scientists discovered that increased levels of CO_2 and reduced levels of O_2 retarded catabolic reactions in respiring foods and inhibited the growth of aerobic spoilage microorganisms. The potential of CO_2-enriched atmospheres to extend the mold-free shelf life of bakery products was demonstrated in early experiments with bread (Skovholt and Bailey, 1933). They showed that mold growth could be partially inhibited in bread stored under 17% CO_2 and completely inhibited when stored under 50% CO_2. Extensive research on the use of CO_2-enriched atmospheres for shelf life extension of bakery products was done by Seiler at the Flour Milling and Bakery Research Association in England. In detailed studies with bread and cake stored under various concentrations of CO_2 (0–60%) at 21 and 27°C, the mold-free shelf life of products increased as the concentration of CO_2 in the atmosphere increased, and this effect was greater the lower the storage temperature (Seiler, 1966). Furthermore, similar extensions in the mold-free shelf life of both bread and cake were observed, irrespective of the ERH of the products (Seiler, 1966). Substantial increases in shelf life have also been observed for bread and cake packaged in mixtures of CO_2 and N_2 and in 100% N_2 (Bogadtke, 1979). However, if the residual O_2 concentration was >1% in the N_2 packaged bread, mold growth was evident after only 5 d compared to 100 d for bread packaged in an atmosphere of 99% CO_2 and 1% O_2 (Bogadtke, 1979).

More recently, the effect of different concentrations of CO_2 (0–100%) on the shelf life of a variety of bakery products (Madeira cake, crumpets, par-baked bread, rye bread and fruit pies) stored at 21 and 27°C was investigated (Seiler, 1989). Once again, the mold-free shelf life increased with increasing concentrations of CO_2 in the package headspace. However, contrary to previous observations, increases in shelf life were similar at both storage temperatures. Furthermore, the mold-free shelf life of products of lower a_w was greater than those of higher a_w products, such as crumpets, fruit pies and bread, particularly at higher concentrations of CO_2 (80–100%) (Seiler, 1989). Seiler concluded that the difference in shelf life extension was due to the types of molds present, rather than the enhanced antimicrobial effect of CO_2 at reduced relative humidities. In low moisture ($a_w < 0.85$) products, *Eurotium* species pre-dominated while in higher a_w products (0.92–0.96), e.g., crumpets and fruit pies, *Penicillium* species were more common (Seiler, 1989).

Extensive studies have been done to investigate the suitability of gas packaging involving a CO_2:N_2 (60:40) gas mixture to extend the shelf life of a variety of bakery products other than bread and cake, and to evaluate secondary problems which might be encountered in these products (Smith et al., 1983, 1986; Ooraikul, 1991). Examples of the products investigated are shown in Table 9.5. These products were selected because they were representative of products with different formulations and hence, different physical and chemical characteristics. Preliminary studies indicated that gas packaging was an effective means of extending the physical, chemical and microbiological shelf life of most of these products. However, both minor and major secondary spoilage problems were encountered in certain gas packaged bakery products.

Minor problems encountered were related to the texture, colour and taste of the gas packaged products. A "stale" taste was encountered in crumpets and waffles after 3–4 weeks storage at ambient temperature. They also became slightly discoloured at the end of the storage period. However, both of these problems were overcome by toasting the products

Table 9.5. Gas packaging of bakery products and consequent problems.

Product	a_w	pH	Moisture %	Swelling*	Mold**	Yeast**	Bacteria**	Summary of Problems†
Dough or batter:								
Cake doughnut	0.82	6.60	17.85	–	–	–	+	T,C
Crumpet	0.97	6.00	47–52.41	+ + +	–	–	+ + +	B,S
Crusty roll (yeast)	0.95	5.60	29.28	–	+ + +	–	+ +	M (14 d)
Yeast doughnut	0.91	6.40	27.98	+	–	+	+ + +	B,Y,S (14 d)
Waffle	0.94	7.20	65.79	–	–	–	+ +	T,C
Cake/pastry:								
Chocolate danishes	0.83	6.28	22.54	–	–	–	–	None
Carrot muffin	0.91	8.70	35.14	–	–	–	+	None
Butter tart	0.78	5.70	19.95	–	–	–	+	None
Cake (layer):								
Strawberry layer cake	0.90	6.66	37.24	+ +	–	+ +	+ +	Y,B,S (21 d)
Cherry cream cheese cake	0.94	4.51	49.90	+ +	–	+ +	–	Y,S (14 d)

Table 9.5. (continued).

Product	a_w	pH	Moisture %	Swelling*	Mold**	Yeast**	Bacteria**	Summary of Problems†
Pie								
Mini blueberry pie	0.94	3.78	40.20	–	–	–	–	T,C
Apple turnover	0.94	4.60	35.12	+ + +	–	+ + +	+ + +	Y,B,S (14 d)
Apple pie baked	0.95	4.21	47.82	–	–			B,Y,T,C
Apple pie raw	0.96	4.25	54.75	+ + +	–	+ + +	+ + +	Y,B,S (7 d)

*Extent of swelling of package during storage due to in-package CO_2 production.

**Product shelf life terminated by visible appearance of molds and/or increase in yeast and bacterial counts.

†Summary of Problems: T = Texture, C = Colour, B = Bacteria, Y = Yeast, M = Mold, S = Swelling (of package). Code: Absent; + = Slight; + + = Medium; + + + = Extensive.

Adapted from Ooraikul (1991).

prior to consumption. The shell of butter tart, mini-blueberry pies and apple pies became crumbly during extended storage (Ooraikul, 1991). A different type of shortening was recommended to overcome this textural problem in these gas packaged products. Cake doughnuts also became sticky during storage, probably due to moisture migration from the inside of the doughnut. This problem could be overcome by using a less hygroscopic sugar or sweetening agent or an anti-caking agent (Ooraikul, 1991).

Major problems encountered in some products were due to either mold, yeast or bacterial growth resulting in off-odours and in certain cases, swelling of packages. Mold growth limited the shelf life of gas packaged crusty rolls to 14 d at ambient storage temperature (Table 9.5). This problem was due to the inability of the packaging system to reduce the headspace O_2 levels to $<0.6\%$ and to maintain it at this level (Smith et al., 1986). Studies have shown that *Aspergillus* species and *Penicillium* species, the common mold contaminants of crusty rolls, can tolerate and grow in the presence of a low concentration of O_2 even in the presence of high concentrations of CO_2 (Smith et al., 1986). The mold problem was resolved using an Ageless O_2 absorbent (Mitsubishi Gas Chemical Co.), to ensure that the residual O_2 was completely "scavenged" from the package headspace.

By contrast, the shelf life of certain products, specifically crumpets, apple turnovers, strawberry layer cake, yeast doughnut and cherry cream cheese cake was limited by in-package production of CO_2, resulting in all packages having a blown appearance. The shelf life of crumpets was terminated after 14 d due to the growth and heterofermentative activity of lactic acid bacteria, specifically, *L. mesenteroides*, while the shelf life of apple turnovers and cheesecakes was terminated after ~ 16 d due to CO_2 production by the yeast *Saccharomyces cerevisiae* (Smith et al., 1983, 1987). Subsequent studies were directed at further extending the shelf life of these products. A process optimization technique, called Response Surface Methodology, was used to determine the levels of a_w, pH, and CO_2 which could be used to control both mold spoilage and gas production by *L. mesenteroides* in gas packaged crumpets. Subsequent reformulation of the product to the appropriate levels of a_w, pH and packaging under a suitable CO_2 concentration inhibited both mold and bacterial problems, and the ambient storage life of crumpets could be extended beyond 21 d (Smith et al., 1988). Similar extensions in shelf life were possible for apple turnovers and cream filled products using ethanol vapour generators (Ooraikul, 1991). These methods of additional control for shelf life extension of specific bakery products using O_2 absorbent technology and ethanol vapour generators will be discussed in more detail in a subsequent chapter.

Examples of other possible gas mixtures to extend shelf life of selected bakery products are shown in Table 9.6. The optimum blend of gases for a specific product cannot simply be determined by trial and error but only through a detailed, systematic study of the variables influencing product shelf life. Using the appropriate CO_2:N_2 mixture, the mold-free shelf life of baked products can be extended for upwards of 3 weeks to 3 months at room temperature (Table 9.7). Although microbial quality is usually the primary concern of the processor, there are many chemical changes which can take place over time which adversely affect colour, freshness, flavour and texture.

Effect of Gas Packaging on Staling of Bakery Products

To date, most studies have focused on the use of MAP to extend the mold-free shelf life of products. A few studies, however, have been done to determine the anti-staling effect of CO_2 with conflicting results

Table 9.6. Typical gas mixtures used in gas packaged bakery products.

	Gas % (v/v)		
Product	CO_2	N_2	O_2
Sliced bread	100	—	—
Rye bread	100	—	—
Buns	100	—	—
Brioches	100	—	—
Cakes	100	—	—
Madeira cake	65	35	—
Madeira cake	80	20	—
Tea cakes	50	50	—
Danish pastries	50	50	—
Crepes	60	40	—
Croissants	100	—	—
Crumpets	100	—	—
Crumpets	60	40	—
Pita bread	99	1	—
Pita bread	73	27	—

Adapted from Goodburn and Halligan (1988) and Ooraikul (1991).

Table 9.7. Typical shelf life for gas packaged bakery products at ambient temperature.

Product	Shelf Life
Bread	1–6 weeks
Cakes	3–9 months
Croissants	15–25 days
Doughnuts	up to 25 days
English muffins	up to 3 weeks
Pastries	up to 45 days
Pizza crusts	1–2 months
Mini muffins	28 days
Nut loaves	3 months
Cinnamon rolls	30 days

Adapted from Goodburn and Halligan (1988), Ooraikul (1991), and American Institute of Baking (Personal Communication).

(Doerry, 1985; Knorr and Tomlins, 1985; Knorr, 1987). The staling rate of white and whole wheat bread and biscuits was significantly reduced when packaged in 100% CO_2 compared to products packaged in 100% N_2 and in air (Knorr and Tomlins, 1985). Subsequent studies confirmed the anti-staling effect of pure CO_2 atmospheres for French bread and white bread compared to air stored samples (Knorr, 1987). Fermentation and baking of enriched white bread dough under CO_2 modified atmosphere also resulted in softer breads than air processed samples (Knorr, 1987). Other studies have also shown that CO_2 has an anti-staling effect on bread (Avital et al., 1990). They concluded that this effect was possibly due to the change in the water relations of bread caused by a blockage of water binding sites in the amylopectin fraction by CO_2 (Avital et al., 1990).

In contrast, no difference was found in the rate of crumb firming between white pan bread, pound cake and sponge cake stored in 100% CO_2, 100% N_2 and in air (Doerry, 1985). Other studies have confirmed that packaging in air or a CO_2-enriched atmosphere had no effect on the staling rate of white bread (Seiler, 1988).

Use of Gas Packaging for Shelf Life Extension of Fresh Pasta Products

The production of fresh pasta products generally involves kneading of the dough followed by extrusion or lamination, cutting or shaping with or without special prepared fillings such as meat or vegetable and cheese mixture, steam heating or pre-cooking, partial surface drying and packaging in air. Fresh and fresh filled pasta products are not usually subjected to pasteurization. Therefore, such high-moisture, high-a_w products, particularly stuffed products such as ravioli, tortellini and cannelloni are highly perishable and have a relatively short shelf life at room and refrigerated temperatures. Attempts at extending the shelf life of these products using food preservation techniques such as canning, freezing or low-dose irradiation have either proved impractical from a textural viewpoint or uneconomical.

Recently, "fresh" or high moisture pastas packaged in barrier trays under a CO_2:N_2 atmosphere have been successfully introduced into the marketplace. These MAP pasta products provide consumers with ready-to-eat, minimally processed products with extended shelf life yet retaining "fresh-like" characteristics. Examples of the gas combinations commonly used to give a 14−21 d shelf life for fresh pastas are shown in Table 9.8 (Castelvetri, 1988). However, a major concern about these non-pasteurized MAP pasta products is their microbiological quality. This depends on the quality of the raw ingredients used in their formulation, as well as good manufacturing practices employed during all stages of processing, packaging and distribution of products. Furthermore, there is concern that these products, if subjected to temperature abuse, could pose a public health hazard. This concern is justified in view of the many pathogenic bacteria such as *Listeria monocytogenes*, *Salmonella* species, *C. botulinum*, *Yersinia enterocolitica* and *S. aureus* which may be tolerant of, and may even be stimulated by high concentrations of CO_2 in the package headspace and can grow at $< 15°C$, i.e., mild temperature abuse storage conditions (Hintlian and Hotchkiss, 1986).

Table 9.8. Examples of gas mixtures to extend the shelf life of fresh pasta.

Product	Gas % (v/v) CO_2	Gas % (v/v) N_2	Storage Temperature (°C)	Shelf Life (days)
Pasta	80	20	4	14
Pasta with meat	80	20	4	14
Pasta with meat	50	50	4	14
Ravioli	20	80	2	21
Lasagne	70	30	2	15

Adapted from Castelvetri (1988) and Goodburn and Halligan (1988).

In a recent survey of fresh pasta products packaged under a CO_2:N_2 (20:80) mixture from five processors for staphylococci and their enterotoxins, ~12% of fresh pasta products were contaminated with *S. aureus*. At 5°C, the number of *S. aureus* decreased with storage time and no enterotoxin was detected in any gas packaged sample (Park et al., 1988). However, when the product was temperature abused, ~22% of samples had *S. aureus* at levels ranging from 10^3 to 10^7 cfu/g and 8% contained staphylococcal enterotoxin after 4 weeks at 16°C (Park et al., 1988). These results clearly show that proper refrigeration is essential to ensure the safety of MAP pastas.

Longer extensions in shelf life of fresh MAP pasta products are possible if the products are pasteurized prior to or immediately after packaging in a modified atmosphere. A shelf life of ~120 d for fresh pasta (fettucine) was obtained using a combination treatment of steam pasteurization and gas packaging (McGuire et al., 1989). The pasta was prepared from a uniform blend of flour and whole egg with a moisture content of 30%. The dough was conditioned, subjected to heat to dry the external surfaces and then cut into shapes. The product was then steam pasteurized for ~4 min to achieve a product center temperature of 180°C, cooled and then packaged in a modified atmosphere (McGuire et al., 1989). The gas mixture used for this product was 80% CO_2:20% N_2. Although high levels of CO_2 are preferred to control microbial growth, elevated levels can cause product blistering. This defect, however, can be prevented by reducing the temperature of the pasteurized product to ~ −4 to 2°C prior to gas packaging. Product shelf life ranged from 50−120 d at 10−4°C (McGuire et al., 1989). During the 14 week storage period, staphylococci, mold, yeast, lactic acid bacteria and psychrotrophic counts remained constant (McGuire et al., 1989). The standard plate count, however, increased slightly after 2−3 weeks then decreased and increased again after week 10 to 11 at 10°C (McGuire et al., 1989).

Several companies are also pasteurizing fresh pasta products after packaging. Products are vacuum or gas packaged in either rigid or semi-rigid containers, and then pasteurized using hot air or microwave energy (Castelvetri, 1988). One company is also using O_2 absorbent technology in conjunction with vacuum/gas packaging to ensure the complete removal of headspace O_2. Additional "hurdles" or "barriers" such as a_w and pH reduction can also be incorporated into the products to extend shelf life. The combined effects of MAP, a_w, pH, pasteurization and storage temperature on the shelf life and microbiological stability of fresh pasta are summarized in Table 9.9 (Castelvetri, 1988).

Microbiological Safety of Minimally Processed Pasta Products

Fresh, pasteurized pasta products fall under the category of REPFEDs, i.e., refrigerated, processed foods with extended durability (Notermans et al., 1990). If the pasteurization process is carried out adequately, all vegetative bacteria, molds and yeasts present are killed. However, spores of many bacteria, specifically *C. botulinum*, can survive the heating process and may pose a public health concern if products are temperature abused at some stage during distribution and storage in the retail or domestic environment, prior to consumption. In recent challenge studies with *C. botulinum* types B, E and F spores in minimally processed pasta products which did not contain intrinsic safety factors, toxin production occurred within 3 weeks in products stored at 8°C, i.e., a temperature at which refrigerated foods are often exposed to during and after retail sale (Notermans et al., 1990). Furthermore, pre-storage at 3°C for 2 to 4 weeks stimulated toxigenesis of *C. botulinum* type B on subsequent storage at 8°C (Notermans et al., 1990). Microwave heating at 700 W for 5 min did not completely inactivate the pre-formed toxin in tortellini, cannelloni and lasagna. Complete inactivation could only be achieved by heating products for 10 min (Notermans et al., 1990). However, a longer heating time adversely affected the ap-

Table 9.9. Combined effect of several "hurdles" on the shelf life and safety of MAP fresh pasta.

MAP	pH	a_w	Pasteurization	Storage Temperature (°C)	Shelf Life (days)
+	>5	>0.95	–	4	<15
+	>5	>0.95	+	4	~30
+	>5	<0.95	–	4	~30
+	>5	<0.95	+	4	90
+	<5	<0.95	–	4	<30
+	<5	<0.95	+	RT*	~90
+	<4.5	•0.89	+	RT	Shelf-stable

*RT = Room temperature.
Adapted from Castelvetri (1988).

pearance of the products. Therefore, microwave heating of REPFEDs prior to consumption cannot be regarded as an additional safety barrier to reduce the botulinum risk. This can only be achieved by storing REPFEDs at <3.3°C. If this storage temperature cannot be guaranteed, the storage time has to be limited (Notermans et al., 1990).

Another strategy to increase the microbiological safety of REPFEDs would be to incorporate additional barriers, such as reduced pH or a_w into the minimally processed products. It is evident from Table 9.9 that reformulation of pasteurized-MAP pasta products to a pH < 5 and an $a_w < 0.95$ can have a pronounced effect on product shelf life at ambient storage temperatures. Such combination treatments in conjunction with refrigerated storage temperature would maximize the safety of minimally processed, MAP pasta products (Castelvetri, 1988).

Advantages and Disadvantages of Gas Packaging Bakery and Pasta Products

The main benefits associated with gas packaging of bakery and pasta products are extended product shelf life and associated increase in market area, improved product presentation and a reduction in processing, transportation and storage costs of MAP products compared to freezing storage costs (Table 9.10). For example, the production, storage and transportation costs of frozen crumpets was 45% more energy intensive than MAP crumpets (Aboagye, 1986). This was reduced to an overall cost advantage of 14% over freezing, due mainly to the higher cost of packaging materials used in gas packaging. There-

Table 9.10. Advantages and disadvantages of gas packaging of food.

Advantages	Disadvantages
Increased shelf life	Initial high cost of packaging equipment, films, etc.
Increased market area	Discolouration of pigments
Reduction in production costs	Leakage
Improved presentation	Secondary fermentation and swelling of products
Fresh appearance	Potential growth of organisms of public health concern
Clear view of product	Increased package volume and storage volumes
Easy separation of slices	
Favorable consumer perception of gas packaging	

fore, for a bakery with an annual production of 1,000,000 kg of crumpets, a saving of $34,730 could be achieved by using MAP instead of freezing (Aboagye, 1986).

Presently, more than 150 European bakery firms are using gas packaging technology to extend the mold-free shelf life and keeping quality of rolls, cakes, pizza, baguettes and sliced bread. In addition, several bakery companies in the U.S. are now reaping the benefits of extended product shelf life through MAP technology (American Institute of Baking, Personal Communication).

Sales of MAP pasta products are also increasing and are anticipated to reach ~$200 million per year. To date, more than forty companies are marketing MAP pastas in refrigerated cases at the retail level. This growth has been catalyzed by more healthy eating habits of consumers and the large variety of pasta products now available on the marketplace, such as protein-enriched pastas, vegetable flavoured and fruit flavoured pastas.

Some of the disadvantages of MAP technology (Table 9.10) include initial cost of packaging equipment, slower throughput of product, secondary fermentation problems caused by CO_2 resistant microorganisms and the potential growth of organisms of public health concern, particularly in meat- and cream-filled products. While this latter topic has been the subject of several investigations in muscle foods packaged under MAP conditions, there is little conclusive evidence that MAP represents a significantly greater hazard than packaging in air, particularly under conditions of temperature abuse where CO_2 is less effective. It is commonly believed that the inclusion of O_2 in the package headspace may prevent the growth of *C. botulinum* in products which may be susceptible to contamination by this pathogen. However, recent studies have refuted this claim and have shown that the inclusion of O_2 in the package headspace offers no additional protection against *C. botulinum* type A and B spores and may even enhance toxin production by this pathogen (Lambert et al., 1991).

CONCLUSION

Some exciting possibilities and challenges face the North American bakery and pasta industry in their search to achieve longer extensions of product shelf life. Gas packaging is now being employed by several bakery companies to extend the mold-free shelf life of high moisture baked products such as crumpets, waffles and muffins. However, with relatively porous products, e.g., crusty rolls or products containing fruit or similar fillings, e.g., cherry cream cheese cake, secondary spoilage problems can occur in the gas packaged products. Additional modifications to product formulation may be required if the desired shelf life is to be achieved under gas packaging conditions.

Gas packaging is also being used by manufacturers of fresh pasta products. This trend is likely to continue as more consumers demand fresh, nutritious and additive-free products on supermarket shelves. However, the success of MAP pasta products depends on their microbiological quality and safety. While this can be achieved by strict temperature control, the use of additional barriers such as reduced a_w and pH may ensure their safety, particularly at mild temperature abuse storage conditions.

In conclusion, the use of MAP technology for the distribution of products is still in its infancy in North America compared to Europe. However, as more and more North American companies become aware of the economic advantages of the technology, MAP will gradually emerge as the preservation technology of the next decade and propel the North American bakery and pasta industries into a new generation of products, distribution and marketing.

REFERENCES

Aboagye, N.Y. 1986. *Food Engineering and Process Applications, Vol 2. Unit Operations.* Elsevier Applied Science Publishers, New York, NY, pp. 417–425.

Avital, Y., Mannheim, C. H. and Miltz, J. 1990. "Effect of carbon dioxide atmosphere on staling and water relations in bread," *J. Food Science*, 55(2):413–416, 461.

Bogadtke, B. 1979. "Use of CO_2 in packaging foods," *Ernahrungswirtschaft*, 7/8:33.

Boyle, P. J. and Hebeda, R. E. 1990. "Anti-staling enzyme for baked goods," *Food Technology*, 44(6):129.

Castelvetri, F. 1988. *Proceedings of the Fourth International Conference on Controlled/Modified/Vacuum Packaging*, December 2–6, 1988, Long Island, New York, pp. 43–62.

Coyne, F. P. 1933. "The effect of carbon dioxide on bacterial growth with special reference to the preservation of fish, Part 1," *J. Society Chemical Industry*, 51:119T–121T.

Daniels, J. A., Krishnamurthi, R. and Rizvi, S. H. 1985. "A review of the effects of carbon dioxide on microbial growth and food quality," *J. Food Protection*, 48:532–537.

Doerry, W. T. 1985. "Packaging bakery products in controlled atmospheres," *American Institute Bakery Research Department Technical Bulletin*, 7(4):1–8.

Doerry, W. T. 1990. "Water activity and safety of bakery products," *American Institute Bakery Research Department Technical Bulletin*, 12(6):1–6.

Donnelly, B. J. 1991. *Pasta: Raw Materials and Processing*. Marcel Dekker Inc., New York, NY, pp. 187–197.

Ellis, W. O., Smith, J. P., Simpson, B. K., Khanizadeh, S. and Oldham, J. H. 1993. "Control of growth and aflatoxin production by *Aspergillus flavus* under modified atmosphere packaging (MAP) conditions," *Food Microbiology*, 10:9–21.

Enfors, S. O. and Molin, G. 1978. "The influence of high concentrations of carbon dioxide on the germination of bacterial spores," *J. Applied Bacteriology*, 4:279–285.

Ernst, L. 1989. "The impact of materials and gas on chilled foods," *Food Engineering*, 10:62–63.

Goodburn, K. E. and Halligan, A. C. 1988. "Modified atmosphere packaging–A technology guide," *Publication of the British Food Manufacturing Association*, Leatherhead, UK, pp. 1–44.

Hintlian, C. B. and Hotchkiss, J. H. 1986. "The safety of modified atmosphere packaging: A review," *Food Technology*, 40(12):70–76.

Hunt, L. and Robbins, L. 1989. "Food expenditure patterns of Canadian Consumers," *Food Market Commentary*, 11(3):42–51.

Jarvis, B. 1972. "Mould spoilage of foods," *Process Biochemistry*, 7:11–14.

King, A. D. and Nagel, C. W. 1975. "Influence of carbon dioxide upon the metabolism of *Pseudomonas aeruginosa*, *J. Food Science*, 40:362–366.

Knorr, D. 1987. "Compressibility of baked goods after carbon dioxide atmosphere processing and storage," *Cereal Chemistry*, 64(3):150–153.

Knorr, D. and Tomlins, R. I. 1985. "Effect of carbon dioxide modified atmosphere on the compressibility of stored baked goods," *J. Food Science*, 50:1172–1176.

Kulp, K. 1979. "Staling of bread," *American Institute Bakery Research Department Technical Bulletin*, 1(8):1–7.

Labuza, T. P., ed. 1982. *Shelf-Life Dating of Foods*. Food and Nutrition Press, Inc., Westport, CT, pp. 119–127.

Lambert, A. D., Smith, J. P. and Dodds, K. L. 1991. "Combined effects of modified atmosphere packaging and low-dose irradiation on toxin production by *Clostridium botulinum* in fresh pork," *J. Food Protection*, 54(2):97–104.

Legan, J. D. and Voysey, P. A. 1991. "Yeast spoilage of bakery products and ingredients," *J. Applied Bacteriology*, 70:361–371.

Malkki, Y. and Rauha. O. 1978. "Mould inhibition by aerosols," *Baker's Digest*, 52:47–50.

McGuire, M., DiGiacomo, R., Palmer, M. and Liggett, L. 1989. "Method for preparing and preserving fresh pasta," U.S. Patent 4,876,104.

Notermans, S., Dufrenne, J. and Lund, B. M. 1990. "Botulism risk of refrigerated, processed foods of extended durability," *J. Food Protection*, 53(12):1020–1024.

Ooraikul, B. 1982. "Gas packaging for a bakery product," *Canadian Institute Food Science Technology Journal*, 15:313–317.

Ooraikul, B., ed. 1991. *Modified Atmosphere Packaging of Food*. Ellis Horwood, New York, pp. 49–117.

Palumbo, S. A. 1986. "Is refrigeration enough to restrain food borne pathogens?" *J. Food Protection*, 49:1003–1009.

Park, C. E., Szabo, R. and Jean, A. 1988. "A survey of wet pasta packaged under a

CO_2:N_2 (20:80) mixture for staphylococci and their enterotoxins," *Canadian Institute Food Science Technology Journal*, 21(1):109–111.

Pomeranz, Y. and Shellenberger, J. A. eds. 1971. *Bread Science and Technology.* AVI Publishing Co., Westport, CT, pp. 169–189.

Robertson, G. L. ed., 1993. *Food Packaging–Principles and Practice*, Marcel Dekker, Inc., New York, NY, pp. 550–588.

Sears, D. F. and Eisenberg, D. F. 1961. "A model representing a physiological role of carbon dioxide at the cell membrane," *J. General Physiology*, 44:869–877.

Seiler, D. A. L. 1966. "Gas packing of cakes in carbon dioxide," *Food Trade Review*, 36:51–56.

Seiler, D. A. L. 1976. *Intermediate Moisture Foods.* Applied Science Publ., London, UK, pp. 166–181.

Seiler, D. A. L. 1978. "The microbiology of cake and its ingredients," *Food Trade Review*, 48:339–344.

Seiler, D. A. L. 1988. "Microbiological problems associated with cereal based foods," *Food Science and Technology Today*, 2:(1)37–41.

Seiler, D. A. L. 1989. *Controlled/Modified Atmosphere/Vacuum Packaging of Foods.* Food and Nutrition Press, Trumbell, CT, pp. 119–133.

Skovholt, O. and Bailey, C. H. 1933. "Effect of carbon dioxide on mold growth in bread," *Cereal Chemistry*, 10:446–451.

Smith, J. P. 1992. *MAP Packaging of Food–Principles and Applications.* Academic and Professional, London, UK, pp. 134–169.

Smith, J. P., Jackson, E. D. and Ooraikul, B. 1983. "Microbiological studies on gas packaged crumpets," *J. Food Protection*, 46(4):279–283.

Smith, J. P., Khanizadeh, S., van de Voort, F. R., Hardin, B., Ooraikul, B. and Jackson, E. D. 1988. "Use of response surface methodology in shelf life extension of a bakery product," *Food Microbiology*, 5:83–97.

Smith, J. P., Ooraikul, B., Koersen, W. J., and Jackson, E. D. 1986. "Novel approach to oxygen control in modified atmosphere packaging of bakery products," *Food Microbiology*, 3:315–320.

Smith, J. P., Ooraikul, B., Koersen, W. J., van de Voort, F. R., Jackson, E. D., and Lawrence, R. A. 1987. "Shelf life extension of a bakery product using ethanol vapor," *Food Microbiology*, 4:329–337.

Smith, J. P., Ramaswamy, H., and Simpson, B. K. 1990. "Developments in food packaging technology. Part 2: Storage aspects," *Trends Food Science and Technology*, 1:112–119.

Spicher, G. 1984. "Die erreger der schimmelbildung bei backwaren 3. Mitteilung: einige beobachtungen uber die biologie der erreger der kreidekranheit des brotes," *Getreide Mehl und Brot*, 38:77–80.

Todd, E. C. D., Jarvis, G. A., Weiss, K. F., Reidel, G. W. and Charbonneau, S. 1983. "Microbiological quality of frozen cream-type pies sold in Canada," *J. Food Protection*, 46:34–40.

Valley, G. and Rettger, L. F. 1927. "The influence of carbon dioxide on bacteria," *J. Bacteriology*, 14:101–137.

Wolfe, S. K. 1980. "Use of CO and CO_2-enriched atmospheres for meats, fish and produce," *Food Technology*, 34:55–58, 63.

Young, L. L., Reviere, R. D. and Cole, A. B. 1988. "Fresh red meats: A place to apply modified atmospheres," *Food Technology*, 42(9):65–69.

CHAPTER 10

Sous Vide: Past, Present, and Future

J. D. BAILEY—*Vie de France, U.S.A.*

SINCE its inception and initial applications during the early 1960s, "sous vide" has attracted steadily increasing interest from the foodservice industry, the media, as well as academic and government institutions associated with foodservice. This interest has certainly been appropriate, as sous vide technology has been of considerable assistance in addressing the sagging bottom lines of foodservice and catering operations suffering from poor quality, inconsistency, waste, portion-control problems, high labour costs, and the inability to find and keep adequately skilled labour.

Sous vide is a French term which literally translated into English means "under vacuum." The term *sous vide* takes its name from the hermetically sealed packaging in which the products are processed and stored. The technology is not constrained to one particular process, but comprises a group of processes which include vacuum packaging and some form of post- and/or pre-vacuum packaging thermal processing. These processes fall into three categories: "la cuisson sous vide" (cooking under vacuum); "cook-chill," or cooking by traditional means followed by vacuum packaging; and "hot fill."

SOUS VIDE DEFINED

Category One: La Cuisson Sous Vide

The first category, "la cuisson sous vide" (cooking under vacuum), and "la cuisson en papillote sous vide" (cooking in a vacuum cocoon), which is normally referred to when speaking of "sous vide," involves cooking a food product inside a hermetically sealed vacuum package. The cooking-under-vacuum process is generally used for protein-based items such as meat, fish, and poultry for two basic reasons.

First, cooking food in water or steam while packaged inside a hermetically sealed pouch is much more efficient than cooking the same product conventionally, such as by baking, broiling, grilling, etc., primarily because thermal energy is transferred more efficiently to the product through a liquid or steam cooking medium. These media completely envelop the product, and liquid media hold more heat energy (i.e., has more heat energy to transfer to the product) per volume than a gas medium such as air. The result is a better-quality product. Also, because the temperature of a liquid cooking medium is easier to control, quality is more consistent from batch to batch.

Second, the flexible or rigid film in which the sous vide product is hermetically sealed is virtually impervious to the external environment. Thus, the thermal process serves as a pasteurization, reducing the microbial populations in the product and rendering the product safer to eat for a longer period of time (i.e., a longer shelf life). Because the product is stored inside its original packaging, it is less likely to be contaminated by an external source. The product has a longer refrigerated shelf life because the reduction in the microbial population retards spoilage.

In addition to protein-based items, other types of foods, including vegetables, pasta, and legumes, can be prepared using this technique. This is advantageous in applications where the vegetables, rice, etc., are processed with herbs, spices or other aromatics, because the aromatic characteristics of herbs and spices are concentrated when a product is cooked under vacuum.

As illustrated in Figure 10.1, the cooking-under-vacuum processes involve the following steps.

Raw-Material Preparation

The broadest step of the process, raw-material preparation, includes all activities that are normally undertaken to prepare a food product for cooking, including unpackaging, defrosting, seasoning, marinating, tumbling, slicing, portioning, etc.

Branding, Marking, Blanching (optional step)

After initial preparation, the food product undergoes branding, marking or blanching. This step of the process is not used for poached products, which are vacuum packaged immediately after being por-

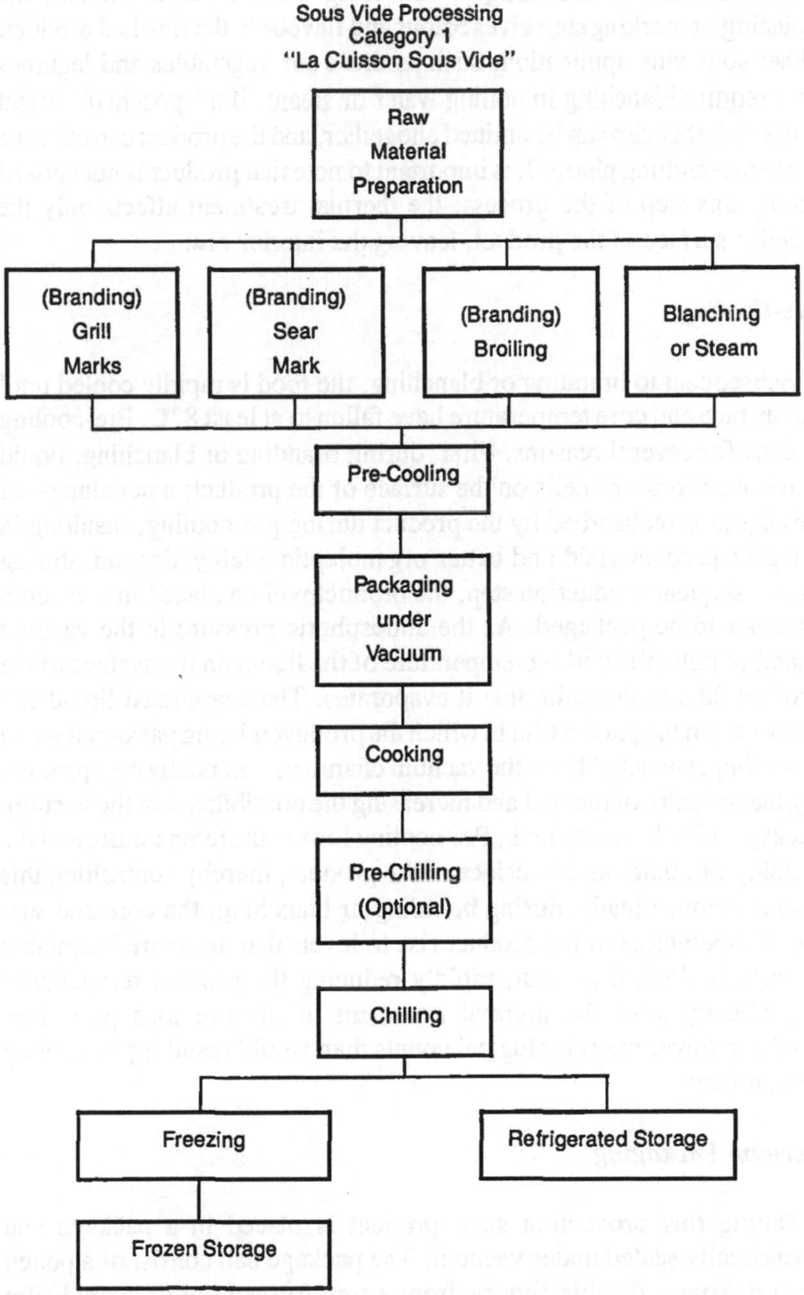

FIGURE 10.1 An illustration of the different phases of "la cuisson sous vide" or cooking under vacuum processing category.

tioned. Similar to browning or rendering a roast before baking, the branding or marking step gives colour and flavour to the finished product. Other sous vide applications such as those for vegetables and legumes may require blanching in boiling water or steam. If a "poached" effect is desired, this step can be omitted altogether, and the product can advance to the pre-chilling phase. It is important to note that product is not cooked during this step of the process; the thermal treatment affects only the exterior surface of the product, leaving the interior raw.

Pre-Cooling

Subsequent to branding or blanching, the food is rapidly cooled until the surface and core temperature have fallen to at least 8°C. Pre-cooling is done for several reasons. First, during branding or blanching, liquid is released from the cells on the surface of the product; a percentage of this liquid is reabsorbed by the product during pre-cooling, resulting in a higher product yield and better organoleptic quality. Second, during the subsequent production step, the product will be placed in a vacuum chamber to be packaged. As the atmospheric pressure in the vacuum chamber falls, the boiling temperature of the liquid on the surface of the product falls to the point that it evaporates. The evaporated liquid can condense on the plastic film in which the product is being packaged while it is being evacuated from the vacuum chamber, potentially compromising the integrity of the seal and increasing the possibility that the vacuum package will ultimately leak. Pre-cooling lowers the temperature and the quantity of liquid on the surface of the product, thereby controlling this phenomenon. Finally, during branding or blanching, the core and surface temperatures of the product rise to levels that are more hospitable to microbiological growth; rapidly reducing the product temperature immediately after the thermal treatment of the previous procedure results in lower microbiological counts than would result if pre-cooling was omitted.

Vacuum Packaging

During this production step, product is placed in a package and hermetically sealed under vacuum. The package can consist of a pouch formed from a flexible film or from a tray formed out of a rigid film covered by a flexible film. There are many types of machines used for vacuum packaging, the most common of which are horizontal form, fill, and seal machines, and single-chamber machines. The product can be vacuum-packaged by itself or combined with other ingredients, including sauces, marinades and garnishes. An adequate vacuum inside the packaging is critical for all sous vide processing, because atmosphere in the packaging creates a barrier between the cooking medium and the product, resulting in an uneven and more inconsistent thermal process. Also, if the product is to be stored at refrigerated temperatures subsequent to processing (i.e., as opposed to being frozen), sufficient vacuum is critical to inhibiting the growth of aerobic organisms.

Cooking

During the cooking phase, the vacuum-packaged, raw product is thermally processed by immersion in water, steam or under running or dripping water. This phase is the most critical of the sous vide process, as it will determine the organoleptic properties and quantity of microorganisms in the finished product.

The thermal treatments or cooking processes to which the vacuum packaged products are subjected are categorized into three basic types:

(1) Type 1: A high cooking temperature (above 70°C) is used, and a high internal temperature (above 70°C) is attained.
(2) Type 2: A high cooking temperature (above 70°C) is used, but a low internal temperature (below 70°C) is attained.
(3) Type 3: A low cooking temperature is used and a low internal temperature (both below 70°C) is attained.

These thermal treatments can be used individually, or combined in cook cycles with multiple phases. For example, some applications call for an introductory cook phase in which high-temperature water or steam is used to rapidly coagulate the proteins on the surface of the product (i.e., the first phase of the cook would be Type 2). After this phase has been achieved, the high-temperature water is drained and replaced with lower-temperature water until the cook has been completed.

Pre-Chilling

After being cooked, the product must be chilled to a temperature at which the growth of microorganisms is slowed. The rate at which the product is chilled is important for several reasons. First, the product must be cooled through the temperature zone at which most microorganisms grow and thrive (50°C to 10°C) as rapidly as possible. However, the product must not be cooled too rapidly, because during the cooling cycle some of the liquid given off by the product during the cook is reabsorbed. The reabsorption is especially facilitated in the cooking process by the vacuum packaging, which forms a second skin around the perimeter of the product, holding the liquid given off during the cooking cycle immediately next to the product.

The pre-chilling phase is optional in the cooling cycle. The product is cooled from the cooking temperature to 25–40°C. The pre-chilling cycle is normally done by exposing the product to a temperature 5 to 10°C cooler than the desired internal temperature.

For example, if an internal temperature of 35°C is desired at the end of the pre-chilling phase, then an ambient temperature of 25–30°C would be used.

The use of a pre-chilling phase serves two primary functions: First, and most importantly, it provides symmetry to the cooking curve (see Figure 10.2) and, as described above, it facilitates the absorption of liquid given off during the cooking cycle.

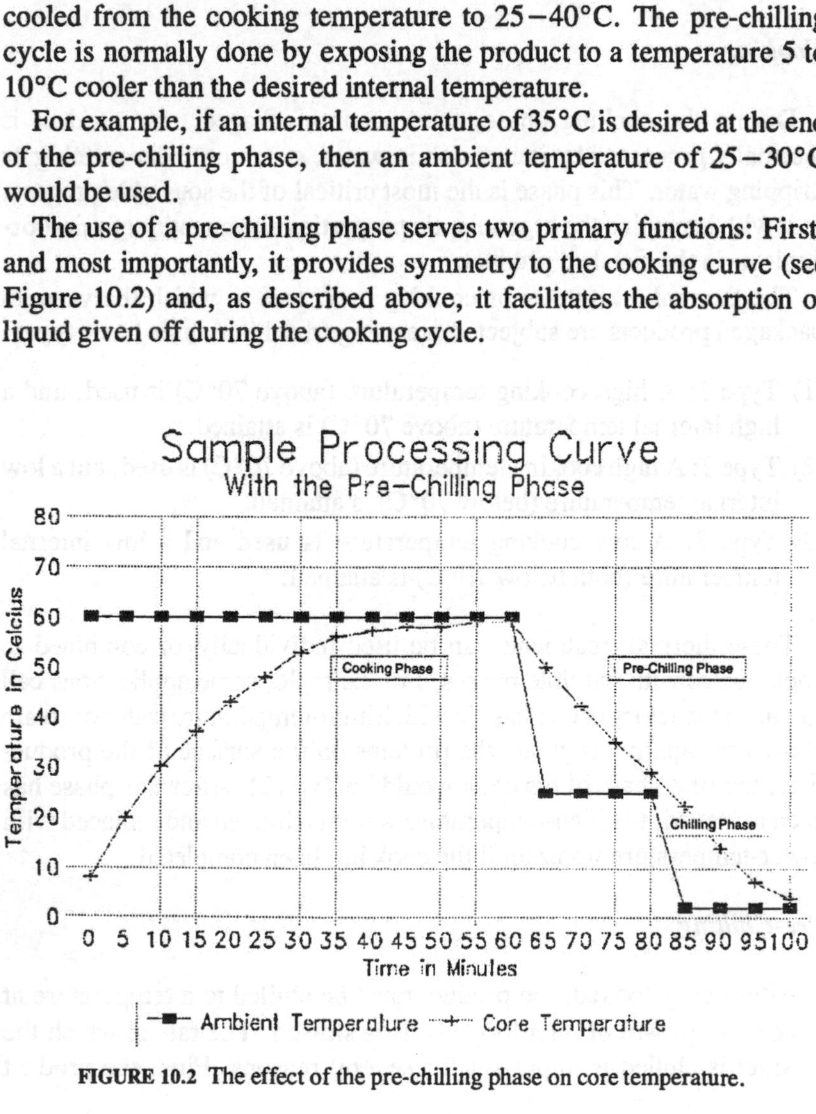

FIGURE 10.2 The effect of the pre-chilling phase on core temperature.

Second, the pre-chilling phase reduces the drain on the medium (water in most cases) that is used to cool the product. That is, depending on the mass of product that needs to be cooled, it requires much more cooling capacity to maintain a water temperature of 2°C when cooling from a cooking temperature of 60°C, than it does to cool from a temperature of 30°C. Albeit the same amount of thermal energy needs to be extracted with or without a pre-chilling phase, the pre-chilling phase allows the water to recover (come back down in temperature) while the cooling phase is underway.

Pre-chilling is often omitted in small batch quantities or in production environments where it is not available or cost-effective.

Chilling

In the chilling phase which follows the cook or pre-chilling phases, the product is chilled to below 5°C. Cooking curves illustrating the effects of pre-chilling and no pre-chilling are shown in Figures 10.2 and 10.3; Figure 10.2 illustrates the most desirable curve. The symmetry of the curve has an impact on the quality and yield of the product; the more symmetrical the curve, the higher the yield and the better the product's organoleptic quality. Figure 10.3 depicts a process curve in which the pre-chilling phase is omitted. As shown by the curve, the decrease in the core temperature of the product is more rapid than that shown in Figure 10.2. The result is a lower product yield and a drier-tasting product.

Category Two: Cook-Chill

The second category of sous vide processes involves vacuum packaging a product that has already been fully cooked. After vacuum packaging, the product is subjected to a thermal treatment that serves as a pasteurization. This process is illustrated in Figure 10.4. As the figure shows, the primary difference between "cook chill" and "cooking under vacuum" is the "cook" phase prior to vacuum packaging. This process is often used for soups, sauces and stews that require cooking by traditional means. It is also used when several products are combined into a single package such as an entrée and other garnished or side dishes. The steps involved in this process are identical to those in "cooking under vacuum."

The "cook" in this category is actually a misnomer; the product is already fully cooked. The cook is a secondary thermal treatment that

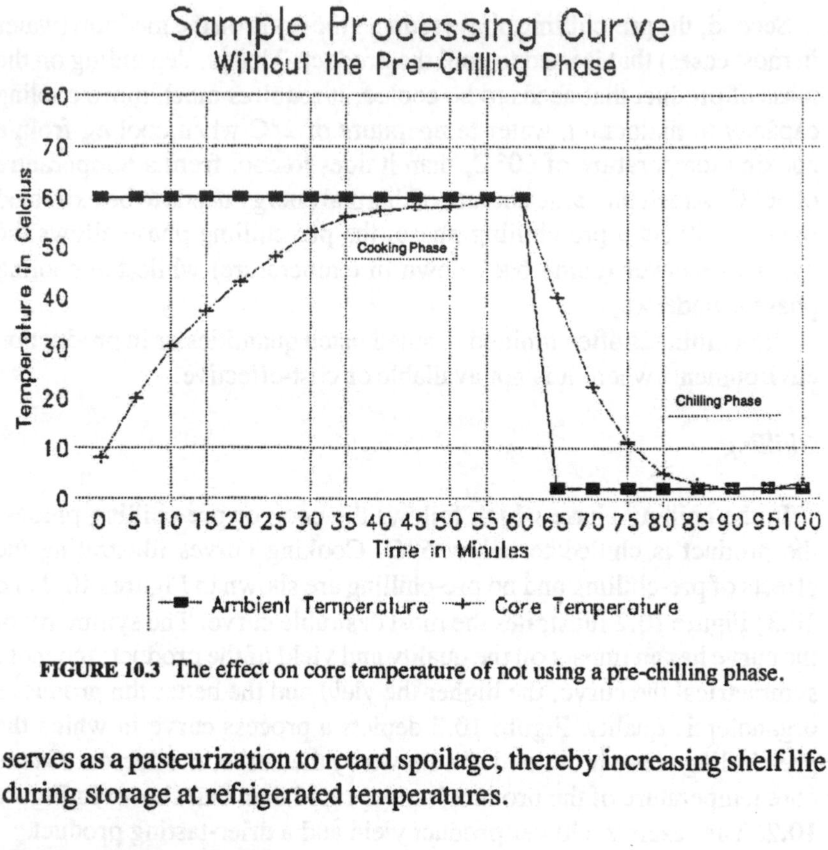

FIGURE 10.3 The effect on core temperature of not using a pre-chilling phase.

serves as a pasteurization to retard spoilage, thereby increasing shelf life during storage at refrigerated temperatures.

Category Three: Hot Fill

The third sous vide category involves vacuum packaging a cooked product while it is still hot, commonly referred to as "hot fill" (Figure 10.5). Subsequent to packaging, the product is rapidly chilled, as in the previous two categories. It is commonly used in the processing of liquid products such as soups, stews, and sauces. The same effect is achieved when hot filling a product as when cook-chilling a product; the shelf life of the refrigerated product is extended.

A BRIEF HISTORY OF SOUS VIDE

The Precursor of Modern Sous Vide Processing

The precursor of the sous vide technology was first discussed by E. F. Kohman in a 1960 *Food Technology* article (Kohman, 1960). The

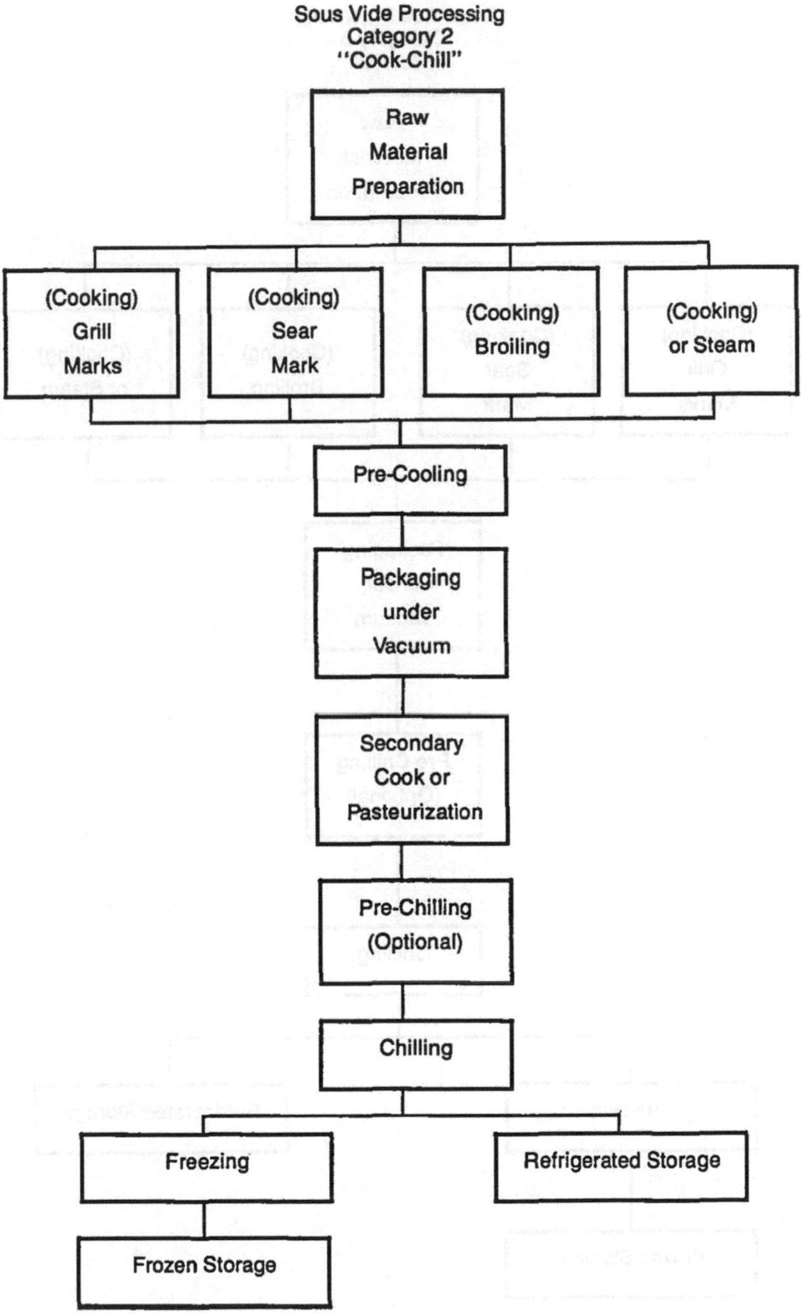

FIGURE 10.4 An illustration of the different phases of the cook-chill processing category.

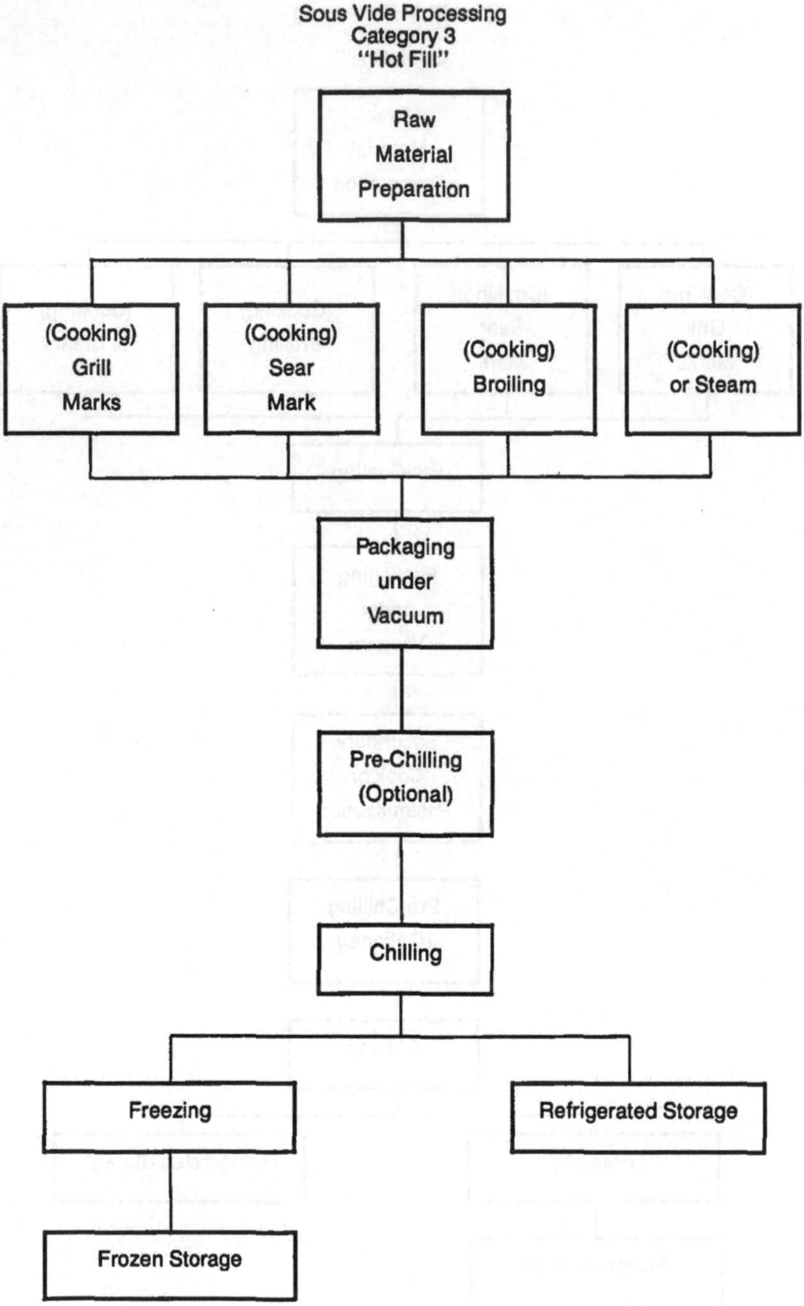

FIGURE 10.5 An illustration of the different phases of the hot-fill processing category.

new process, termed "Frigi-Canning," described shelf life extension and food preservation techniques by the thermal processing of product that had been hermetically sealed in cans or jars. The process was sufficient to kill the vegetative forms of microorganisms present in the food. However, because the thermal processing was not sufficient to destroy spore-forming pathogenic organisms, rapid chilling subsequent to processing and refrigeration at temperatures as cold as possible, without freezing, were required.

Kohman (1960) cites two primary factors that led to the development of Frigi-Canning: the relatively poor quality of canned foods sterilized at high temperatures for short periods of time, and the use of low-temperature pasteurization and subsequent refrigerated storage to retard the spoilage of milk, thereby allowing for its mass distribution.

Kohman points out that reactions between the various organic compounds inside a can of fruit or vegetables that has been subjected to enough thermal energy to inactivate all enzymes, are dramatically slowed down if the can is stored at refrigerated temperatures. In fact, the reactions are slowed by one-half for every decrease of 18°F.

On the subject of milk, Kohman discusses how formaldehyde was used in early milk distribution to prevent spoilage. After preservatives were made illegal, many people thought distribution of fresh milk would be impossible. However, the development of pasteurization and refrigerated storage, neither of which alone is sufficient to adequately prevent spoilage, made such distribution feasible.

Frigi-Canning was developed in connection with home gardening. Kohman notes that after he harvested his garden, he had a surplus of vegetables, the majority of which went to waste. Using the reasoning applied to preserving milk, Kohman discovered that he was able to preserve excess vegetables for extended periods of time with Frigi-Canning.

It Started in Sweden

The first formal application of the sous vide technology was developed and tested at two hospitals in Sweden from 1960 to 1965. The process was termed "The Nacka System," from one of the hospitals in which it was developed (Bjorkman and Delphin, 1966). The Nacka System was developed by a committee appointed by the Stockholm City Council to study different food processing technologies that would allow for the centralized production and distribution of meals to various hospital locations. Nacka was selected over other processes because it allowed for extended shelf life while maintaining quality at acceptable levels.

The Nacka System differed from "true" sous vide processing because the products were not cooked in their packages. Food produced according to the Nacka technique was fully cooked by conventional means (broiling, frying, grilling, etc.) prior to being vacuum packaged; the food was then packaged while it was still hot. Employing the hot fill technique, Nacka required a fill temperature of 80°C.

Because it was difficult to insure that all parts of the product had reached the required fill temperature, or that the surface of the product had not cooled while waiting to be vacuum packaged, products were subjected to a thermal treatment in boiling water after vacuum packaging. The duration of the heat treatment was generally 3 min, but could be extended up to 10 min, depending on the fill temperature of the product.

The program in which the Nacka System was implemented, having been designed for institutional foodservice (where it was necessary to serve many people in a relatively short period of time), called for pouches which contained five portions, minimizing the number of pouches that would need to be opened during the meal service. Only entrée or main-dish items were prepared using the Nacka technique; side dishes were prepared by conventional means immediately prior to being served. The Nacka products were immersed in boiling water for 30 min prior to being removed from their pouches and served.

During a five-year study of the Nacka System, more than 5 million portions were produced, from which more than 10,000 microbiological analyses were performed as part of the Nacka Quality Assurance program (Bjorkman and Delphin, 1966), the results of which indicate that the Nacka System produced "an entirely satisfactory product" from a food safety standpoint. The system was evaluated by the Swedish National Veterinary Board, which stated, "this method of food preparation meets the highest hygienic requirements in a satisfactory manner."

In addition to its safety record and standards, the Nacka System was also evaluated by the Stockholm School of Economics from a business standpoint. The school concluded that, through economies of scale, there were substantial savings realized by the centralized production and distribution of food products, as compared to the decentralized production in hospitals. Although it was difficult to calculate the exact amount of savings that were realized through the use of the Nacka System as compared to a standard foodservice operation, it was stated that the equivalent of $6 million would be saved during a 10-year continuous operation. In the economic context of the 1960s, this represented a substantial figure (Bjorkman and Delphin, 1966).

Made in the U.S.A.

During the late 1960s a refined form of Nacka – the AGS System – was developed and tested at three hospitals in the U.S. (McGuckian, 1969). Taking its name from the hospitals, AGS set out to improve the organoleptic quality of the food products pioneered by Nacka. The project was undertaken by the hospitals in coordination with Groen, a process-equipment design and manufacturing company, and the Cryovac Division of W.R. Grace, a plastic film design and manufacturing company.

Quality and wholesomeness being the top priorities, A. T. McGuckian, the AGS project manager, set out to evaluate traditional methods of convenience-food preparation in terms of overall advantages and disadvantages. The results of his analysis concluded there were obvious advantages to centralizing the food-preparation operations of several satellites into one location. However, food prepared and stored by conventional means would require such frequent transportation that a centralized food-production operation could not be profitable.

All things considered, AGS's overseers decided that a foodservice system similar to that used by the Nacka hospital in Sweden should be employed. The benefit of extended shelf life resulting from products that have been pasteurized subsequent to being packaged would provide the needed flexibility that is required for standardized production and distribution over a relatively large urban area. Moreover, the burden of seven-day-a-week, 24-hour-a-day foodservice operations in the various satellite locations could be converted to a five-day, 40-hour week in a single location.

Like the Nacka System, the initial tests were done with product that was fully cooked prior to being vacuum packaged. However, after taste panelists rated ten of the twelve initial products as "barely satisfactory," describing them as "tired" because they were overcooked, it was decided that a new approach was needed. McGuckian, although somewhat discouraged by the results from the taste panels, realized that the problem lay in processing fully-cooked food. He discovered that by vacuum packaging partially cooked, or in some cases raw food, the cooking process would occur or be completed in the pouch. The new technique was vastly superior to the Nacka System in terms of organoleptic quality because the product was only cooked a single time. The single cook, occurring in a vacuum pouch, allowed the product to retain its natural flavour and juices.

Unlike the Nacka System, AGS did not confine itself to main dish or entrée items. All types of food products that employed the cooking-under-vacuum approach, including vegetables and side dishes, were developed and produced. The only products which did not lend themselves well to the technique were some egg-based products and fried foods. Partially fried or deep-fried foods lose their crispiness after being thermally processed inside a vacuum pouch. The AGS System was successful in its goal to develop a foodservice system that would provide products of superior quality and nutritional value to its predecessors. AGS products consistently scored higher in taste tests than product that had been prepared according to conventional means (in the hospital kitchens).

The taste tests further concluded that product which had been properly prepared, stored, and distributed tasted as good after it had been reheated as it had at packaging.

Despite its success in producing consistent-quality products, AGS, like the Nacka System, was discontinued because of distribution problems and a reluctance by hospitals outside the three in the original test group to purchase the products.

Building on the work done in the Nacka and AGS programs, Groen and Cryovac formed a contractual agreement to develop a cook-chill system that could be marketed and sold commercially. Known as "CAP-KOLD," for "Controlled Atmosphere Packaging done Cold," the project was started in the late 1960s. The contract between Groen and Cryovac provided for an agreement in which Groen would design, build, and market the equipment required for the process, and Cryovac would do the same for the plastic film that would be used to package the product. The program met with great success and is still used in hundreds of applications throughout the world today.

While having some similarities to AGS, the CAPKOLD system more closely resembles the process done at Nacka. The process requires pre-cooking the product, and packaging it under a "controlled atmosphere" while it is still hot. Most applications call for a fill temperature of 180°F (82°C).

The French Influence

Sous vide technology appeared in France in 1972 (Goussalt, 1987). The technology was first used in the processing of hams in what were called the Gatineau and Soplaril processes. The initial efforts in France met with limited success and were later discontinued. During the mid 1970s, through cooperative efforts between chef Albert Roux, Cryovac, and Groen in England, and separately by chef George Pralus in France, sous vide was adapted to the hotel and restaurant segment of the catering industry.

The first large-scale application of modern sous vide technology was undertaken in 1985 by the French national railway company, SNCF (Gaussalt, 1987). SNCF was internationally known in previous eras for its high levels of culinary artistry and gastronomic exploits on trains such as the Orient Express.

In an effort to revitalize its once-legendary culinary program, SNCF contracted world-renowned French food critic Henri Gault to spearhead a program that would provide "haute cuisine" meal service in the first class section of the trains termed "Nouvelle Premiere" (New First Class). The initial program was to be tested on the trains travelling between Paris and Strasbourg. After serious consideration, Gault decided that the sous vide technology was the only method that would allow for the centralized production and distribution of the ultra-high quality meals that would be served on the trains.

Gault worked in cooperation with chef Joel Robuchon to supervise the creation of the concepts and recipes for the new program. In an effort to insure the best-possible product quality and wholesomeness, the Institute Superior Alimentaire (ISA), under the direction of Bruno Goussault, was contracted to provide quality assurance and technical support services.

Quality being of utmost importance, the initial research and development on the recipes for the project focused on cooking temperatures and durations, and their impact on product quality.

The initial research concluded that the use of low-cook temperatures (between 54 and 68°C) yielded the best organoleptic quality when compared to identical raw materials processed at higher temperatures. However, the use of low temperatures posed risks in the areas of wholesomeness and premature spoilage. To address these problems, Goussalt and ISA undertook a study to determine the risks of processing sous vide products at lower temperatures.

The results of Goussalt's study, in accordance with previously conducted research, concluded that temperature is not the sole critical factor in the destruction of microorganisms – the duration of exposure to the cooking temperature is equally important (Goussalt, 1987). Thus, the destructiveness of a thermal process for a product cooked at 80°C could be equivalent to a product processed at 58°C, providing the latter was held at that temperature for a sufficient period of time.

The research done by Goussalt and ISA was instrumental in the development of modern sous vide technology, as it provided the data that was necessary to determine the cooking parameters for many products, and more importantly, it provided vital information on the relationship between time/temperature thermal processing and shelf life.

SOUS VIDE IN THE 1990S

Ranging from institutional foodservice to in-home convenience foods, the applications for sous vide in the 1990s are extremely varied. The needs that sous vide products fill are similar in both markets. These needs are described next.

The needs that sous vide responds to in the retail market are as follows:

- The value of time is increasing, thus the increasing market for foods that require less time to prepare.
- The mean disposable income of many target market segments is increasing, so people are willing to pay additional money for better-quality food products.
- Sous vide products are not cooked in fats or oils, and are not prepared with preservatives such as salt and nitrites. In addition, because they are cooked at low temperatures inside their final packaging, they retain most or all of the nutrients; thus, they respond to the trend toward healthier foods.

Sous vide's advantages in the wholesale (hotel/restaurant/institutional) markets are as follows:

- Sous vide products need only to be reheated; therefore, they reduce the need for skilled labour in the food preparation facility.
- Sous vide products increase the flexibility in the foodservice program by allowing the facility to respond to changes on shorter notice. For example, a banquet manager in a hotel could accept an additional 100 diners without a great deal of additional (meal preparation) work or planning with sous vide products available.
- The quality and freshness of sous vide products is more consistent than products served in most restaurants. A sous vide chicken breast tastes the same after a week in the refrigerator; a raw chicken breast tastes much worse.

- Sous vide reduces waste because of its extended shelf life.
- Sous vide eliminates portion-control problems because the products are generally proportioned.

For the above reasons, the number of sous vide producers in the U.S. is steadily increasing. This follows the trend in Europe, which started about ten years ago. The number of new European firms that are entering the industry is slowing, however, because the demand has slowed and recent economic problems in Europe have caused many sous vide firms to go out of business. Only the stronger firms have survived.

U.S. firms getting into the sous vide business face problems of their own—problems that don't exist in Europe. First, the geography and population density in the U. S. is much different than Europe. The lower population densities in many areas of the U.S. and the vast distance between many U.S. cities have made distribution of sous vide products extremely challenging.

Because of the buying habits of Americans and the problems associated with distribution of fresh, refrigerated, products in the U.S., the majority of the current U.S. sous vide manufacturers produce frozen sous vide products. The term "sous vide" is mistakenly thought of by many as referring only to refrigerated finished product, and it is often confused with the new wave of chilled foods that has recently been introduced in the U.S.

In a paper submitted at the Eastern Food Science Conference VII Symposium, a microbiologist for the U.S. Food and Drug Administration stated, "Sous vide is a process whereby foods are vacuum packaged and then cooked, chilled, and stored refrigerated" (Rhodehamel, 1992). The microbiologist, or his source of information, voiced one of the more common fallacies relating to sous vide; he assumed that all sous vide products are stored at refrigerated temperatures subsequent to processing when, in fact, two of the largest producers of sous vide in the U.S. specialize in frozen sous vide products.

Chilled foods or foods stored at refrigerated temperatures are not necessarily sous vide, and sous vide products are not necessarily stored at refrigerated temperatures. As previously stated, sous vide refers to the various processes by which the products are prepared.

Quality: Frozen Sous Vide Stakes Its Claim

With regard to quality, many varieties of frozen sous vide products have been preferred in taste panels over sous vide products stored at refrigerated temperatures. The primary factor contributing to these superior organoleptic characteristics is the duration of the thermal processing—the cook time. Sous vide products held at refrigerated temperatures must be processed at higher temperatures and/or for longer periods of time. The quality of frozen product is dependent on the product being promptly and properly frozen and stored at temperatures below 0°F (−18°C).

The longer thermal-process times and higher temperatures required for fresh sous vide products are necessary to retard spoilage by reducing the population of microorganisms. As microbial spoilage is not a factor in frozen sous vide products, the cooking temperature at the core of the product in processing need not be maintained longer than necessary to render the product cooked. Increased thermal processing time results in varying degrees of lower organoleptic quality, depending on the type of product.

Regarding shelf life for refrigerated sous vide products, it is widely believed in the U.S. that all fresh sous vide products can be stored at refrigerated temperatures for up to 21 d. In fact, no blanket shelf life can be placed on these products. The shelf life for a refrigerated product is determined by several factors, including the thermal treatment the product receives, pH, available water (a_w), NaCl concentration, and the presence of any ingredients or additives that may have bacteriocidal or bacteriostatic properties.

The fallacy of a "blanket" 21-d shelf life did not come about arbitrarily. In France, many manufacturers of refrigerated sous vide products put a 21 d shelf life on retail and foodservice products. However, French federal legislation strictly determines and enforces the requirements for the three-week shelf life. Prior to placing such a shelf life on any product (meat, poultry, fish, sauce, etc.), a detailed product-information package must be submitted to the French veterinary service for approval.

The information contained in the package must include all the product ingredients and the thermal-processing parameters expressed in terms of the pasteurization, or F value. F value is a number that quantifies the destructiveness of the thermal energy (transferred from the cooking medium to the product) to a specific reference bacterium. French regulations state that for a shelf life of 21 d, a sous vide product must be exposed to a thermal process equal to or greater than a pasteurization or F value of 100. Similarly, French regulations allow a 42-d shelf life for all products that have an F value of 1,000. The regulations also state that a refrigerated product must be held at or below 3°C.

Sous Vide and HACCP

In an effort to maintain the highest levels of safety and product quality, the following key elements are currently being adopted by sous vide manufacturers as industry standards:

- A mandatory Hazard Analysis Critical Control Points (HACCP) program in all sous vide operations that covers raw materials suppliers, processing facilities, distribution and retail locations.
- Government regulations that restrict the production of sous vide products to qualified facilities.
- The mandatory use of "super-chill" equipment in storage facilities, distribution and retail facilities.
- The mandatory use of time-temperature indicators (TTIs) suitable to the different shelf lives of products.
- Government regulation based on thermal-processing parameters or F values that controls the length of shelf life that can be placed on a product.

Without question, quality assurance and HACCP will continue to play extremely important roles in the manufacturing of sous vide products.

REFERENCES

Adams, C. E. 1991. "Applying HACCP to sous vide products," *Food Technol.*, 45(4):148–151.

Bjorkman, A. and Delphin, K. A. 1966. "Sweden's Nacka Hospital food system centralizes preparation and distribution," *The Cornell H. R. A. Quarterly*, 7:84–87.

Folkenberg, J. 1989. "The big chill, how safe are "sous vide" refrigerated foods," *FDA Consumer*, 23(7):33–36.

Goussalt, B. 1987. *La Cuisson Sous Vide, Utilization Des Cuissons A Basse Temperature*, Paris, France.

Kalish, F. 1991. "Extending the HACCP concept to product distribution," *Food Technol.*, 45(6):119–120.

Kohman, E. F. 1960. "Frigi-Canning," *Food Technol.*, 14:254–256.

McGuckian, A. T. 1969. "The A. G. S. food system—Chilled pasteurized food," *The Cornell H. R. A. Quarterly*, 10:87–92.

Rhodehamel, J. E. 1992. "FDA's concerns with sous vide processing," *Food Technol.*, 46(12):73–76.

Schehter, M. 1990. "A revolution in a pouch," *Food Management*, 25(10):88, 92, 96.

Smith, J. P., Toupin, C., Gagnon, B., Voyer, R., Fiset, P. P. and Simpson, M. V. 1990. "A hazard analysis critical control approach (HACCP) to ensure the microbiological safety of sous vide processed meat/pasta," *Food Microbiol.*, 7:177–198.

Vie de France. 1989–1990. *Evaluation of Organoleptic Difference Between Fresh and Frozen Sous Vide Products*. Alexandria, Virginia.

CHAPTER 11

The Potential Use of Additional Hurdles to Increase the Microbiological Safety of MAP and Sous Vide Products

I. R. GRANT—*The Queen's University of Belfast, Northern Ireland*
M. F. PATTERSON—*The Queen's University of Belfast, Northern Ireland*

INTRODUCTION TO HURDLE CONCEPT

THERE are a wide variety of preservation methods available to the food industry. Only a few of these methods act by destroying all the microorganisms likely to be present in foods, e.g., heat and irradiation sterilization. Such methods usually require an extreme use of the treatment and this can result in deterioration of product quality. The majority of food preservation techniques, e.g., modified atmosphere packaging (MAP), vacuum packaging and addition of chemical preservatives, act by inhibiting the activity of microorganisms to varying degrees. The use of these inhibitory factors in combination can be advantageous as they permit less extreme use of any single treatment and can result in improved product quality (Gould and Jones, 1989). In many foods more than one factor is relied upon to control microbial growth and therefore prevent spoilage and/or foodborne disease. This approach of superimposing a number of limiting factors has been described as the Hurdle Concept (Leistner, 1985). It was originally used in the formulation of a number of shelf-stable meat products, but is increasingly being applied to a wide range of minimally processed foods which consumers perceive as being more "natural" and "healthy." The hurdle concept is illustrated in Figure 11.1. Figure 11.1(a) illustrates the principle and represents a food which contains six hurdles (heating, chilling, water activity, acidification, redox potential and preservatives). The microorganisms present cannot overcome (jump over) these hurdles and so the product is microbiologically stable and safe. However, this example is only a theoretical case as all the hurdles are the same height, i.e., of equal intensity. A more realistic example is a food product in which the hurdles are of different intensity [Figure 11.1(b)].

The hurdle concept applied to MAP products is illustrated in Figure 11.2. The majority of MAP products are stored under refrigerated

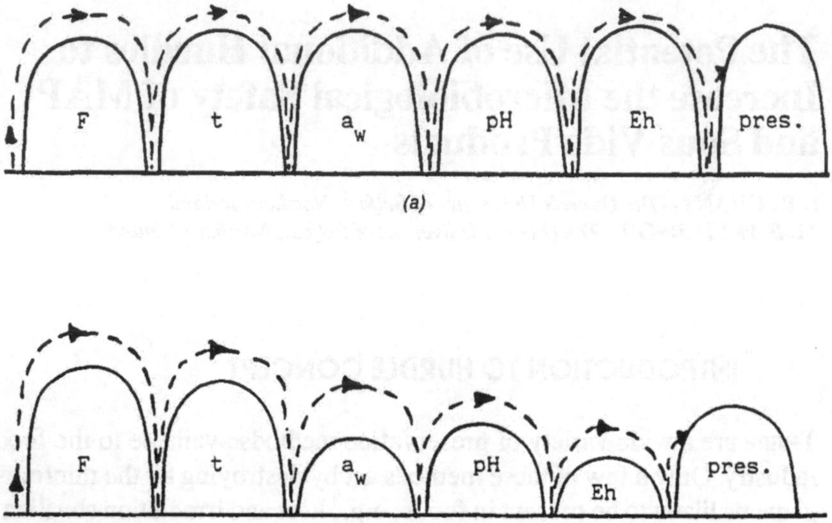

FIGURE 11.1 Illustration of the hurdle concept [reprinted from Leistner (1992) with permission from Elsevier on behalf of the Canadian Institute of Food Science and Technology].

conditions and so temperature is usually included as the primary hurdle. Any alteration of the packaging atmosphere will have a selective effect on the microorganisms growing in the food. A lack of O_2 will inhibit the growth of strict aerobes, such as the pseudomonads, and so spoilage is delayed. In addition, the growth rates and yields of many facultative anaerobes may be lower due to less energy being available from fermentative rather than oxidative metabolism. Thus O_2-free packaging in combination with other hurdles that place extra energy demands on the

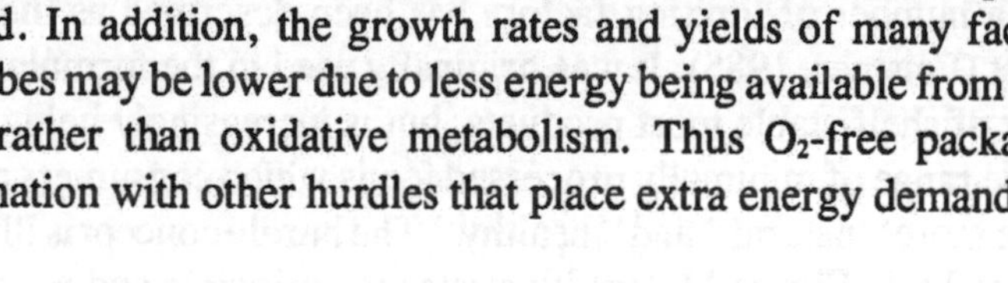

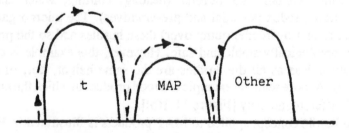

FIGURE 11.2 Illustration of the hurdles in MAP products.

cell can be particularly beneficial (Gould and Jones, 1989). Such treatments might include a reduction in pH value by addition of weak organic acid or co-inoculation with lactic acid bacteria. In these cases the hurdles act synergistically as they have different targets within the microbial cell. Sublethally damaged cells, for example the organisms which survive low-dose irradiation, may be more susceptible to inhibition by the packaging atmosphere.

Sous vide products possess the additional hurdle of a mild heat treatment (Figure 11.3) which inactivates the majority of vegetative microorganisms but may permit the survival of spore-forming organisms. Refrigeration is also an important hurdle in sous vide products. Non-sporogenic psychrotrophic pathogens, e.g., *Listeria monocytogenes*, can be a potential hazard, even under proper chill storage, if the product is underprocessed. The vacuum packaging restricts the growth of aerobic sporeformers, so the organisms of main concern are the clostridia, especially *Clostridium botulinum* (Chapter 5). It should be noted that the initial microbial quality of the food is important. If the microbial load is too high, then the heat treatment will be less effective and undesirable organisms may survive this and subsequent hurdles and so cause spoilage or food poisoning.

Some of the additional hurdles which may be used in combination with MAP or sous vide packaging will be described in the following sections.

IRRADIATION

Low-dose irradiation can improve the microbiological quality of foods by reducing the number of spoilage organisms and potential

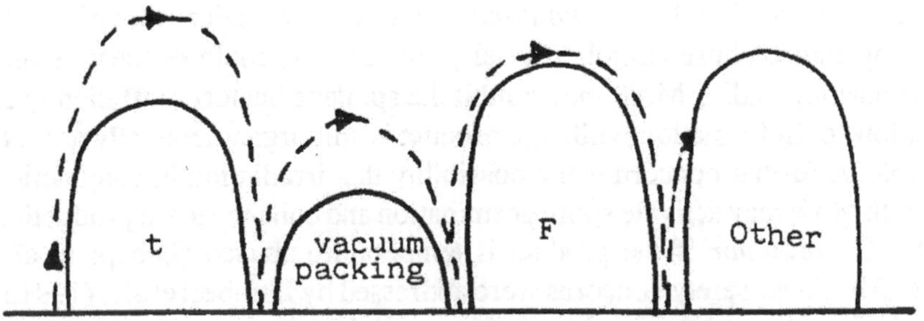

FIGURE 11.3 Illustration of the hurdles in sous vide products.

pathogens. However, the process is not a panacea for all food problems. It is likely to be used commercially only when it can offer a technological advantage over other existing techniques, for example in improving food safety, and be economically viable (Stevenson, 1991). Some would argue that it is in the area of combination treatments that the major benefits of food irradiation will eventually be seen.

The amount of irradiation which can be applied to a particular food, and hence the extent of microbial kill, is limited by undesirable changes in flavour, appearance and texture which may occur. The "threshold dose," above which these organoleptic changes can be detected, varies with different foods. By using low-dose irradiation in combination with other treatments, the desired microbial effect can be achieved without significant loss of organoleptic quality. The combination of low-dose irradiation and MAP or sous vide packaging has potential as it could be possible to: (a) reduce the numbers of spoilage and pathogenic organisms by irradiation, and (b) suppress the growth of surviving microorganisms during storage by MAP or sous vide packaging without significantly affecting organoleptic quality. Organoleptic problems, such as production of off-odours and changes in colour and flavour, encountered in irradiated meats (Lea et al., 1960) may also be overcome by the use of MAP or vacuum packaging (Dempster, 1985; Egan and Wills, 1985). The combination of irradiation and MAP has successfully been applied to beef (Niemand et al., 1981), pork (Mattison et al., 1986; Ehioba et al., 1987; Grant and Patterson, 1991a, 1991b; Lambert et al., 1991a, 1991b, 1992) and fresh fish (Licciardello et al., 1984; Lescano et al., 1990) to improve their microbiological and sensory quality.

The microbiological safety of MAP foods has been the subject of much debate (Genigeorgis, 1985; Hintlian and Hotchkiss, 1986; Reddy et al., 1992). There have been concerns expressed that MAP alone, or in combination with low-dose irradiation, might favour the growth of and toxin production by *C. botulinum* particularly under conditions of temperature abuse (Lambert et al., 1991c). The main concern is that irradiation and/or MAP may inhibit the spoilage bacteria sufficiently to allow toxin formation while the product is still organoleptically acceptable. A further concern is the possibility that irradiation in conjunction with MAP may activate spore germination and enhance toxin production by *C. botulinum* if the product is temperature abused (Lebepe et al., 1990). These safety concerns were addressed by Lambert et al., (1991a) who investigated the combined effects of initial O_2 concentration (0, 10 or 20% O_2, balance N_2), irradiation dose (0, 0.5 or 1.0 kGy) and storage temperature (5, 15 or 25°C) on toxin production by *C. botulinum* in MAP pork. Temperature appeared to have the greatest effect on toxin production. No toxin was detected in any product stored at 5°C, indicating the importance of proper refrigerated storage to prevent the growth of proteolytic strains of *C. botulinum*. All samples were toxic after 2 days at 25°C, irrespective of headspace atmosphere or irradiation treatment. When storage was at a less abusive temperature (15°C), irradiation significantly delayed toxin production in samples packaged under anaerobic conditions, but when O_2 was initially present, irradiation had no effect on toxigenesis. It was suggested that initial packaging of the product with O_2 enhanced toxin production at 15°C due to increased production of CO_2, which has been shown to encourage spore germination (Foegeding and Busta, 1983). This theory was investigated in a subsequent study (Lambert et al., 1991b). As before, low-dose irradiation was found to delay toxin production in samples packed anaerobically. Toxin production occurred faster in packs containing 15 or 30% CO_2 than in packs containing higher levels of CO_2. However, contrary to expectations, inclusion of a CO_2 absorbent in the package also enhanced toxin production by *C. botulinum*. This was attributed to the production of H_2 by the CO_2 absorbent, possibly resulting in a decrease in the redox potential of the meat.

Non-proteolytic *C. botulinum* strains, capable of growing and producing toxin at refrigeration temperatures without exhibiting any signs of spoilage, could have particular significance in MAP and vacuum-packaged products, especially if the competing microorganisms were reduced by irradiation. Firstenberg-Eden et al., (1982, 1983) inoculated vacuum-packed chicken skins with *C. botulinum* type E spores, before irradiating with a total dose of 3 kGy and storage at 10 and 30°C. At both temperatures, spoilage was found to precede toxin production. The results indicated that an irradiation dose of 3 kGy (but not 5 kGy) left enough viable microorganisms to compete with *C. botulinum* type E on severely temperature abused (30°C) chicken skins. Therefore, the concern about toxin production occurring before spoilage may only be valid if doses above 3 kGy are used. Similar results were obtained in a study using *C. botulinum* types A and B inoculated onto chicken skins irradiated at 3 kGy and stored aerobically or vacuum packed at 30°C. However, during storage at 10°C, the irradiated and non-irradiated *C. botulinum* spores failed to grow or produce toxin (Dezfulian and Bartlett, 1987).

Several workers have reported the successful use of irradiation with vacuum packaging or MAP to improve the shelf life of fish such as hake and cod (Licciardello et al., 1984; Przybylski et al., 1989; Lescano et al., 1990). Problems of development of rancidity in fatty fish, such as petrale sole, have been overcome by exclusion of O_2 during irradiation and subsequent storage. The most obvious safety concern with MAP or vacuum-packaged fish is the potential for *C. botulinum* type E growth and toxin production. The evidence for the safety of these products is conflicting. Some workers have reported that spoilage occurs before toxin is detectable in MAP fish (Stier et al., 1981), whereas others have reported that toxin was detectable before the MAP fish was visibly spoiled (Post et al., 1985). The potential hazard from *C. botulinum* type E in MAP fish depends on a number of factors including: chemical composition of the fish; initial spore inoculum levels; storage temperature and methods used to estimate shelf life (Reddy et al., 1992).

The dose used to irradiate MAP fish is important. Doses in the order of 1.0 kGy have been proposed for fish products (Eklund, 1982; Urbain, 1986) so that a sufficient number of spoilage microorganisms survive irradiation to compete with *C. botulinum* during subsequent storage. Storage at or below 3.3°C is still necessary to prevent botulinum toxin production, irrespective of packaging atmosphere or irradiation treatment. However, there is some evidence that in certain vacuum-packed fish such as trout, irradiation to 2 kGy and storage at higher temperatures (5°C but not 10°C) is possible with a good margin of safety between spoilage and toxin production (Hussain et al., 1977).

The requirements for clostridial spore germination, growth and toxin production, and their interactions with atmosphere and temperature are obviously complex. The additional factor of irradiation injury also needs to be considered when MAP products are irradiated.

With respect to other foodborne pathogens, research indicates that MAP storage does not increase the hazards from *Salmonella*, *Staphylococcus aureus*, *Yersinia enterocolitica* and *L. monocytogenes* but there is evidence that MAP will not completely inhibit the growth of these pathogens, particularly if temperature abuse occurs (Hintlian and Hotchkiss, 1986; Gill and Reichel, 1989). The survival and growth of some of these pathogens in irradiated, MAP (25% CO_2:75% N_2) minced pork stored at abuse temperatures (10 or 15°C) was assessed by Grant and Patterson (1991b). An earlier study had shown that packaging in 25% CO_2:75% N_2 combined with irradiation (1.75 kGy) significantly improved the microbiological and sensory quality of pork chops (Grant and Patterson, 1991a). *Salmonella typhimurium*, *L. monocytogenes*, *Escherichia coli*, *Y. enterocolitica* or *C. perfringens* were inoculated onto the surface of pork mince at two levels: 10^3 cells/g and 10^6 cells/g, the latter representing an extreme level of contamination. Irradiation to a dose of 1.75 kGy reduced pathogen numbers by 1 to >5 $\log_{10}$ cycles depending on the strain (Table 11.1). *C. perfringens* was the most radiation-resistant and *Y. enterocolitica* the most radiation-sensitive of the pathogens studied. It was found that, with the exception of *C. perfringens*, pathogens present in low numbers (10^3 cells/g) prior to irradiation were either not detected, or only detected after an extended lag period. When present in high numbers (10^6 cells/g) before irradiation, low numbers of the pathogens survived but during storage they did not grow to reach the levels attained in the unirradiated MAP pork. *Y. enterocolitica* was not detected even when inoculated at high levels (10^6 cells/g), as it is very sensitive to irradiation. When the more radiation-resistant *C. perfringens* was inoculated onto the pork in low numbers (10^3 cells/g), clostridial numbers in the unirradiated and irradiated pork did not increase over the storage period (7 d). When inoculated at a high level (10^6 cells/g) clostridial counts did not increase in the unirradiated samples, but increased by two $\log_{10}$ cycles in the irradiated pork stored at 15°C after an initial lag period of 4 d. It has been shown that the growth of *C. perfringens* on anaerobically packaged raw beef and vacuum-packaged frankfurters was inhibited by the growth of lactic acid bacteria (LAB) (Goepfert and Kim, 1975; Nielsen and Zeuthen, 1985). In the MAP pork, the LAB were able to outgrow the clostridia when low numbers of the pathogen were present prior to irradiation. However *C. perfringens*, when present at high levels prior to irradiation, was able to grow on the irradiated pork. This would suggest that the pathogen was able to compete more successfully when fewer spoilage organisms were present. It was concluded that the microbiological safety of MAP pork was improved by irradiation to a dose of 1.75 kGy. However, only foods

Table 11.1. Reduction in pathogen numbers achieved by irradiating MAP pork mince to a dose of 1.75 kGy (Grant and Patterson, 1991b).

Pathogen	$\log_{10}$ Reduction
Yersinia enterocolitica	>5
Escherichia coli	4
Salmonella typhimurium	2–3
Listeria monocytogenes	2–3
Clostridium perfringens	1

of good microbiological quality should be treated by irradiation and MAP if the safety of the product is to be assured.

In several of the above studies it was also noted that LAB form a significant part of the microbial population of irradiated MAP and vacuum-packaged meats during storage (Niemand et al., 1981; Grant and Patterson, 1991a; Lambert et al., 1992). As many LAB have been shown to be antagonistic towards food poisoning bacteria (Schaack and Marth, 1988; Gill and Reichel, 1989; Spelhaug and Harlander, 1989), growth of LAB in irradiated MAP meats during storage could confer additional safety to these products.

The effect of irradiation in combination with sous vide processing on the survival and growth of *C. botulinum* has not been addressed. However, it has been shown that *C. botulinum* spores can be sublethally injured by irradiation (Rowley et al., 1983), and that an irradiation dose of 2.32 kGy can sensitize clostridial spores to subsequent heat treatment (Morgan and Reed, 1954). It is likely that such a combination would be beneficial in reducing spore numbers and the risk of toxin production in sous vide products, although this would need to be investigated by a series of inoculation studies.

The combined effect of electron beam irradiation (2.9 kGy) and sous vide treatment (cooking to an internal temperature of 65.6°C) on the survival and growth of *L. monocytogenes* in chicken breast meat has been investigated (Shamsuzzaman et al., 1992). The initial inoculum level of *L. monocytogenes* used was 10^5 cfu/g. The sous vide treatment alone had little effect on the numbers of the pathogen while the radiation treatment caused a 4.6 $\log_{10}$ reduction. However, the radiation and heat treatments acted synergistically and the organism remained undetectable during 8 weeks storage at 2°C. Additional experiments indicated that the combination of irradiation with sous vide treatment extended shelf life and had no significant effect on odour or flavour of the reheated samples. It was concluded that irradiation could greatly enhance the microbial safety and shelf life of sous vide chicken breast meat.

BIOPRESERVATION

During the past few years there has been an increasing interest in biocontrol preservation of foods (also referred to as biopreservation), and much research has been initiated making use of microbial antagonism in the preservation of fermented and non-fermented foods.

Biopreservation may be described as the use of non-pathogenic LAB, or their metabolic products (e.g., lactic acid, nisin), to control the proliferation of pathogenic bacteria in food products (Hanlin and Evancho, 1992). Rather than relying on the natural development of LAB within a food during storage, appropriate "starter" cultures could be added to certain foods (Mossel and Struijk, 1991). Numerous studies have demonstrated the antagonistic action of streptococci (Saleh and Ordal, 1955; Gilliland and Speck, 1972; Wang et al., 1986), pediococci (Daly et al., 1973; Tanaka et al., 1980; Hutton et al., 1991), lactobacilli (Daly et al., 1973; Nielsen and Zeuthen, 1985; Raccach et al., 1989) and *Carnobacterium piscicola* (Buchanan and Klawitter, 1992) against a number of food pathogens including *E. coli, Salmonella* spp., *C. botulinum, S. aureus* and *L. monocytogenes*. The antimicrobial activity of LAB has been attributed to: (a) lowering of pH by production of lactic and other acids, (b) competition for nutrients, (c) generation of hydrogen peroxide, and (d) production of bacteriocins (Gibbs, 1987; Gould, 1992).

Use of Starter Cultures

The best example of the use of starter cultures to preserve a refrigerated food is the "Wisconsin Process" which was developed in the early 1980s by scientists at the University of Wisconsin's Food Research Institute. Tanaka et al. (1980) evaluated the antibotulinal effects of *Lactobacillus plantarum*, as a producer of lactic acid, and sucrose, as a fermentable carbohydrate, for use in lowering the amount of, or eliminating, sodium nitrite in bacon. These workers postulated, and eventually showed (Tanaka et al., 1980, 1985b), that botulinal toxigenesis was inhibited if a controlled level of *L. plantarum* or *Pediococcus acidilactici* (4×10^6 cells/g) and sucrose (0.9%) were added to bacon. Only one of forty-nine samples became toxic when bacon was formulated with *L. plantarum* and sucrose, but no nitrite. When nitrite (40 ppm), *L. plantarum* and sucrose were added to the bacon, none of the thirty samples became toxic. Tanaka et al. (1980, 1985b) found that sufficient acid was produced and the pH of the bacon reduced fast enough to prevent botulinal toxigenesis under temperature abuse (27°C) conditions. However, if the bacon was not temperature abused there was no significant growth or acid production by the starter culture and cooked bacon was as organoleptically acceptable as traditionally processed bacon (Tanaka et al., 1985a). The "Wisconsin

Process'' was approved by the USDA in 1986 (9 CFR Part 318.7) as an alternative method for processing bacon.

The principles of the ''Wisconsin Process'' have also been applied to egg and chicken salads with some success (Hutton et al., 1991). Application to sous vide and MAP products may be possible but a number of factors which affect the efficacy of this preservation technology must be taken into consideration: 1) type and concentration of fermentable carbohydrate, 2) species/strain and concentration of starter culture, 3) initial pH and buffering capacity of the food, 4) nutritional profile of the product, 5) presence of factors inhibitory to starter culture growth, and 6) growth characteristics (e.g., acid production) of starter cultures over a wide temperature range (Gombas, 1989).

An alternative approach is to use starter cultures of bacteriocin-producing strains of LAB, rather than strong acid producers (Wang et al., 1986; Buchanan and Klawitter, 1992). Bacteriocins are biologically active proteins displaying bactericidal activity towards other bacteria (Tagg et al., 1976). Evidence for the in-situ production of the bacteriocin nisin by *Lactococcus lactis* was provided in a study by Wang et al. (1986) who inoculated frankfurters during manufacture with 10^9 *L. lactis*/g and found that growth of spoilage microorganisms was inhibited during storage. A more recent study by Buchanan and Klawitter (1992) utilized a bacteriocin-producing strain of *C. piscicola* to control the growth of *L. monocytogenes* in a number of meat and fish products at refrigeration temperatures. *L. monocytogenes* did not grow on the surface of frankfurters at either 5 or 19°C, when cultured in the presence of *C. piscicola*.

Where reliance is placed on the production of bacteriocins in food, the extent of biosynthesis of these substances in the food under ''real life'' conditions, and diffusion of the bacteriocins within the food, need to be accurately determined (Mossel and Struijk, 1991).

Use of Metabolic Products

In many food applications production of high levels of acid by LAB is undesirable and therefore inhibition of pathogenic organisms must be achieved by other means. Direct application of bacteriocins produced by LAB may be an option in these circumstances. Many bacteriocins have been identified and new ones are continually being isolated. The variety of bacteriocins known was reviewed recently by Lücke and Earnshaw (1991). To date only nisin, produced by strains of *L. lactis*, has been approved by the Food and Drug Administration for use in the U.S. It possesses antimicrobial activity against *Bacillus* and *Clostridium* spp., *S. aureus* and *L. monocytogenes* (Delves-Broughton, 1990). Few other bacteriocins have enjoyed substantial commercial use.

Where bacteriocins are used as additives, binding to and partial inactivation by food constituents, as well as possible inactivation by enzymes produced by organisms resistant to the bacteriocins, must be taken into account (Mossel and Struijk, 1991). For example, it was reported by Scott and Taylor (1981) that in meat systems the binding of nisin to meat particles decreases its activity sufficiently to allow outgrowth of *C. botulinum*. Consequently, the use of nisin in meat preservation may be limited to its use in combination with other preservatives.

The combination of nisin and nitrite in cured meat products has received some attention (Rayman et al., 1981; Calderon et al., 1985; Taylor et al., 1985). Taylor et al. (1985) showed that botulinum toxin formation in chicken frankfurter emulsions at 27°C was slowed by the addition of nitrite and/or nisin. In the absence of any antibotulinal agent, toxin was detected in all samples tested after only one week of storage. Inhibition of toxin formation for 5–6 weeks was achieved by addition of 100 or 250 ppm nisin, and 120 ppm nitrite. When even higher levels of nisin (500 ppm) were added to the meat, the corresponding level of nitrite required for inhibition of toxin formation fell to 40 ppm. However, the addition of such high levels of nisin is considered uneconomic (Calderon et al., 1985).

A large number of LAB have been shown to produce bacteriocins inhibitory towards *L. monocytogenes* in laboratory trials (Harris et al., 1989; Raccach et al., 1989; Schillinger and Lücke, 1989). As yet, direct addition of these bacteriocins to MAP and/or sous vide food systems to control the growth of this pathogen has not received much attention. Mossel and Struijk (1991) suggest that, even though the direct use of bacteriocins has potential, the practical applicability of bacteriocins in foods requires further investigation.

CHEMICAL PRESERVATIVES

In recent years the trend in food preservation has been to reduce the use of chemical preservatives in food products, principally in response to consumer pressure for more "additive-free" foods. Consumers fail to realise that removing all chemical preservatives from food would prove detrimental to the food supply, rather than beneficial. Therefore, in order to maintain the range, quality, safety and convenience of the food supply, there is still a vital role for chemical preservation. However, it may be possible to reduce the levels of certain chemical preservatives by application of the hurdle technology. The main chemical preservatives which can be used as additional hurdles in MAP or sous vide products include: organic acids, fatty acids, salts and nitrite. The potential of each of these preservatives to improve the safety of MAP and sous vide products will be discussed.

Organic Acids

The organic acids and their derivatives which are principally used as antimicrobial agents in foods are acetic, lactic, benzoic, propionic and sorbic acids (Wagner and Moberg, 1989). The organic acids have been studied mostly as antimicrobial agents in fresh meats (Woolthius and Smulders, 1985; Bell et al., 1986). Their use in sous vide and MAP meats has been investigated more recently.

The potential for clostridial growth in refrigerated sous vide pork chops was demonstrated by Prabhu et al. (1988). These workers investigated the potential to improve the microbiological safety of cooked pork chops by dipping them in 1% acetic acid prior to vacuum packaging and recooking-in-the-bag to 66°C (150°F). The acetic acid dip effectively prevented growth of surface-inoculated *C. sporogenes* during exposure of the sous vide product to simulated mishandling (48 h at 24–25°C).

In a subsequent study to investigate anti-clostridial and anti-listerial effects of acetic acid, Unda et al. (1991) incorporated 1% acetic acid into a brine which was pumped into beef roasts prior to surface inoculation with *C. sporogenes* and *L. monocytogenes* (10^3 spores or cells per roast), vacuum packaging and cooking-in-the-bag to 62.8°C. Although acetic acid exhibited some ability to delay growth of clostridia in beef roasts during simulated mishandling (25°C for up to 48 h after storage at 2–4°C for two weeks), it did not prevent greater survival of *C. sporogenes* spores than any of the other antimicrobials tested (glycerol monolaurin, sodium lactate and potassium sorbate). The same study also showed that a combination of the polyphosphate blend Brifisol 414™ and 1% acetic acid nearly eliminated listeriae in refrigerated (but not temperature-abused) sous vide beef roasts.

The efficacy of lactic acid as a meat decontaminant is well established

(Smulders et al., 1986). Lactic acid decontamination of poultry and meat products has been studied (Smulders, 1987; Cudjoe, 1988; Zeitoun and Debevere, 1990). Zeitoun and Debevere (1991) investigated the effect of buffered 2%, 5% and 10% lactic acid treatment and modified atmosphere packaging (90% CO_2:10% O_2) on the inhibition, survival and growth of *L. monocytogenes* on chicken legs. A combined treatment of 10% lactic acid/sodium lactate (pH 3.0) and MAP was also included among the treatments. *L. monocytogenes* numbers increased by just 0.56 cfu/cm^2 during 17 d storage at 6°C on the legs treated with 10% lactic acid/sodium lactate combined with MAP, whereas *L. monocytogenes* numbers increased by 1.80 cfu/cm^2 and 2.93 cfu/cm^2 on 10% buffered lactic acid treated and untreated MAP samples, respectively. Zeitoun and Debevere (1991) concluded that the antimicrobial effect of lactic acid/sodium lactate buffer systems increased with increasing concentration of lactic acid, but there seemed to be a synergistic effect on the inhibition of *L. monocytogenes* between MAP (90% CO_2:10% O_2) and treatment with 10% lactic acid/sodium lactate buffer (pH 3.0).

The incorporation of sodium lactate into sous vide turkey products and beef roasts has also been studied (Maas et al., 1989; Unda et al., 1991). Both of these studies were carried out to investigate the possibility of controlling the growth of *C. botulinum* under conditions of temperature abuse. Maas et al. (1989) incorporated 2%, 2.5%, 3% or 3.5% sodium lactate as a 60% aqueous solution into comminuted turkey meat, which was subsequently vacuum packaged and cooked to an internal temperature of 71.1°C by submersion in an 88°C water bath, prior to storage at 27°C for up to 10 days. This study showed that sodium lactate exhibited an antibotulinal effect which was concentration dependent (Table 11.2).

Unda et al. (1991) pumped 2% sodium lactate into beef roasts in brine form, prior to vacuum packaging and cooking to 62.8°C. The brine also included 1% sodium chloride and 0.3% Brifisol 414™. Lowest survival of viable clostridial spores during low-temperature storage (2−7°C) was evident in roasts pumped with brines containing sodium lactate compared with brines containing a number of other antimicrobials. Sodium lactate also proved effective as an antibotulinal agent in temperature-abused samples (24−25°C after initial storage at 2−4°C for two weeks). The same study also showed that the greatest inactivation of *L. monocytogenes* in sous vide beef roasts was obtained using brines containing Brifisol 414™ and 2% sodium lactate.

Sorbate has traditionally been used as an antifungal agent in a wide range of food products (Sofos and Busta, 1981). In the U.S. it is a

Table 11.2. Effect of sodium lactate concentration on toxin production by Clostridium botulinum *in cook-in-bag turkey products (Maas et al., 1989).*

% Sodium Lactate	Time to Toxin Detection (d)
0 (control)	3
2.0	4–5
2.5	4–5
3.0	7
3.5	7–8

Generally Regarded as Safe (GRAS) preservative. Levels of incorporation into food products vary from 0.01–0.30% sorbic acid. Prabhu et al. (1988) and Unda et al. (1991) investigated the effect of adding 2.5% potassium sorbate into sous vide pork chops and beef roasts, respectively, on the microbiological safety of these products. Neither study showed potassium sorbate to exhibit much anti-clostridial activity when compared with the effects of a number of other chemical preservatives (including glycerol monolaurin, sodium lactate and acetic acid).

The combination of potassium sorbate and MAP for fresh fish and fish products has been found to be beneficial. Studies have reported that a combination of 1% potassium sorbate ice and MAP (Fey and Regenstein, 1982), or dipping in 0.1–1.2% potassium sorbate solution and MAP (Bremner and Statham, 1983; Statham et al. 1985) protects against spoilage and pathogenic organisms, and inhibits the growth of trimethylamine-producing bacteria in fresh fish stored at 0–1°C. It is generally thought that a potassium sorbate pre-treatment is necessary to minimize safety risks associated with MAP fish (Reddy et al., 1992). Seward (1982) found that dipping fish fillets in 5% potassium sorbate/10% tripolyphosphate solution and storing them under elevated CO_2 atmosphere delayed growth and toxin production by *C. botulinum* longer than did the use of CO_2 atmosphere alone.

Fatty Acids

The fatty acids capric (C10:0), lauric (C12:0) and myristic (C14:0) are the most biocidal of the saturated fatty acids, while palmitoleic (C16:1) is the most active unsaturated fatty acid (Kabara, 1981). Fatty acids esterified to polyhydric alcohols are used in the food industry for their emulsifying properties, the most important being the monoglycerides. One of these monoglycerides, monolaurin, has been shown to be inhibitory towards a number of gram-positive bacteria. Notermans and Dufrenne (1981) showed that when glyceryl monolaurin was added to a meat slurry (5 g per kg), toxin production by *C. botulinum* types A, B and E was inhibited during storage at 30°C for 7–8 d. Bacteriostatic and bacteriocidal effects of monolaurin on *S. aureus* in a model agar-meat system were reported by Kabara (1984).

Potential anti-clostridial and anti-listerial effects of 0.25% glyceryl monolaurin were investigated by Unda et al. (1991) who incorporated monolaurin into a NaCl-phosphate brine used to pump beef roasts prior to vacuum packaging and cooking. Glyceryl monolaurin was found to reduce survival of clostridial spores, and to result in the greatest inactivation of *L. monocytogenes* in sous vide beef roasts cooked to 62.8°C in a water bath. Wang and Johnston (1991) also found monolaurin and a number of other fatty acids to be inhibitory towards *L. monocytogenes* in milk.

Salts

The two major salts which have been used as hurdles in the preservation of MAP and sous vide products are sodium chloride and polyphosphates. In recent years the antimicrobial importance of sodium chloride (NaCl) has been reduced, chiefly because additional processing factors and preservatives are now employed in combination to preserve foods (Sofos, 1984). The antimicrobial contribution of NaCl in a food system may also be influenced by the presence of other preservatives. Synergistic interactions of NaCl with benzoate, sorbate, phosphates, antioxidants (BHA), spices and liquid smoke have been reported.

The primary role of phosphates or polyphosphates in foods is to confer specific functional properties. In meat products, polyphosphates may buffer pH, sequester metal ions and influence the system through their polyanionic properties. They can improve the stability of cured meat colour, decrease oxidative rancidity and improve cook yields, texture and tensile strength of meat products (Ellinger, 1972). A review by Tompkin (1984) of the antimicrobial effects of phosphates in foods and laboratory media indicated that under certain conditions phosphates exhibit antimicrobial activity. A number of phosphates have been examined for their anti-botulinal and anti-listerial properties including: sodium acid pyrophosphate (SAPP), sodium polyphosphate glassy

(SPG), tetrasodium pyrophosphate (TSPP) and sodium tripolyphosphate (STPP) (Molins et al., 1985, 1986).

Molins et al. (1985) added 0.5% SAPP, STPP, TSPP or SPG to bratwurst sausages inoculated with *C. sporogenes* prior to cooking and vacuum packaging. No significant bacterial inhibition by any phosphate was observed during refrigerated (5°C) storage. However, SAPP caused significant inhibition of *C. sporogenes* during temperature abuse, followed in effectiveness by TSPP and STPP. Cooking to 65.5°C was found to help retain antimicrobial properties of phosphates to some extent. It is known that meat contains phosphatases, which, if not inactivated, could hydrolyse added phosphates and render them useless (Sutton, 1973). Molins et al. (1985) stressed that cooking should take piace as soon as possible after phosphate addition in order to retain maximal antimicrobial effects.

In most applications, phosphates are not used alone but in combination with one or more other chemical preservatives. A study by Molins et al. (1986) examined the effect of 0.5% SAPP or 0.5% STPP in the presence or absence of added sodium nitrite (50 or 100 ppm) on the inhibition of *C. sporogenes* inoculated into cooked vacuum-packaged bratwurst. Phosphates alone, or combined with nitrite, were found to result in reduced clostridial counts at 5°C. Upon temperature abuse, inhibition of *C. sporogenes* for up to 48 h was significant, and greatly enhanced by phosphates combined with 100 ppm, but not 50 ppm, sodium nitrite.

A number of other chemical combinations incorporating phosphates were examined by Prabhu et al. (1988) and Unda et al. (1991) in sous vide meat products. Meat pumped with brines containing only salt and phosphates (STPP or Brifisol 414™) did not prevent clostridial growth during 24 h of temperature abuse, whereas the inclusion of acetic acid, sodium lactate, monolaurin or potassium sorbate in the brines improved the microbiological safety of sous vide pork chops and beef roasts (as described previously).

Nitrite

Traditionally, sodium nitrite has been added to cured meats, such as ham, bacon, bologna and frankfurters, to confer a distinct colour and flavour, and retard *C. botulinum* growth and toxin production (Sofos et al., 1979a). Protection against *C. botulinum* in sous vide meat products is essential as conditions within this type of product are potentially conducive to the growth of this pathogen. However, nitrite cannot be used to provide botulism control in uncured sous vide products because of the inevitability of cured meat colour development. Extensive research has been undertaken to find other antimicrobial agents that might be used to replace nitrite, totally or partially, as antibotulinal agents in meats (Sofos and Busta, 1980). Where it is desirable to confer cured meat colour to a sous vide product the incorporation of reduced levels of sodium nitrite in conjunction with potassium sorbate shows promise. Research carried out by Sofos and coworkers showed that sorbic acid (0.2%) inhibited botulinal spore germination in mechanically-deboned chicken meat, beef and pork frankfurter emulsions temperature abused at 27°C. Added nitrite (20–156 μg/g) did not affect spore germination, but at high levels (156 μg/g) it delayed toxin production by *C. botulinum* (Sofos et al., 1979b). These workers also found that when 0.2% sorbic acid was added along with nitrite, nitrite depletion from the product was slower during temperature abuse and toxin was not produced until nitrite levels decreased to 5–15 μg/g. One possible explanation for the combined effectiveness of sodium nitrite and sorbic acid could be that the two react and form a potent inhibitor(s) (Sofos et al., 1979a).

Buchanan et al. (1989) investigated the effects and interactions of temperature, pH, presence and absence of O_2, sodium chloride and sodium nitrite on the growth of *L. monocytogenes* Scott A in a model broth system. Data obtained by these workers suggest that sodium nitrite can have significant bacteriostatic activity against *L. monocytogenes* and may provide cured meats with a degree of protection against this pathogen, particularly if employed in conjunction with acidic pH, vacuum packaging, high salt concentration and adequate refrigeration. Clearly these findings would have some application to sous vide and vacuum-packaged cured meat products.

Liquid Smoke

Smoke components such as formaldehyde, acetic acid, creosote and high-boiling phenols exhibit bactericidal properties and also confer a desirable flavour and colour to processed meats. Several commercially available smoke concentrates (CharSol-10, Aro-Smoke P-50, CharOil Hickory, CharSol PN-9 and CharDex Hickory) prepared as 0.25% and 0.5% solutions have been shown to exhibit considerable antimicrobial activity against *L. monocytogenes* (Messina et al., 1988). These workers employed one liquid smoke preparation (CharSol-10) as a dip for beef frankfurters prior to vacuum packaging. In untreated frankfurters *L.*

monocytogenes numbers remained unchanged after 72 h at 4 °C, whereas frankfurters dipped in CharSol-10 liquid smoke exhibited a greater than 99.9% reduction in *L. monocytogenes* numbers after 72 h at 4 °C.

Application of liquid smoke preparations to fresh fish fillets prior to MAP or vacuum packaging may also have potential, although to our knowledge these combinations have not been investigated.

CONCLUSIONS

In the past the microbiological safety of MAP and sous vide products has relied almost exclusively on refrigeration. However, in recent years it has become clear that refrigeration alone is no longer adequate to prevent the growth of psychrotrophic food poisoning bacteria such as *L. monocytogenes*, *Y. enterocolitica* and the non-proteolytic clostridia. Fortunately, a number of additional hurdles exist which could potentially be applied to safeguard MAP and sous vide products. These have been discussed in this chapter.

With the exception of irradiation which inactivates microorganisms, the other hurdles described inhibit microbial growth, but are not lethal. It is unfortunate that consumer resistance to irradiation has prevented the widespread use of this technology to eliminate food poisoning bacteria from food products. Therefore, the potential microbiological benefits of low-dose irradiation as applied to MAP and sous vide may not be achieved for some time.

Consumers are increasingly demanding foods which are fresh, natural and more "preservative-free." In light of this fact, the use of chemical preservatives in MAP or sous vide products may not be pertinent at this point in time, although the anti-botulinal effectiveness of nitrite, for example, may be difficult to equal by other means.

Of all the potential hurdles which might be applied to MAP/sous vide products, biopreservation would appear to be the most promising. It is a more acceptable means of preservation than any of the others in the sense that the inhibitory action of LAB towards some food poisoning bacteria occurs naturally in the environment. Therefore, the addition of LAB to foods may be perceived by the consumer as being natural, and hence acceptable. However, biopreservation would need to be evaluated on a product to product basis in order to determine the appropriate LAB strain, inoculum levels, and growth characteristics of the strain under the conditions prevailing in the food product.

The commercial potential of each of the hurdles discussed in this chapter has yet to be fully evaluated as there have been few practical applications to date. Consequently, considerable research is still required on the application of hurdle technology to particular MAP and sous vide products.

REFERENCES

Bell, M. F., Marshall, R. T. and Anderson, M. E. 1986. "Microbiological and sensory tests of beef treated with acetic and formic acids," *J. Food Protection*, 49:207–210.

Bremner, H. A. and Statham, J. A. 1983. "Effect of potassium sorbate on refrigerated storage of vacuum-packed scallops," *J. Food Science*, 48:1042–1047.

Buchanan, R. L. and Klawitter, L. A. 1992. "Effectiveness of *Carnobacterium piscicola* LK5 for controlling the growth of *Listeria monocytogenes* Scott A in refrigerated foods," *J. Food Safety*, 12:219–236.

Buchanan, R. L., Stahl, H. G. and Whiting, R. C. 1989. "Effects and interactions of temperature, pH, atmosphere, sodium chloride and sodium nitrite on the growth of *Listeria monocytogenes*," *J. Food Protection*, 52:844–851.

Calderon, C., Collins-Thompson, D. L. and Usborne, W. R. 1985. "Shelf life studies of vacuum-packaged bacon treated with nisin," *J. Food Protection*, 48(4): 330–333.

Cudjoe, K. S. 1988. "The effect of lactic acid sprays on the keeping qualities of meat during storage," *Int. J. Food Microbiology*, 7:1–7.

Daly, C., La Chance, M., Sandine, W. E. and Elliker, P. R. 1973. "Control of *Staphylococcus aureus* in sausage by starter cultures and chemical acidulation," *J. Food Science*, 38:426–430.

Delves-Broughton, J. 1990. "Nisin and its use as a food preservative," *Food Technology* 44(11):100–117.

Dempster, J. F. 1985. "Radiation preservation of meat and meat products: A review," *Meat Science*, 12:61–89.

Dezfulian, M. and Bartlett, J. G. 1987. "Effects of irradiation on growth and toxigenicity of *Clostridium botulinum* types A and B inoculated onto chicken skins," *Applied and Environmental Microbiology*, 53:201–203.

Egan, A. F. and Wills, P. A. 1985. "The preservation of meats using irradiation," *CSIRO Food Research Quarterly*, 45:49–54.

Ehioba, R. M., Kraft, A. A., Molins, R. A., Walker, H. W., Olson, D. G., Subbaraman, G. and Skowronski, R. P. 1987. "Effect of low-dose (100 krad) gamma radiation on the microflora of vacuum-packaged ground pork with and without added sodium phosphates," *J. Food Science*, 52:1477–1480, 1505.

Eklund, M. W. 1982. "Significance of *Clostridium botulinum* in fishery products preserved short of sterilization," *Food Technology*, 36:107–112, 115.

Ellinger, R. H., ed. 1972. Phosphate applications as microbial inhibitors, In *Phosphates as Food Ingredients*, CRC Press, Cleveland, Ohio, pp. 147–151.

Fey, M. S. and Regenstein, J. M. 1982. "Extending shelf life of fresh wet red-hake and salmon using CO_2-O_2 modified atmosphere and potassium sorbate ice at 1°C," *J. Food Science*, 47:1048–1054.

Firstenberg-Eden, R., Rowley, D. B. and Shattuck, G. E. 1982. "Factors affecting growth and toxin production by *Clostridium botulinum* type E on irradiated (0.3 Mrad) chicken skins," *J. Food Science*, 47:867–870.

Firstenberg-Eden, R., Rowley, D. B. and Shattuck, G. E. 1983. "Competitive growth of chicken skin microflora and *Clostridium botulinum* type E after an irradiation dose of 0.3 Mrad," *J. Food Protection*, 46:12–15.

Foegeding, P. M. and Busta, F. F. 1983. "Effect of carbon dioxide, nitrogen and hydrogen gases on germination of *Clostridium botulinum* spores," *J. Food Protection*, 46:987–989.

Genigeorgis, C. A. 1985. "Microbial and safety implications of the use of modified atmospheres to extend the storage life of fresh meat and fish: A review," *Int. J. Food Microbiology*, 1:237–251.

Gibbs, P. A. 1987. "Novel uses for lactic acid fermentation in food preservation," *Society for Applied Bacteriology Symposium Series Number 16*, C. S. Gutteridge and J. R. Norris, eds., Oxford, Blackwell Scientific Publications, pp. 51S–58S.

Gill, C. O. and Reichel, M. P. 1989. "Growth of cold-tolerant pathogens *Yersinia enterocolitica*, *Aeromonas hydrophilia* and *Listeria monocytogenes* on high pH beef packaged under vacuum and carbon dioxide." *Food Microbiology* 6: 223–230.

Gilliland, S. E. and Speck, M. L. 1972. "Interactions of food starter cultures and food-borne pathogens: lactic streptococci versus staphylococci and salmonellae," *J. Milk Food Technology*, 35:307–310.

Goepfert, J. M. and Kim, H. U. 1975. "Behaviour of selected foodborne pathogens in raw ground beef," *J. Milk Food Technology*, 38:449–452.

Gombas, D. E. 1989. "Biological competition as a preserving mechanism," *J. Food Safety*, 10:107–117.

Gould, G. W. 1992. "Ecosystem approaches to food preservation," *Society for Applied Bacteriology Symposium Series Number 21*, Blackwell Scientific Publications, Oxford, pp. 58S–68S.

Gould, G. W. and Jones, M. V. 1989. Combination and synergistic effects, In *Mechanisms of Action of Food Preservation Procedures*, G. W. Gould, ed., Elsevier Applied Science Publishers, Amsterdam, pp. 401–421.

Grant, I. R. and Patterson, M. F. 1991a. "Effect of irradiation and modified atmosphere packaging on the microbiological and sensory quality of pork stored at refrigeration temperatures," *Int. J. Food Science and Technology*, 26:507–519.

Grant, I. R. and Patterson, M. F. 1991b. "Effect of irradiation and modified atmosphere packaging on the microbiological safety of minced pork stored under temperature abuse conditions," *Int. J. Food Science and Technology*, 26:521–533.

Hanlin, J. H. and Evancho, G. M. 1992. The beneficial role of microorganisms in the safety and stability of refrigerated foods, In *Chilled Foods: A Comprehensive Guide*, C. Dennis and M. Stringer, eds., Ellis Horwood Ltd., Chichester, England, pp. 229–259.

Harris, L. J., Daeschel, M. A., Stiles, M. E. and Klaenhammer, T. R. 1989. "Antimicrobial activity of lactic acid bacteria against *Listeria monocytogenes*," *J. Food Protection*, 52(6):384–387.

Hintlian, C. B. and Hotchkiss, J. H. 1986. "The safety of modified atmosphere packaging: A review," *Food Technology*, 40:70–76.

Hussain, A. M., Ehlermann, D. and Diehl, J. F. 1977. "Comparison of toxin production by *Clostridium botulinum* type E in irradiated and unirradiated vacuum-packed trout (*Salmo gairdneri*)," *Archiv für Lebensmittelhygiene*, 28:23–27.

Hutton, M. T., Chehak, P. A. and Hanlin, J. H. 1991. "Inhibition of botulinum toxin production by *Pediococcus acidilactici* in temperature abused refrigerated foods," *J. Food Safety*, 11:255–267.

Kabara, J. J. 1981. "Food grade chemicals for use in designing food preservative systems," *J. Food Protection*, 44(8):633–647.

Kabara, J. J. 1984. "Inhibition of *Staphylococcus aureus* in a model agar-meat system by monolaurin: A research note," *J. Food Safety*, 6:197–201.

Lambert, A. D., Smith, J. P. and Dodds, K. L. 1991a. "Combined effect of modified atmosphere packaging and low-dose irradiation on toxin production by *Clostridium botulinum* in fresh pork," *J. Food Protection*, 54:94–101.

Lambert, A. D., Smith, J. P. and Dodds, K. L. 1991b. "Effect of headspace CO_2 concentration on toxin production by *Clostridium botulinum* in MAP irradiated fresh pork," *J. Food Protection*, 54:588–592.

Lambert, A. D., Smith, J. P. and Dodds, K. L. 1991c. "Shelf life extension and microbiological safety of fresh meat–A review," *Food Microbiology*, 8:267–297.

Lambert, A. D., Smith, J. P. and Dodds, K. L. and Charbonneau, R. 1992. "Microbiological changes and shelf life of MAP, irradiated fresh pork," *Food Microbiology*, 9:231–244.

Lea, C. H., Macfarlane, J. J. and Parr, L. J. 1960. "Treatment of meats with ionizing radiations. V. Radiation pasteurisation of beef for chilled storage," *J. Science Food and Agriculture*, 11:690–694.

Lebepe, S., Molins, R. A., Charoen, S. P., Farrar IV, H. and Skowronski, R. P. 1990. "Changes in the microflora and other characteristics of vacuum-packaged pork loins irradiated at 3.0 kGy," *J. Food Science*, 55:918–924.

Leistner, L. 1985. Hurdle technology applied to meat products of the shelf stable product and intermediate moisture food types, In *Properties of Water in Foods*, D. Simatos and J. L. Multon, eds., Martinus Nijhoff, Dordrecht, pp. 309–329.

Leistner, L. 1992. "Food preservation by combined methods," *Food Research International*, 25:151–158.

Lescano, G., Kairiyama, E., Narvaiz, P. and Kaupert, N. 1990. "Studies on quality of radurized (refrigerated) and non-radurized (frozen) hake (*Merluccius merluccius hubbsi*)," *Lebensmittel-Wissenshaft und Technologie*, 23:317–321.

Licciardello, J. J., Ravesi, E. M., Tuhkunen, B. E. and Racicot, L. D. 1984. "Effect of some potentially synergistic treatments in combination with 100 krad irradiation on the iced shelf life of cod fillets," *J. Food Science* 49:1341–1346, 1375.

Lücke, F-K. and Earnshaw, R. G. 1991. Starter cultures, In *Food Preservatives*, N. J. Russell and G. W. Gould, eds., Blackie & Son, Glasgow, pp. 215–234.

Maas, M. R., Glass, K. H. and Doyle, M. P. 1989. "Sodium lactate delays toxin production by *Clostridium botulinum* in cook-in-bag turkey products," *Applied and Environmental Microbiology*, 55(9):2226–2229.

Mattison, M. L., Kraft, A. A., Olson, D. G., Walker, H. W., Rust, R. E. and James, D. B. 1986. "Effect of low dose irradiation of pork loins on the microflora, sensory characteristics and fat stability," *J. Food Science*, 51:284–287.

Messina, M. C., Ahmad, H. A., Marchello, J. A., Gerba, C. P. and Paquette, M. W. 1988. "The effect of liquid smoke on *Listeria monocytogenes*," *J. Food Protection*, 51:629–631.

Molins, R. A., Kraft, A. A., Walker, H. W. and Olson, D. G. 1985. "Effect of poly- and pyro-phosphates on the natural bacterial flora and inoculated *Clostridium sporogenes* PA 3679 in cooked, vacuum-packaged bratwurst," *J. Food Science*, 50:876–880.

Molins, R. A., Kraft, A. A., Olson, D. G., Walker, H. W. and Hotchkiss, D. K. 1986. "Inhibition of *Clostridium sporogenes* PA 3679 and natural bacterial flora of cooked, vacuum-packaged bratwurst by sodium acid pyrophosphate and sodium tripolyphosphate with or without added sodium nitrite," *J. Food Science*, 51:726–730.

Morgan, B. H. and Reed, J. M. 1954. "Resistance of bacterial spores to gamma irradiation," *Food Research*, 19:357–366.

Mossel, D. A. A. and Struijk, C. B. 1991. "Public health implications of refrigerated, pasteurised ("sous vide") foods." *Int. J. Food Microbiology*, 13:187–206.

Nielsen, H.-J. S. and Zeuthen, P. 1985. "Influence of lactic acid bacteria and the overall flora on development of pathogenic bacteria in vacuum-packed, cooked emulsion-style sausage," *J. Food Protection*, 48(1):28–34.

Niemand, J. G., Van Der Linde, H. J. and Holzapfel, W. H. 1981. "Radurization of prime beef cuts," *J. Food Protection*, 44:677–681.

Notermans, S. and Dufrenne, J. 1981. "Effect of glyceryl monolaurate on toxin production by *Clostridium botulinum* in meat slurry," *J. Food Safety*," 3:83–88.

Post, L. S., Lee, D. A., Solberg, M., Furang, D., Specchio, J. and Graham, C. 1985. "Development of botulinal toxin and sensory deterioration during storage of vacuum and modified atmosphere packaged fish fillets," *J. Food Science*, 50:990–996.

Prabhu, G. A., Molins, R. A., Kraft, A. A., Sebranek, J. G. and Walker, H. W. 1988. "Effect of heat treatment and selected antimicrobials on the shelf life and safety of cooked, vacuum-packaged, refrigerated pork chops," *J. Food Science*, 53(5):1270–1272, 1326.

Przybylski, L. A., Finerty, R. M., Grodner, R. M. and Gerdes, D. L. 1989. "Extension of shelf life of iced fresh channel catfish fillets using modified atmosphere packaging and low-dose irradiation," *J. Food Science*, 54:269–273.

Raccach, M., McGrath, R. and Daftarian, H. 1989. "Antibiosis of some lactic acid bacteria including *Lactobacillus acidophilus* towards *Listeria monocytogenes*," *Int. J. Food Microbiology*, 9:25–32.

Rayman, M. K., Aris, B. and Hurst, A. 1981. "Nisin: A possible alternative or adjunct to nitrite in the preservation of meats," *Applied and Environmental Microbiology*, 41(2):375–380.

Reddy, N. R., Armstrong, D. J., Rhodehamel, E. J. and Kautter, D. A. 1992. "Shelf life extension and safety concerns about fresh fishery products packaged under modified atmospheres: A review," *J. Food Safety*, 12:87–118.

Rowley, D. B., Firstenberg-Eden, R., Powers, E. M., Shattuck, G. E., Wasserman, A. E. and Wierbicki, E. 1983. "Effect of irradiation on the inhibition of *Clostridium botulinum* toxin production and the microbial flora in bacon," *J. Food Science*, 48:1016–1021, 1030.

Saleh, M. A. and Ordal, Z. J. 1955. "Studies on growth and toxin production of *Clostridium botulinum* in a pre-cooked frozen food. II. Inhibition by lactic acid bacteria," *Food Research*, 20:340–350.

Schaack, M. M. and Marth, E. H. 1988. "Interaction between lactic acid bacteria and some foodborne pathogens: A review," *Cultured Dairy Products J.*, 23:17–18, 20.

Schillinger, U. and Lücke, F.-K. 1989. "Antibacterial activity of *Lactobacillus saké*, isolated from meat," *Applied and Environmental Microbiology*, 55(8): 1901–1906.

Scott, V. N. and Taylor, S. L. 1981. Effect of nisin on the outgrowth of *Clostridium botulinum* spores," *J. Food Science*, 46:117–126.

Seward, R. A. 1982. Efficacy of potassium sorbate and other preservatives in preventing toxigenisis by *Clostridium botulinum* in modified atmosphere packaged fresh fish. Ph.D. Thesis, University of Wisconsin, Madison.

Shamsuzzaman, K., Chuaqui-Offermanns, N., Lucht, L., McDougall, T. and Borsa, J. 1992. "Microbiological and other characteristics of chicken breast meat following electron-beam and sous-vide treatments," *J. Food Protection*, 55:528–533.

Smulders, F. J. M. 1987. Prospectives for microbial decontamination of meat and poultry by organic acids with special reference to lactic acid, In *Elimination of Pathogenic Organisms from Meat and Poultry*, F. J. M. Smulders, ed., Elsevier Applied Science Publishers, Amsterdam, pp. 319–344.

Smulders, F. J. M., Barendsen, P., Van Logestijn, J. G., Mossel, D. A. A. and Van der Marel, G. M. 1986. "Lactic acid; considerations in favour of its acceptance as a meat decontaminant," *J. Food Technology*, 21:419–436.

Sofos, J. N. 1984. "Antimicrobial effects of sodium and other ions in foods: A review," *J. Food Safety*, 6:45–78.

Sofos, J. N., Busta, F. F. and Allen, C. E. 1979a. "Botulism control by nitrite and sorbate in cured meats: A review," *J. Food Protection*, 42:739–770.

Sofos, J. N., Busta, F. F. and Allen, C. E. 1979b. "Sodium nitrite and sorbic acid effects on *Clostridium botulinum* spore germination and total microbial growth in chicken frankfurter emulsions during temperature abuse," *Applied and Environmental Microbiology*, 37:1103–1109.

Sofos, J. N. and Busta, F. F. 1980. "Alternatives to the use of nitrite as an antibotulinal agent," *Food Technology*, 34:244–251.

Sofos, J. N. and Busta, F. F. 1981. "Antimicrobial activity of sorbate," *J. Food Protection*, 44(8):614–622.

Spelhaug, S. R. and Harlander, S. K. 1989. "Inhibition of foodborne bacterial pathogens by bacteriocins from *Lactococcus lactis* and *Pediococcus pentosaceus*," *J. Food Protection*, 52:856–862.

Statham, J. A., Bremner, H. A. and Quarmby, A. R. 1985. "Storage of Morwong (*Normadactylus macropterus*) in combinations of potassium sorbate, polyphosphate and carbon dioxide at 4.0°C," *J. Food Science*, 50:1580–1585.

Stevenson, M. H. 1991. An overview of food irradiation, In *New Developments in Fundamental and Applied Radiobiology*, C. B. Seymour and C. Mothersill, eds., Taylor and Francis Ltd., London, pp. 427–436.

Stier, R. F., Bell, L., Ito, K. A., Shafer, B. D., Brown, L. A., Seeger, M. L., Allen, B. H., Porcuna, M. N. and Lerke, P. A. 1981. "Effect of modified atmosphere storage on *Clostridium botulinum* toxigenesis and the spoilage microflora of salmon fillets, *J. Food Science*, 46:1639–1642.

Sutton, A. H. 1973. "The hydrolysis of sodium triphosphate in cod and beef muscle," *J. Food Technology*, 8:185–195.

Tagg, J. R., Dajani, A. S. and Wannamaker, L. W. 1976. "Bacteriocins of Gram-positive bacteria," *Bacteriological Reviews*, 40:722–756.

Tanaka, N., Traisman, E., Lee, M. H., Cassens, R. G. and Foster, E. M. 1980. "Inhibition of botulinal toxin formation in bacon by acid development," *J. Food Protection*, 43:450–457.

Tanaka, N., Gordon, N. M., Lindsay, R. C., Meske, L. M., Doyle, M. P. and

Traisman, E. 1985a. "Sensory characteristics of reduced nitrite bacon manufactured by the Wisconsin Process," *J. Food Protection*, 48:687–692.

Tanaka, N., Meske, L., Doyle, M. P., Traisman, E., Thayer, D. W. and Johnston, R. W. 1985b. "Plant trials of bacon made with lactic acid bacteria, sucrose and lowered sodium nitrite," *J. Food Protection*, 48:679–686.

Taylor, S. L., Somers, E. B. and Krueger, L. A. 1985. "Antibotulinal effectiveness of nisin-nitrite combinations in culture medium and chicken frankfurter emulsions," *J. Food Protection*, 48(3):234–239.

Tompkin, R. B. 1984. "Indirect antimicrobial effects in foods: phosphates," *J. Food Safety*, 6:13–27.

Unda, J. R., Molins, R. A. and Walker, H. W. 1991. "*Clostridium sporogenes* and *Listeria monocytogenes* survival and inhibition in microwave-ready beef roasts containing selected antimicrobials," *J. Food Science*, 56(1):198–205, 219.

Urbain, W. M., ed. 1986. Marine and freshwater animal foods, In *Food Irradiation*, Academic Press, London, pp. 145–169.

Wagner, M. K. and Moberg, L. J. 1989. "Present and future use of traditional antimicrobials," *Food Technology*, 43(1):143–147, 155.

Wang, L. L. and Johnston, E. A. 1991. "Inhibition of *Listeria monocytogenes* by fatty acids, "*J. Food Protection*, 54:817.

Wang, S. Y., Dockerty, T. R., Ledford, R. A. and Stouffer, J. R. 1986. "Shelf life extension of vacuum-packaged frankfurters made from beef inoculated with *Streptococcus lactis*," *J. Food Protection* 49(2):130–134.

Woolthius, C. H. J. and Smulders, F. J. M. 1985. "Microbial decontamination of calf carcasses by lactic acid sprays," *J. Food Protection*, 48:832–837.

Zeitoun, A. A. M. and Debevere, J. M. 1990. "The effect of treatment with buffered lactic acid on microbial decontamination and on shelf life of poultry," *Int. J. Food Microbiology*, 11:305–312.

Zeitoun, A. A. M. and Debevere, J. M. 1991. "Inhibition, survival and growth of *Listeria monocytogenes* on poultry as influenced by buffered lactic acid and modified atmosphere packaging," *Int J. Food Microbiology*, 14:161–170.

CHAPTER 12

Modified Atmosphere Packaging—Present and Future Uses of Gas Absorbents and Generators

J. P. SMITH—*McGill University, Quebec, Canada*
Y. ABE—*Mitsubishi Gas Chemical Company, Japan*
J. HOSHINO—*Mitsubishi Gas Chemical Company, Japan*

INTRODUCTION

OVER the past decade, there has been a tremendous growth in Modified Atmosphere Packaging (MAP) for shelf life extension of food. MAP can simply be defined as a gas atmosphere within a packaged food whose composition differs from that of air. The growth in MAP technology has resulted from advances in packaging technology, the food industry's need for less energy intensive forms of food preservation other than drying, freezing or thermal processing, and consumer needs for fresh, convenience foods with extended shelf life.

While the atmosphere in MAP foods has usually been modified by vacuum or gas packaging (discussed in detail in other chapters), novel methods of atmosphere modification have been developed, primarily by the Japanese. These include O_2/CO_2 absorbents, O_2 absorbents/CO_2 generators, ethylene absorbents and ethanol vapour generators. This technology involves the use of sachets which can be placed alongside the food and actively modify the package headspace, thereby extending product shelf life.

This chapter will review the various types of absorbent/generator sachets available on the marketplace, the methods by which these sachets actively modify the gas atmosphere in the packaged product, and their uses, advantages and disadvantages for shelf life extension of food.

O_2 Control

Although vacuum/gas packaging can be used to extend the shelf life and keeping quality of food, aerobic spoilage can still occur in these packaged products depending on the level of residual O_2 in the package headspace. The level of residual O_2 in vacuum/gas packaged products could be affected by a number of factors such as (1) O_2 permeability of the packaging material; (2) ability of the food to trap air; (3) leakage of air through poor sealing; and (4) inadequate evacuation and/or gas flushing (Smith et al., 1986). Tomkins (1932) observed that the inhibitory effect of CO_2 was, within wide limits, independent of the O_2 concentration. He showed that the extent of mold growth reduction in 20, 40 and 60% CO_2 was the same whether the O_2 concentration was 20% or 5%. Dallyn and Everton (1969), reported that *Xeromyces bisporus*, a xerophilic mold which causes spoilage of food with low water activity, could grow in an atmosphere of 95% CO_2 and 1% O_2. Studies in our laboratory (Smith et al., 1986) have shown that *Aspergillus* and *Penicillium* mold species, the major spoilage isolates of gas-packaged crusty rolls, can tolerate and grow in <1% (v/v) headspace O_2. These studies clearly demonstrate that aerobic microorganisms can tolerate very low concentrations of headspace O_2, even in the presence of elevated levels of CO_2, and that additional control measures are necessary to completely inhibit aerobic spoilage microorganisms in vacuum/gas packaged products.

O_2 ABSORBENTS

One novel and innovative method of O_2 control and of atmosphere modification involves the use of O_2 absorbents. O_2 absorbents can be defined as "a range of chemical compounds introduced into the MAP package (not the product) to alter the atmosphere within the package" (Agriculture and Agri-Food Canada, Personal Communication). Developed in Japan in 1976, O_2 absorbents were first marketed by the Mitsubishi Gas Chemical Co., under the trade name Ageless. Several other Japanese companies also produce O_2 absorbents, with the best known being the Toppan Printing Company, which produces a range of O_2 absorbents under the label Freshilizer Series (Smith et al., 1990). In 1989, almost 7000 million sachets were sold in Japan with sales of absorbents growing at a rate of 20% annually. The Mitsubishi Gas Chemical Co. dominates the O_2 absorbent market (73%) while the Toppan Printing Co. has ~11% of the market, with 9 other companies sharing the remaining 16% of the market (Figure 12.1). O_2 absorbent technology has been successful in Japan for a variety of reasons, including a hot and humid climate during the summer months which is conducive to mold spoilage of food products. Another important factor for their success is that Japanese consumers are

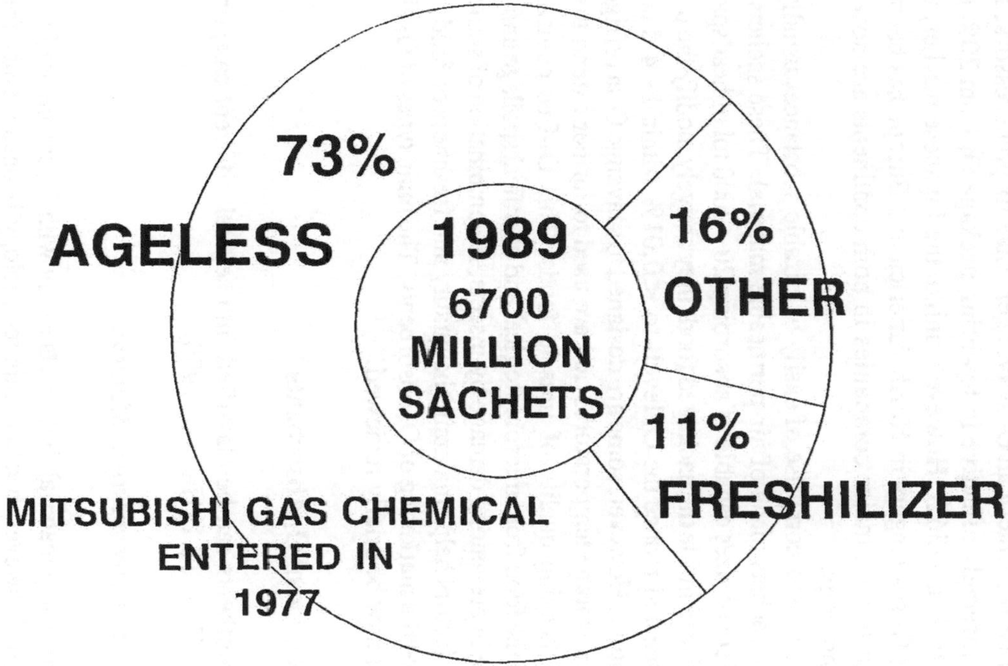

FIGURE 12.1 Market share of O_2 absorbents in Japan (courtesy of Mitsubishi Gas Chemical Co., Japan).

willing to pay higher prices for preservative-free products with an increased shelf life. Several years after the Mitsubishi Gas Chemical Company successfully launched O_2 absorbent technology onto the marketplace, Multiform Desiccants in the U.S. introduced FreshPax, the first North American O_2 scavenger. Like its predecessors, FreshPax is extremely effective in lowering package O_2 from 20% to less than 0.05% in ~36 h. However, unlike the Japanese market, acceptance of O_2 scavengers in North America and Europe has been slow, although several major companies in both continents are now using this technology.

O_2 absorbents comprise of easily oxidizable substances usually contained in sachets made of air-permeable material. These sachets come in a variety of sizes capable of absorbing 20–2000 ml of headspace O_2. When placed inside the packaged food, they actively modify the package headspace and reduce the O_2 levels to <0.01% within 1–4 d at room temperature. However, some are designed to scavenge O_2 at refrigerated or frozen storage temperatures and are used to further extend the shelf life and keeping quality of muscle foods. This O_2-free environment protects the food from microbiological and chemical spoilage and is also effective in preventing damage by insects. A combination of some or all of these factors helps maintain the quality and freshness of food, which facilitates the marketing of O_2 absorbents. The main types of O_2 absorbents will now be briefly reviewed.

Classification of O_2 Absorbents

O_2 absorbents can be classified into several different categories as shown in Table 12.1 (Harima, 1990).

Classification According to Material

In theory, any material that reacts easily with O_2 can be used as an O_2 scavenger. However, because this technology is used mainly for food preservation, the material used inside the sachet must meet the following criteria prior to approval by regulatory agencies:

(1) It must be safe.
(2) It must be handled easily.
(3) It must not produce toxic substances or offensive odours/gases.
(4) It must be compact in size.

Table 12.1. Classification of O_2 absorbents (Harima, 1990).

A. CLASSIFICATION ACCORDING TO MATERIAL
1. Inorganic-iron powder
2. Organic-ascorbic acid, catechol

B. CLASSIFICATION ACCORDING TO REACTION STYLE
1. Self-reaction type
2. Moisture-dependent type

C. CLASSIFICATION ACCORDING TO REACTION SPEED
1. Immediate effect type
2. General type
3. Slow effect type

D. CLASSIFICATION ACCORDING TO USE
1. For very moist food
2. For moderately moist food
3. For low water food
4. For extra dry food

E. CLASSIFICATION ACCORDING TO FUNCTION
1. Single function type—O_2 absorption only
2. Composite function type
   a. O_2 absorbtion—CO_2 generation
   b. O_2 absorbtion—CO_2 absorbtion
   c. O_2 absorbtion and ethanol generation
   d. O_2 absorbtion and other gases

(5) It must absorb a large amount of O_2.
(6) It must have an appropriate O_2 absorption speed.
(7) It must be economically priced (Harima, 1990).

Iron powder and ascorbic acid are commonly used in existing O_2 absorbers, with powdered iron being most frequently used either alone, or in conjunction with other specific chemical compounds in dual function absorbents.

Classification According to Reaction Style

Water is essential for O_2 absorbents to function. In the self-reaction type, water required for the chemical reaction is added and these absorbents must be handled carefully as the O_2 absorbing reaction commences as soon as the self-reacting absorbent is exposed to air. In the moisture-dependent type, the O_2 absorption reaction only takes place after moisture has been absorbed from the food. These type of absorbents are easier to handle as they do not react immediately upon exposure to air. However, they absorb O_2 quickly after sealing and O_2 can be absorbed within 0.5 to 1 d in certain products.

Classification According to Reaction Speed

O_2 absorbents can be classified as immediate effect type, general effect and slow effect type (Harima, 1990). The average time for O_2 absorption is 0.5 to 1 d for the immediate type, 1 to 4 d for the general type, and 4 to 6 d for the slow reacting type (Harima, 1990). The reaction time depends on the storage temperature and water activity (a_w) of the food. Most O_2 absorbents are used with foods stored at ambient temperature. However, some absorbents can now be used with foods stored under refrigerated and frozen storage conditions.

Classification According to Use

O_2 absorbents can be used for a variety of foods with different moisture contents ranging from very moist to very dry food products. In general, foods with a high moisture content are more susceptible to mold spoilage. Therefore, an immediate effect absorbent would be used with these products to rapidly absorb O_2 and extend the mold-free shelf life of the product. General type absorbents are used with intermediate moisture food products where a moderate speed of O_2 absorption is required. For low moisture food products, which are not susceptible to microbial spoilage but are to chemical spoilage, a slow effect O_2 absorbent can be used. The relationship between the types of absorbents, reaction speeds and their end uses is shown in Table 12.2.

Classification According to Function

The majority of absorbents have only one function—absorption of O_2. However, dual functional absorbents have been developed for use in specific products. These include O_2-CO_2 absorbents for use in coffee, and O_2 absorbents-CO_2 generators. These sachets absorb O_2 and generate the same amount of CO_2 as that of absorbed O_2. They are mainly used in products where package volume and appearance is critical e.g., packaged peanuts. These sachets contain iron car-

Table 12.2. Relationship between type of O_2 absorbent, reaction speed and end use (adapted from Harima, 1990).

Material	Reaction Type	Reaction Speed	Use
Inorganic	Self-reaction	Immediate	High moisture food
		General	Intermediate moisture food
		Slow	Low moisture food
Inorganic	Moisture dependent	Immediate	High moisture food
		General	High moisture food

bonate and ascorbic acid as the reactants. Another dual functional absorbent scavenges O_2 and releases alcohol vapour. These absorbents could be used to control the growth of facultative bacteria and yeasts which can grow under reduced O_2 tensions. However, this type of dual functional absorber is not widely used in the Japanese market (Harima, 1990).

Types of O_2 Absorbents

Ageless

Ageless, made by the Mitsubishi Gas Chemical Co., consists of a range of gas scavenger products designed to reduce O_2 levels to less than 100 ppm in the package headspacc. While both organic types (based on ascorbic acid) and inorganic types (based on iron powder) are available, the inorganic types are most commonly used in the Japanese market. The basic system is made up of finely divided powdered iron which, under appropriate humidity conditions, uses up residual O_2 to form non-toxic iron oxide, i.e., it rusts. The oxidation mechanism can be expressed as follows (Smith et al., 1990; Smith, 1992):

$$Fe \rightarrow Fe^{2+} + 2e$$

$$1/2O_2 + H_2O + 2e \rightarrow 2OH^-$$

$$Fe^{2+} + 2(OH)^- \rightarrow Fe(OH)_2$$

$$Fe(OH)_2 + 1/2O_2 + 1/2H_2O \rightarrow Fe(OH_3)$$

To prevent the iron powder from imparting colour to the food, the iron is contained in a sachet (like a desiccant). The sachet material is highly permeable to O_2 and, in some cases, to water vapour.

Since Ageless relies on a chemical reaction and not on the physical displacement of oxygen as in gas packaging, it completely removes all traces of residual O_2 and protects the packaged food from aerobic spoilage and quality changes. Several types and sizes of Ageless sachets are commercially available and are applicable to many types of foods including those with a high moisture content, intermediate moisture products, low moisture product foods and foods containing or treated with oil. The main types of absorbents are type Z, S, FX, E and G. Three new types of Ageless products are commercially available in Japan. These are Ageless type SS, type FM and type SE (Table 12.3).

Type Z is designed for food products with a_w values less than 0.65 and reduces residual headspace O_2 to 100 ppm in 1–3 d. It is available in sizes that can "scavenge" 20–2000 ml of O_2 (an air volume of 100–10,000 ml).

Two other types of Ageless (FX and S) work best at a higher a_w and have a faster reaction rate (0.5 to 2 d). They have the same O_2 scavenging

Table 12.3. Types and properties of Ageless O_2 absorbents.

Type	Function	Moisture Status	Water Activity	Absorption Speed (Day)
Z	Decreases O_2	Self reacting	<0.65	1–3
S	Decreases O_2	Self reacting	>0.65	0.5–2
SS	Decreases O_2	Self reacting	>0.85	2–3 (0–4°C) 10 (−25°C)
FX	Decreases O_2	Moisture dependent	>0.85	0.5–1
FM	Decreases O_2	Moisture dependent	>0.85	0.5–1
E	Decreases O_2	Self reacting	<0.3	3–8
G	Decreases CO_2 Decreases O_2 Increases CO_2	Self reacting	0.3–0.5	1–4
SE	Decreases O_2 Increases ethanol	Self reacting	>0.85	1–2

Courtesy of Mitsubishi Gas Chemical Co., Tokyo, Japan.

capacity as above. Type FX is moisture dependent and does not absorb O_2 until it is exposed to an a_w >0.85. Thus, it can be easily handled if kept dry. Type S, on the other hand contains moisture in the sachet and is a self-working type. This type of absorbent requires careful handling since it begins to react immediately on exposure to O_2. Absorbent type SS is similar to type S. However, it has the ability to rapidly scavenge O_2 under refrigerated and frozen storage conditions. These absorbents (Ageless type SS) are widely used to extend the shelf life of muscle foods such as fresh meat, fish and poultry. Yet another new absorbent is type FM which can be used with microwaveable products (Table 12.3).

A commonly used absorbent is type E which also contains $Ca(OH)_2$ in addition to iron powder. Type E scavenges CO_2 as well as O_2. It is used for ground coffee, where CO_2 removal reduces the chance of the package bursting. Marketed under the brand name "Fresh Lock," it was used in Maxwell House ground coffee cans (Table 12.3).

Two other types commonly used in the Japanese market are type G and type SE. Type G is a self working type and absorbs O_2 and generates an equal volume of CO_2. It is used mainly with snack food products, such as nuts, to maintain the package volume and hence appearance of the product. Another new innovation is Ageless type SE. This absorbent is self-reacting and absorbs O_2 and generates ethanol vapour. It is commonly used to extend the mold-free shelf life of bakery products in Japan. The various types of Ageless and their characteristics are summarized in Table 12.3.

Freshilizer Series

The Freshilizer Series of "Freshness Keeping Agents" made by Toppan Printing Co., Japan, consists of the F series and the C series (Table 12.4). Series F Freshilizers use mainly ferrous metal and absorb only O_2. Three types are commercially available–type FD, FH and FT. Type FD is designed for use in food products with a_w values less than 0.8 (nuts, tea, chocolate), while type FH is suitable for use in products with a range of aw values ranging from 0.6–0.9, and is used mainly with beef jerky and salami to maintain the colour of these products. Type FT works best in foods with a_w values >0.8, such as pizza crusts. Series F absorbents can absorb 20–300 ml O_2, corresponding to a package volume of 100–1500 ml of air (Toppan Technical Information, 1989).

Table 12.4. Types and properties of Freshilizer O_2 absorbents.

Type	Function	Moisture Status	Water Activity	Absorption Speed (Day)
F Series (ferrous metal)				
FD	Decreases O_2	Self-reacting	<0.8	1–3
FH	Decreases O_2	Self-reacting	0.6–0.9	0.5–1
FT	Decreases O_2	Moisture dependent	>0.8	0.5–1
C Series (non-ferrous metal)				
C	Decreases O_2 Increases CO_2	Self-reacting	<0.8	3–5
CW	Decreases O_2 Increases CO_2	Self-reacting	0.8–0.9	2–3
CV	Decreases O_2 Decreases CO_2	Self-reacting	<0.3	1–4

Courtesy of Toppan Printing Co., Tokyo, Japan.

The Freshilizer C series of absorbents consists of types C, CW and CV. These sachets consist of non-ferrous metal particles and can therefore be used in products which must pass through a metal detector. Types C and CW absorb O_2 and generate an equal volume of CO_2, thereby preventing package collapse. Type C is used in foods with an a_w of 0.8 or less (nuts) while type CW is suitable for foods with higher a_w values, i.e., (>0.8). Type CW is commonly used to prevent mold growth in sponge cakes. Type CV absorbs both O_2 and CO_2 and was developed for use with roasted or ground coffee (Table 12.4; Toppan Technical Information, 1989).

FreshPax

FreshPax is a patented O2 absorbing system developed by Multiform Desiccants, a leading manufacturer of desiccants and other protection products for more than 30 years. Manufactured in the U.S., FreshPax, like all other O_2 absorbent technology, provides an alternative to gas/vacuum packaging as a means of improving shelf life and product quality, while simultaneously cutting down on costs and increasing profitability. Produced in sachet form, FreshPax absorbs headspace O_2 to $<0.1\%$ using safe, non-toxic ingredients that rapidly absorb O_2 before the food deterioration process begins. Four main types

Table 12.5. Types and properties of FreshPax™ O_2 absorbents.

Type	Function	Moisture Status	Water Activity	Absorption Speed (Day)
Type B	Decreases O_2	Moisture dependent	>0.65	0.5–2
Type D	Decreases O_2 at 2° to −20°C	Self reacting	<0.7	0.5–4
Type R	Decreases O_2	Self reacting	ALL	0.5–1 (depending on temperature)
Type M	Decreases O_2 Increases CO_2	Moisture dependent	>0.65	0.5–2

Courtesy of Multiform Desiccants, Buffalo, NY.

of FreshPax are commonly available—type B, type D, type R and type M (Table 12.5). Type B is used with moist or semi-moist foods while type D can be used with dehydrated or dried foods. Type R can be used to scavenge O_2 at refrigerated or frozen storage temperature and is similar to Ageless type SS. It is mainly used to extend the shelf life and keeping quality of muscle foods. Type M is used with moist or semi-moist gas-flushed products to maintain package volume and to remove all traces of residual O_2 (Table 12.5; FreshPax, 1994).

Factors Influencing the Choice of O_2 Absorbents

O_2 absorbent sachets come in a range of sizes capable of absorbing 5 to 2000 ml of O_2. Several interrelated factors influence the choice of the type and size of absorbent selected for shelf life extension of food products (Harima, 1990; Smith et al., 1990; Smith, 1992). These are

(1) The nature of the food, i.e., size, shape, weight
(2) The a_w of the food
(3) The amount of dissolved O_2 in the food
(4) The desired shelf life of the product
(5) The initial level of O_2 in the package headspace
(6) The O_2 permeability of the packaging material

This latter parameter is critically important for the overall performance of the absorbent and shelf life of the product. If films with high O_2 permeabilities are used, e.g., >100 cc/m^2/atm/mil/day, the O_2 concentration in the container will reach zero within a week but then returns to ambient air level after 10 d because the absorbent is saturated. However, if films of low O_2 permeability (<10 cc/m^2/atm/mil/d), such as PVDC coated nylon/LDPE are used, the headspace O_2 will be reduced to 100 ppm within 1–2 d and remain at this level for the duration of the storage period, providing packaging integrity is maintained. Examples of appropriate packaging materials for use with O_2 absorbents and their permeability characteristics are shown in Table 12.6.

A rapid and efficient method of monitoring package integrity throughout the storage period is through the incorporation of a redox indicator, e.g., Ageless Eye. When placed inside the package alongside an Ageless sachet, the colour of the indicator changes from blue (oxidized state) to pink (reduced state) when the O_2 content of the container reaches ~0.1%. If the indicator reverts back to its blue colour, it is indicative of poor packaging integrity caused by: (1) poor sealing of film or (2) minute pin holes in the film. Similar redox indicators are made by the Toppan Printing Company.

Table 12.6. Examples of films used with O_2 absorbents (FreshPax Technical Pamphlet, *1994*).

	Film Laminates Including	OTR (cc/m^2/d)	MVTR (g/m^2/d)
Long-term	Aluminum	<0.6	<0.6
preservation	EVOH	<3	<4
Short-term	PVDC	<15	<8
preservation	Nylon	<16	<40
	PET	<15	<100
Not appropriate	Cellophane	<200	<20
	PP	<2000	<6
	PE	•3000	<5

OTR = O_2 transmission rate.
MVTR = Moisture vapour transmission rate.
EVOH = Ethylene vinyl alcohol.
PVDC = Polyvinylidene chloride.
PET = Polyester.
PP = Polypropylene.
PE = Polyethylene.

Application of O_2 Absorbents for Shelf Life Extension of Food

In Japan

Several studies have reported significant extension in the chemical and microbiological shelf life of low, intermediate and high moisture food products using O_2 absorbent technology (Nakamura and Hoshino, 1983; Abe and Kondoh, 1989; Harima, 1990; Minakuchi and Nakamura, 1990). As a result of these studies, O_2 absorbent technology is used extensively in Japan to prevent discolouration problems in highly pigmented products such as cured meat products and tea, rancidity problems in high fat foods such as peanuts, peanut brittle and dried sardines, and mold spoilage especially in intermediate moisture and high moisture bakery products. Examples of products with extended shelf life found in Japanese supermarkets as a result of O_2 absorbent technology are summarized in Tables 12.7 to 12.10. More detailed information on the use of both Ageless and Freshilizer O_2 absorbents for shelf life extension of food can be found in the excellent articles referenced at the beginning of this section.

In the U.S.

While O_2 absorbents are used extensively in Japan, their use in North America is still in its infancy. Examples of the known U.S. companies currently using O_2 absorbents for shelf life extension of products are shown in Table 12.11. The earliest use of this technology was with Maxwell House coffee, which used a dual function absorbent (Ageless E). This absorbent, marketed under the trade name "Fresh Lock," contains iron powder for absorption of O_2 and calcium hydroxide for scavenging CO_2. The use of this absorbent in coffee delays oxidative flavour changes and absorbs the occluded CO_2 produced in the roasting process which, if not removed, would cause the packages to burst. More recently, a mold-free shelf life of 1 year was achieved for a specialty therapeutic gluten-free bread called Ener-Getic. The bread is packaged in a copolymer film of Mylar/EVOH/Surlyn film with 100% CO_2 and a Freshilizer O_2 absorbent/CO_2 generator (type CW) supplied by the Toppan Printing Company, Japan. The bread remained mold-free for 1 year at room temperature and physico-chemical changes, i.e., staling, were minimal at the end of storage in the gluten-free bread (Anon., 1988).

Table 12.7. Typical shelf lives for bakery products packaged with O_2 absorbents.

Type of Food	Water Activity of Food	Packaging Materials	Type and Size of Ageless	Prevents	Shelf Life @ RT*
Cheesecake	0.88–0.92	KON	S200	Mold growth/ oxidation	1 month
Buttercake	0.9	KON	S100	Oxidation/mold growth	1 month
Plain sponge cake	0.86–0.88	KON	S200	Mold growth	1 month
Egg bread	0.9	KON	S100	Mold growth	1 month
Pumpernickel bread	0.9	KON	FX100	Mold growth	3–4 weeks
Pita bread	0.9	KOP	S100	Mold growth	3–4 weeks
Tortilla chips	<0.6	KON/Box	Z200	Oxidation	6 months
Custard cakes	0.92–0.94	PET/PVDC/PE	S50	Mold growth	7–10 days
Blueberry cheesecake	0.88–0.9	KON	S200	Mold growth	1 month
Soybean pie	0.93	KON	S10–15	Mold growth	1 month
Bean/jam cake	0.85	KON/PS Tray	ZPT100–200	Mold growth	2 months
Steamed bean/jam cake	0.9	PP/EVAL/KON/PE	S200	Mold growth	2–3 weeks

*RT = Room temperature.

PET = polyester; PE = low density polyethylene; PS = polystyrene; AC = acrynonitrile; KON = nylon/ethylene vinyl alcohol/low density polyethylene; KOP = polypropylene/ethylene vinyl alcohol/low density polyethylene; PVDC = polyvinylidene chloride; PP = polypropylene; EVAL = ethylene vinyl alcohol; EVA = ethylene vinyl acetate.

Source: Courtesy of Mitsubishi Gas Chemical Co., Japan.

Table 12.8. Use of Ageless for shelf life extension of fish products.

Type of Food	Water Activity of Food	Packaging Materials	Type and Size of Ageless	Prevents	Shelf Life @ RT
Dried seaweed	<0.3	PET/AC/PE	Z100	Oxidation/Discolouration	1 year
Dried salmon jerky	0.7–0.8	KON	Z50	Oxidation/Discolouration	3 months
Dried sardines	<0.6–0.88	KON	Z100–200	Oxidation	6 months
Dried shark's fin	0.8	KON	S200	Oxidation/Mold growth	1–2 months
Dried rose mackerel	0.7–0.8	KON/PET Tray	S2000	Oxidation	1 month
Dried cod	<0.6	KOP	Z100	Oxidation	6 months
Dried squid	0.7–0.8	KON	Z100	Oxidation/Mold growth	3 months
Fresh yellowtail	0.99	Nylon/PVDC/PP	SS100	Oxidation/Discolouration	1 week at −3°C
Sliced salmon	0.99	KON	FX100	Discolouration	6 months at −20°C
Dried/smoked salmon	0.8	KON	Z100	Mold/Discolouration/ Oxidation	1 month at 10°C
Dried octopus leg	<0.85	KON	FX100	Oxidation/Mold	1 month
Dried bonito	0.6–0.7	KON	Type Z200 + 100% N_2	Oxidation/Discolouration	3–6 months
Salmon roe	0.8–0.9	KON	SS100	Discolouration	6 months at −3°C
Dried squid/ vinegar/ soybean sauce	0.7–0.8	KOP	Z200	Mold/Oxidation	3 months
Sea urchin	0.98	Wood Tray/KON	SS100	Discolouration Bacterial growth	2 weeks at 5°C

Source: Courtesy of Mitsubishi Gas Chemical Co., Japan; see Table 12.7 for list of abbreviations.

Table 12.9. Use of Ageless for shelf life extension of meat and deli products.

Type of Food	Water Activity of Food	Packaging Materials	Type and Size of Ageless	Prevents	Shelf Life
Pizza crust	~0.94–0.95	KON	FX200	Mold growth	2 months at 5°C
Pizza	0.99	KON	S200	Mold growth/ Discolouration	2 weeks at 5°C
Fresh carved sausage	0.99	KON	FX50	Discolouration/ Rancidity	3 weeks at 0–4°C
Pre-cooked hamburgers	0.99	PS Tray/KON	S100	Discolouration/ Rancidity	3 weeks at 0–4°C
Pre-cooked chicken nuggets	0.99	KON	FX100	Flavor discolouration Oxidation	3 weeks at 0–4°C
Salami sticks	0.8–0.85	KON	Z50	Discolouration	1–2 months @RT
Sliced salami	0.8–0.85	KON	S100	Discolouration	1 month @RT

Source: Courtesy of Mitsubishi Gas Chemical Co., Japan; see Table 12.7 for list of abbreviations.

Table 12.10. Use of Ageless for shelf life extension of miscellaneous products.

Type of Food	Water Activity of Food	Packaging Materials	Type and Size of Ageless	Prevents	Shelf Life @RT
Rice crackers/fried bean	<0.3	Tin/Plastic/ Aluminum lid	Z100	Oxidation/Rancidity	1 year
Peanuts	<0.3	KON	G100	Oxidation/Rancidity	6 months
Peanut butter/ chocolate coated	0.5–0.6	KON/Paper box	Z200	Oxidation/Rancidity	9 months
Chocolate peanuts	0.5–0.6	KON	Z30–100	Oxidation/Rancidity	9 months
Candy type cheese	0.9–0.95	KON	FX50 + 100% N_2	Oxidation/Rancidity	1 month at 0–4°C
Smoked cheese	0.9	KON	S30–50	Oxidation/Mold growth	1 month at 0–4°C
Soybean paste	0.8–0.85	Paper/AC/EVA	FX20	Discolouration	3 months
Bakery food (milk powder)	<0.3	Tin plate	Z100	Oxidation/Rancidity	6–12 months
Green tea	<0.3	Tin can	Z50	Oxidation	1 year
Flavoured tea	0.5–0.6	KON	Z50	Oxidation	1 year

Source: Courtesy of Mitsubishi Gas Chemical Co., Japan; see Table 12.7 for list of abbreviations.

Studies by Powers and Berkowitz (1990) at the U.S. Army Natick Research, showed that a FreshPax O_2 scavenger enclosed in a high gas barrier pouch made of polyester/aluminum foil/HDPE with baked, meal, ready-to-eat (MRE) bread prevented mold growth on bread for 13 months at ambient storage temperature. Based on the results of this study, two companies, Sterling Foods and Franz bakery were producing MRE bread and buns with a 6–36 months mold-free shelf life for U.S. Army troops in the Gulf War Crisis (Table 12.11). Staling was again minimal due to recipe reformulation involving higher levels of shortening, which are claimed to delay staling by inhibiting the amylose-gluten interactions which occur during staling (American Institute of Baking, Personal Communication). Another U.S. company using O_2 absorbent technology is the Famous Pacific Dessert Company. This company uses Ageless O_2 absorbents to prevent rancidity and mold problems in packaged tortes (Table 12.11). Clearly, there is an increasing interest by many companies, particularly baking companies, in the use of O_2 absorbent technology as an alternative or adjunct to gas packaging to prevent the perennial problem of mold growth.

Research Activities in O_2 Absorbent Technology

Although the use of O_2 absorbent technology has not been accepted commercially with the same enthusiasm in North America compared to Japan, there is however tremendous interest and ongoing research in this innovative method of atmosphere modification.

Studies in our laboratory have shown O_2 absorbents to be three times more effective than gas packaging for increasing the mold-free shelf life of crusty rolls (Smith et al., 1986). While the mold-free shelf life of crusty rolls can be extended by packaging in 60% CO_2 and 40% N_2, mold growth still occurred after 19 d due to the O_2 permeability of the

Table 12.11. Use of O_2 absorbent technology in the U.S.

Company	Product	Shelf Life (months @RT*)
General Foods	Maxwell House coffee	12–36
Ener-G Foods	Ener-Getic Gluten-	12
Sterling Foods	Free bread	
Franz Bakery	MRE bread	36
The Famous Pacific	MRE bread/buns	6
Dessert Company	Fruit tortes	6–9

*RT = Room temperature.

packaging film. Studies have shown that *Aspergillus* and *Penicillium* species, the major spoilage isolates of crusty rolls, can tolerate and grow in <1% (v/v) headspace O_2, even in the presence of elevated levels of CO_2. However, when Ageless (Type S or Type FX) was packaged alongside crusty rolls, either alone or in conjunction with gas packaging, headspace O_2 never increased beyond 0.05% and the product remained mold-free for >60 d at ambient storage temperature (Smith et al., 1986). While a longer extension in the mold-free shelf life was possible using O_2 absorbents, mold problems occasionally arose in the Ageless packaged product. This was due to absorption of headspace gas and package collapse, resulting in some products being tightly wrapped with the packaging film. This created pockets of localized environments between product surface and the film where the O_2 concentration may have increased to a level sufficient to permit mold growth. This clearly demonstrates the need for a free flow of gas around the product if Ageless is to be totally effective as an O_2 scavenger (Smith et al., 1986).

More recently, Ageless O_2 absorbents have been used in our laboratory to extend the mold-free shelf life of fresh bagels and pizza crusts. Furthermore, the use of specific types of absorbents appeared to have an anti-staling effect on bagels. Further studies are now under way to compare the anti-staling effects of O_2 absorbents with other methods of atmosphere modification for shelf life extension of bakery products.

The use of absorbent technology for shelf life extension of food has also been extensively studied by Cryovac, a world leader in vacuum packaging technology. Cryovac in the U.S., has been given the distribution rights to Ageless in North America by the Mitsubishi Gas Chemical Company. Researchers at Cryovac have found that O_2 absorbent technology can be used to prevent mold growth, discolouration and flavour changes in cooked cured sausage, cooked poultry and smoked poultry chops. Using a combination of O_2 absorbent technology and Cryovac's high barrier shrink film, headspace O_2 (which can be 4–7% due to entrapped O_2 in the foam tray) is reduced to <0.3% before products are exposed to retail display lighting. Thus, O_2 absorbent technology can be used to inhibit photodegradation of meat pigments in cured meat products, thereby enhancing consumer appeal for such products (Cryovac, Personal Communication). O_2 absorbent technology has also been investigated to inhibit mold growth and rancidity problems in fresh pasta and pizza crusts. Other areas of applied research by Cryovac involving O_2 absorbents include (1) prevention of rancidity problems in snack food, (2) prevention of flavour changes in wine stored in O_2 permeable PET bottles using a barrier pouch overwrap with an O_2 absorbent and, (3) master packaging of fresh meat and poultry. This latter concept is in response to consumer concerns about absorbents being visible inside the packaged food product. In the MasterPak, retail cuts of meat/poultry are packaged in trays with an O_2 permeable overwrap. These are placed inside a cardboard carton with a high gas barrier film and absorbents added. The system is then evacuated, gas-flushed and heat sealed, and then distributed under refrigeration to the retailer, where the individual packs are removed from the MasterPak and displayed on retail shelves. Colour or "bloom" returns to the meat/poultry due to the low barrier characteristics of the overwrap.

Extensive research on FreshPax absorbents is being conducted by Multiform Desiccants in conjunction with Dr. Joe Hotchkiss at Cornell University. Studies by Alarcon and Hotchkiss (1993) have shown that FreshPax O_2 absorbents can prevent mold growth in preservative free white bread and low-moisture, low-fat mozzarella cheese for 8 weeks at ambient/refrigerated temperature. Furthermore, O_2 absorbents were also effective in reducing the oxidative formation of *n*-hexanal (an indicator of off-flavour development) in both sunflower seeds and cornchips. Sensory panel analysis also showed that absorbents inhibited the formation of undesirable rancid odours during accelerated storage tests (Alarcon and Hotchkiss, 1993). These authors also reported that FreshPax O_2 absorbents were effective in controlling permeated O_2, and hence mold growth and rancidity problems, in bread and cheese packaged in medium barrier films with an OTR of ~ 32 cc/m²/d. Therefore, a medium barrier film could be substituted for a high barrier more expensive film, if an O_2 absorber pack was included in the package (Alarcon and Hotchkiss, 1993).

Advantages and Disadvantages of O_2 Absorbents

O_2 absorbents have several advantages for the food processor, both from a marketing and food quality viewpoint (Harima, 1990; Smith et al., 1990; Smith, 1992). These are as follows.

(1) Inexpensive and simple to use
(2) Approved by the U.S. Food and Drug Administration as non-toxic and safe to use.
(3) Prevents aerobic microbial growth and extends shelf life of product.
(4) Arrests the development of rancid off-flavour in fats and oils.

(5) Maintains flavour quality by preventing oxidation of the flavour compounds.
(6) Maintains product quality without additives.
(7) Increases product shelf life and distribution radius.
(8) Allows a long time between deliveries and fewer, longer deliveries.
(9) Increases length of time product can stay in the distribution pipeline.
(10) Reduces distribution losses.
(11) Replaces chemical pesticides to prevent insect damage of foods.
(12) Reduces evacuation/gas flushing times in gas-packaged products, thereby increasing product throughput.
(13) Reduces costs required for gas-flushing equipment.

A few of the disadvantages of O_2 absorbents are as follows (Harima, 1990; Smith et al., 1990; Smith, 1992).

(1) There needs to be a free flow of air surrounding the sachet in order to "scavenge" headspace O_2 if an O_2 absorbent is used alone.
(2) It may cause package collapse; this can be overcome by using a O_2 absorbent/CO_2 generator.
(3) O_2 absorbents/CO_2 generators may cause flavour changes in high moisture/fat foods due to the dissolution of CO_2 in the aqueous/fat phase of product.
(4) Cost is approximately 2.5 – 10 cents per sachet depending on size of sachet and volume ordered. Many processors regard O_2 absorbent sachets as too expensive.
(5) There are consumer concerns about sachets inside the packages and possible consumer misuse of sachets.
(6) It may promote growth of potentially harmful anaerobic bacteria.

These two latter concerns will be briefly discussed.

Consumer Resistance to Sachets

The two main consumer concerns about sachets being placed inside the packaged food product are: (1) fear of ingestion even though the contents are safe and the label clearly says "Do not eat" and (2) spillage of sachet contents into food and adulteration of the food product. These concerns have been addressed by the Mitsubishi Gas Chemical Company by producing the deoxidizer in a tablet form or board form and affixing it to the lid or to the bottom of packages. This approach has been successfully used for health foods sold in bottles and with individually wrapped teacakes in the Japanese marketplace. An alternative method is to package the primary packages inside a secondary packaging containing the absorbent. This is the principle behind the MasterPak system i.e., bag-in-box system, designed to hold 12–48 consumer size primary packages made of material with a higher O_2 transmission rate than the secondary container, and discussed in the previous section. The main advantage of this approach is that the consumer never sees the absorbents, thus eliminating the possibility of consumers misusing the packets or reacting negatively to them (Smith, 1992).

Another approach to overcome consumer resistance to sachets is to incorporate the O_2 scavenging capacity in the form of a package label. This approach has been developed by both Mitsubishi Gas Chemical Company and Multiform Desiccants. This latter company currently markets both Type B and Type M absorbents in the form of labels marketed under the trade name FreshMax. This label can be backed with various adhesives for different product applications. The technology has a printed surface and is contact acceptable (oil/grease resistant). The label can be applied at conventional line speeds and is compatible with both moist and dry food products. The prototype FreshMax is 2.5″ × 2.5″ and 15 mil thick. It is capable of absorbing 100 cc of O_2 in a 24 h period and can be made into various shapes. The labels can reduce headspace O_2 to less than 0.01% (100 ppm) and can be used to retard both chemical and microbiological spoilage in stored products. Marks and Spencers Ltd., London are now using FreshMax labels to extend the shelf life and keeping quality of deli meats. The new labels cover only one third of the lid surface providing consumers with a clearer view of the product, yet they absorb O_2 and protect products from the adverse oxidizing effect of light and O_2 on meat pigments (Multiform Desiccants, Personal Communication).

Public Health Concerns

Another consumer and regulatory concern about O_2 absorbent technology is that the anaerobic environment created inside the package may be conducive to the growth of facultative anaerobic or strictly anaerobic food pathogens. These concerns are justified in view of (1) the ability of many food pathogens to grow at refrigeration temperatures, e.g.,

Listeria monocytogenes and non-proteolytic *Clostridium botulinum*; (2) the inhibition of aerobic microorganisms as indicators of spoilage; and (3) the potential for temperature abuse. Recent studies showed that average temperatures of retail cases in supermarkets were ~ 12°C while the temperature of domestic refrigerators ranged from 2–20°C (Palumbo, 1986). Nakamura and Hoshino (1983) reported that an O_2-free environment alone was insufficient to inhibit the growth of *Staphylococcus aureus*, *Vibrio* species, *Escherichia coli*, *Bacillus cereus* and *Enterococcus faecalis* at ambient storage temperature. For complete inhibition of these microorganisms, they recommended combination treatments involving O_2 absorbents with thermally processed food, refrigerated storage, or a CO_2 enriched atmosphere. The authors found that this latter modified atmosphere (i.e., low O_2-high CO_2), inhibited the growth of *S. aureus* and *E. coli* but promoted the growth of *E. faecalis*. While low levels of CO_2 (20%) failed to inhibit the growth of *B. cereus*, higher concentrations (100%) completely inhibited the growth of this microorganism in an O_2-free environment. When *Clostridium sporogenes* was cultivated on agar plates at 37°C for 3 d under various gas atmospheres, some interesting results were obtained as shown in Table 12.12.

These studies demonstrate that an O_2/CO_2 absorbent will inhibit, while an O_2 absorbent-CO_2 generator will enhance the growth of *Clostridium* species, indicating the importance of selecting the correct absorbent to control the growth of *Clostridium* species in MAP food. In challenge studies with proteolytic *C. botulinum*, Lambert et al. (1991a) reported that toxin was detected in fresh pork packaged in 100% N_2 with an O_2

Table 12.12. Effect of various gas atmospheres on the growth of C. sporogenes *(adapted from Nakamura and Hoshino, 1983).*

Packaging Conditions	Growth of *C. sporogenes*
Air	–
Ageless O_2 absorbent generating CO_2	4+
Ageless O_2 absorbent not absorbing CO_2	2+
Ageless O_2 absorbent absorbing CO_2	1+

– = No growth; 1+ = slight growth; 2+ = medium growth; 3+ = heavy growth; 4+ = extensive growth.

absorbent after 21 d at 15°C, while it was detected in all samples packaged in air only after day 14. This was attributed to the growth of aerobic spoilage bacteria in the air packaged products resulting in a reduced O_2 tension and production of CO_2, which may have enhanced toxin production by *C. botulinum* (Lambert et al., 1991a). However, subsequent studies showed that the presence of CO_2 in the package headspace was not a significant factor affecting time until toxin production (Lambert et al., 1991b, 1991c).

In more recent agar plate studies with *L. monocytogenes*, high CO_2 levels (>60%) promoted the growth of this pathogen at 10–15°C (Morris et al., 1994). However, when an O_2-free environment was achieved using Ageless SS, growth of *L. monocytogenes* was completely inhibited, even at mild temperature abuse storage conditions. Further studies are now underway to determine the antimicrobial efficacy of Ageless SS and gas packaging on *L. monocytogenes* in packaged pork.

Effect of O_2 Absorbents on Aflatoxigenic Mold Species

While the use of O_2 absorbent technology may fail to completely inhibit the growth of facultative or strictly anaerobic pathogenic bacteria, it is very effective in controlling the growth of *Aspergillus flavus* and *Aspergillus parasiticus*. In inoculation studies with these aflatoxigenic molds in peanuts packed in air alone and with an O_2 absorbent, mold growth and aflatoxin production were completely inhibited, in absorbent packaged peanuts, while ~1000 ppb (1000 ng) of aflatoxin B_1 was detected in all air packaged samples after only 6 d at room temperature (Ellis et al., 1993). Similar control of *A. parasiticus* has also been reported in inoculation studies with peanuts using O_2 absorbent technology (Ellis et al., 1994). However, control was dependent on both the OTR of the packaging films used and storage temperature. While packaging in high-medium barrier films inhibited aflatoxin B_1 production, aflatoxin was detected in all absorbent packaged peanuts using a low barrier film (OTR ~4000 cc/m²/d). However, when absorbent packaged peanuts were packaged in medium barrier films (OTR ~50-cc/m²/d), aflatoxin production occurred in all peanuts stored at 30°C (Ellis et al., 1994). This was attributed to the greater permeability of the film to O_2 at higher storage temperatures resulting in saturation of the absorbents, a concomitant increase in headspace O_2, and subsequent mold growth and aflatoxin production. Similar results were obtained with a CO_2 generating O_2 absorbent (Ageless G) indicating the importance of packaging film permeability to ensure the efficacy of O_2 absorbents and the public health safety of absorbent packaged peanuts.

ETHYLENE ABSORBENTS

Another type of absorbent sachet is ethylene absorbents. Two types are commercially available—Ethysorb produced by StayFresh Ltd., London, U.K. and Ageless C produced by the Mitsubishi Gas Chemical Co., Japan. Ethysorb is a non-toxic, non-corrosive product composed of aluminum oxide and potassium permanganate. These compounds absorb and oxidize ethylene and other volatile gases, e.g., sulphur dioxide and hydrogen sulphide from the atmosphere surrounding the product. Ageless type C, also known as Freshness Keeping agent, also contains calcium hydroxide to absorb carbon dioxide. Ethylene production in climacteric fruits and vegetables initiates ripening and senescence. Hence, removal of this gas from the headspace environment slows down this process and extends product shelf life. Ethysorb has been reported to extend the shelf life of perishable fruit products by more than 60%, thereby reducing wastage and minimizing spoilage losses (Goodburn and Halligan, 1988).

ETHANOL VAPOUR

The use of ethanol (ethyl alcohol) as an antimicrobial agent is well documented. Alcohol was used by the Arabs over 1000 years ago to preserve fruit from mold spoilage. However, alcohol is most commonly known as a surface sterilant or disinfectant. In high concentrations (60–75% v/v) ethanol acts against vegetative cells of microorganisms by denaturing the protein of the protoplast (Seiler and Russell, 1993). Lower concentrations of alcohol (5–20% v/v) have also been shown to have a preserving action against food spoilage and pathogenic microorganisms in agar model systems (Table 12.13). In tests with surface inoculated agar medium containing ethanol concentrations ranging from 4–12% (v/v), ethanol was shown to be effective in controlling ten species of mold including *Aspergillus* and *Penicillium* species; fifteen species of bacteria including *S. aureus* and *E. coli* and three species of spoilage yeasts (Table 12.13). Most molds were inhibited by 4% ethanol

Table 12.13. Inhibitory effect of ethanol addition on growth of microorganisms (Freund Technical Information, 1985).

Species of Microorganism	Ethanol Content of Agar Medium (%)		
	4	8	12
Bacteria			
Escherichia coli	+	–	–
Bacillus cereus	+	–	–
Bacillus megaterium	+	–	–
Bacillus natto	+	–	–
Staphylococcus aureus	+	+	–
Sarcina lutea	+	–	–
Aerobacter aerogenes	+	–	–
Serratia marcescens	+	–	–
Pseudomonas fluorescens	+	–	–
Salmonella typhimurium	+	–	–
Brevibacterium ammoniagenes	+	–	–
Micrococcus epidermis	+	+	–
Enterococcus faecalis	+	+	+
Lactobacillus plantarum	+	+	–
Lactobacillus sake	+	+	+
Yeast			
Torulopsis utilis	+	–	–
Schizosaccharomyces pombe	+	+	–
Candida albicans	+	–	–
Saccharomyces carlsbergensis	+	+	–
Mycotorula japonica	+	–	–
Endomycopsis fibuliger	+	–	–
Endomyces selsii	–	–	–
Pichia membranaefaciens	+	–	–
Saccharomyces rouxii	+	+	–
Fungi			
Aspergillus awamori	+	–	–
Aspergillus niger	+	–	–
Aspergillus usami	–	–	–
Penicillium chrysogenum	–	–	–
Penicillium notatum	–	–	–
Rhizopus javanicus	+	–	–
Mucor plumbeus	±	–	–
Monilia formosa	+	–	–
Trichoderma viride	–	–	–
Dematium pullulans	–	–	–

Table 12.13. (continued).

Species of Microorganism	Ethanol Content of Agar Medium (%) 4	8	12
Actinomycetes			
Streptomyces albus	+	−	−
Streptomyces grieus	−	−	−
Nocardia garneri	+	−	−

while yeasts and *S. aureus* were more resistant and required 8 and 12% ethanol, respectively, for complete inhibition (Table 12.13). Shapero et al. (1978) reported that the effectiveness of ethanol against *S. aureus* was a function of a_w. They reported that in broth media adjusted to a_w values of 0.99, 0.95 and 0.90 with glycerol, inhibition of *S. aureus* occurred at 9, 7 and 4% ethanol, respectively. A similar synergism between a_w and the antimicrobial efficacy of ethanol against *Aspergillus niger* and *Penicillium notatum* has been observed by Smith et al. (1994). In studies with unadjusted Potato Dextrose Agar (PDA) plates, a concentration of 6% ethanol was required for complete inhibition of these common mold contaminants of bakery products. However, in plates adjusted to a_w values of 0.95 and 0.9 with glycerol, complete inhibition of *A. niger* and *P. notatum* was observed with 4 and 2% ethanol, respectively. These results indicate that in lower a_w foods, lower concentrations of ethanol could be used to extend the mold-free shelf life of these products.

The antimicrobial effect of low concentrations of ethanol for shelf life extension of bakery products has also been demonstrated. Plemons et al. (1976) showed that the mold-free shelf life of pizza crusts could be extended for about 20 weeks at ambient temperature by spraying the crusts with 95% ethanol and overwrapping in a large unsealed polyethylene bag. Seiler (1978, 1988) and Seiler and Russell (1993) also reported a 50–250% increase in the mold-free shelf life of cake and bread sprayed with 95% ethanol to give concentrations ranging from 0.5–1.5% by product weight. Both these studies indicate the potential of ethanol as a vapour phase inhibitor. However, a more practical and safer method of generating ethanol vapour is through the use of ethanol vapour generating sachets.

Ethanol Vapour Generators

A novel and innovative method of generating ethanol vapour, again developed by the Japanese, is the use of ethanol vapour generating sachets or strips. These contain absorbed or encapsulated ethanol in a carrier material, and enclosed in packaging films of selective permeabilities which allow the slow or rapid release of ethanol vapour. Several Japanese companies produce ethanol vapour generators, with the most commercially used system being Ethicap or Antimold 102 produced by the Freund Industrial Co., Ltd., Japan.

Ethicap consists of food grade alcohol (55% by weight) adsorbed on to silicon dioxide powder (35%) and contained in a sachet made of a laminate of paper/ethylene vinyl acetate copolymer. To mask the odour of alcohol, some sachets contain traces of vanilla or other flavours as masking compounds. The sachets are labelled "Do not eat contents" and include a diagram demonstrating this warning. Another type of ethanol vapour generator is Negamold (Table 12.14), which scavenges O_2 as well as generating ethanol vapour. Negamold, like its competitor, Ageless type SE, can be used to extend the shelf life and keeping quality of bakery products. However, Negamold is not widely used since it does not generate as much ethanol vapour as Ethicap (Freund Technical Information, 1985).

Sachet sizes of Ethicap range from 0.6 to 6 g or 0.33 to 3.3 g of ethanol that can be evaporated. The actual size of the sachet used depends on the (1) weight of food, (2) a_w of food, and (3) the desired shelf life of product. This interrelationship is shown in Figure 12.2. For example, if the a_w of the product is 0.95 and a 1–2 week shelf life is desired, then a 2 g size of Ethicap (~1.1 g ethanol) should be used per 100 g of product.

Table 12.14. Types of alcohol generators.

Type	Function	Application	a_w of Products
Ethicap (Antimold 102)	Generates ETOH vapour	Moisture dependent Cakes/bread	>0.85
Negamold	Absorbs O_2 Generates ETOH vapour	Moisture dependent Cakes/bread	>0.85

Source: Courtesy of Freund Technical Co., Japan.

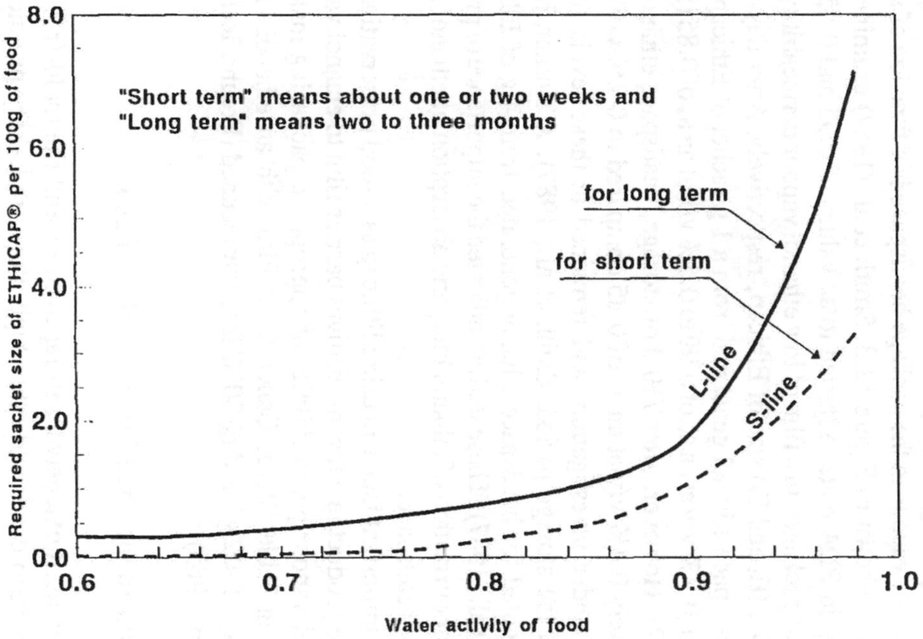

FIGURE 12.2 Relationship between a_w of food and required size of Ethicap sachet (Courtesy of Freund Technical Information, 1985).

However, if a longer shelf life is desired (2–3 months), a 4 g size of Ethicap should be used for the same product (Freund Technical Information, 1985).

When food is packed with a sachet of Ethicap, moisture is absorbed from the food and ethanol vapour is released from encapsulation and permeates the package headspace. However, both the initial and final level of ethanol vapour in the package headspace is a function of sachet size and a_w as shown in Figure 12.3. Smith et al. (1987) examined this relationship in PDA plates adjusted to a_w values of 0.95 and 0.85 with glycerol and packaged in a film of low ethanol vapour permeability with a 1 g and 4 g (E_1 and E_4) size of Ethicap, respectively, After day 1, the level of headspace ethanol generated from a 1 g sachet of Ethicap (E_1), ranged from 0.7% v/v at an a_w of 0.99 to 0.5% v/v at an a_w of 0.85 (Smith et al., 1987). However, after 7–10 d of storage, headspace ethanol was approximately 0.4% v/v at an a_w of 0.85 compared to 0.2% v/v at the highest a_w under investigation, and remained at these levels for the duration of the storage period (Smith et al., 1987). A similar, higher trend was noted for headspace ethanol generated from 4 g of Ethicap, E_4 (Smith et al., 1987). These studies indicated the importance of product a_w on the vapourization of ethanol into, and absorption of ethanol from, the package headspace.

Another important factor in using Ethicap as a food preservative is to package the product in a low or medium barrier film to ethanol vapour. The ethanol vapour permeabilities of appropriate packaging materials are shown in Table 12.15. Generally, a film with an ethanol vapour permeability of <2 g/m²/d @ 30°C is recommended (Freund Technical Information, 1985).

Uses of Ethicap for Shelf Life Extension of Food

Ethicap is used extensively in Japan to extend the mold-free shelf life of high ratio cakes and other high moisture bakery products. A 5–20 times extension in mold-free shelf life has been observed for high ratio cakes depending on size of Ethicap used (Freund Technical Information, 1985). Results also showed that products with sachets did not get as hard as the controls and results were better than those using an O_2 scavenger to inhibit mold growth, indicating that the ethanol vapour also exerts an anti-staling effect (Freund Technical Information, 1985). Ethicap is also widely used in Japan to extend the

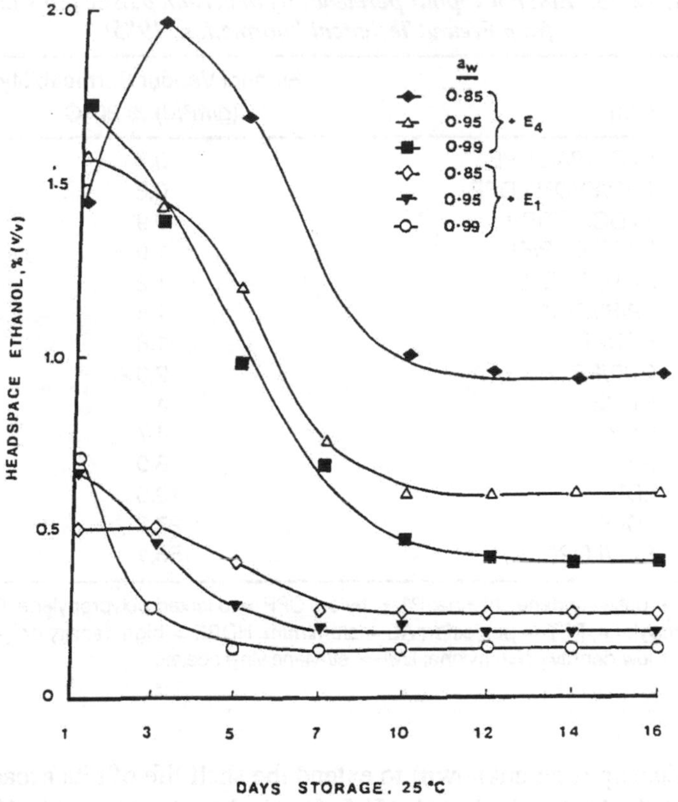

FIGURE 12.3 Effect of water activity (a_W) on generation and absorption of ethanol vapour [from Smith et al. (1990)].

shelf life of semi-moist and dry fish products. Examples of bakery and fish products preserved by Ethicap in the Japanese market are shown in Table 12.16.

Pafumi and Durham (1987) also found that the mold-free shelf life of 200 g Madeira cake could be extended for ~6 weeks at room temperature using a 3 g sachet of Ethicap. However, there was a significant change in quality after 3 weeks characterized by a loss of moistness, firmness of the texture and development of soapy and rancid flavours. Therefore, while Ethicap could be used to extend the mold-free shelf life of the product, the sensory quality of the product was not significantly extended.

Black et al. (1993) examined the combined effect of gas packaging

Table 12.15. Ethanol vapour permeability of certain plastic films (adapted from Freund Technical Information, 1985).

Film	Ethanol Vapour Permeability ($g/m^2/d$) @ 30°C
PVDC/PA/LDPE	0.7
OPP/EVOH/LDPE	0.8
PVDC/PET/PP	0.9
PVDC/OPP/PP	1.0
AL/PET/LDPE	1.2
PP/PVDC/PP	1.5
PET/PP	1.8
OPP/PP	2.0
HDPE	4.1
OPP	4.7
PP	8.0
LDPE	19.0
LDPE	27.3
EVA/LDPE	56.1

PVDC = polyvinylidene chloride; PA = nylon; OPP = oriented polypropylene; PP = cast polypropylene; PET = polyester; AL = aluminum; HDPE = high density polyethylene; LDPE = low density polyethylene; EVA = ethylene vinyl acetate.

and Ethicap (size unknown) to extend the shelf life of pita bread. They reported a 14 d mold-free shelf life for pita bread packaged in 40% CO_2 (balance N_2) or 100% CO_2. However, the shelf life was doubled when an Ethicap sachet was incorporated into the gas packaged products. However, these authors reported that ethanol vapour had little anti-staling effect on pita bread. While the effect of increased firmness due to staling was reversed by microwave heating, it failed to eliminate the stale flavours in the pita bread (Black et al., 1993). The authors observed a slight increase in headspace O_2 in all gas packaged products stored with Ethicap. They hypothesized that the ethanol vapour may have dissolved in the film and acted as a plasticizer, thereby affecting the film's permeability to both O_2 and CO_2.

Ethicap has also been used as a means of further extending the shelf life of gas packaged apple turnovers (Smith et al., 1987). Apple turnovers, with an a_w of 0.93, had a shelf life of 14 d when packaged in a CO_2:N_2 (60:40) gas mixture at ambient temperature due to growth and CO_2 production by *Saccharomyces cerevisiae*. However, when Ethicap (E_4) was incorporated into the packaged product, either alone or in conjunction with gas packaging, yeast growth was completely suppressed and all packages appeared normal at the end of the 21 d storage period.

Ethicap has also proved effective in controlling secondary spoilage problems by *S. cerevisiae* in gas-packaged strawberry and vanilla layer cakes and cherry cream cheese cake. Ooraikul (1993) reported that 5 g sachet of Ethicap inhibited growth of and CO_2 production by *S. cerevisiae* in both strawberry and vanilla layer cakes. This size of Ethicap had no adverse effect on the organoleptic quality of cakes. Indeed, Ooraikul (1993) reported that the aroma of the cake packaged with Ethicap was more pleasant than the fresh cake and the taste also remained excellent. However, a 6 g sachet of Ethicap failed to inhibit yeast spoilage in cherry cream cheese cake (Ooraikul, 1993). He recommended a 7–8 g sachet of Ethicap in conjunction with a preservative, such as ethyl paraben, to control secondary yeast fermentation problems in cherry cream cheese cake (Ooraikul, 1993). These studies show that larger sizes of Ethicap, and hence higher levels of ethanol vapour, are required to inhibit yeast growth compared to mold growth. Indeed, agar model studies have shown that yeasts can grow in media containing 8%

Table 12.16. Use of Ethicap for shelf life extension of food (adapted from Freund Technical Information, 1985).

Food	a_w of Ethicap	Size of Material	Packaging	Shelf Life @RT*
Bakery products				
Bread	0.92	1 G	OPP	1 week
Cupcake	0.85	3 G	OPP/PP	2 months
Jam doughnut	0.83	3 G	OPP	20 days
American cake	0.80	2 G	OPP/PP	6 months
Rice cake	0.76	4 G	OPP/PP	2 months
Chocolate				
Sponge cake	0.72	2 G	OPP/PP	6 months
Fish Products				
Smoked squid	0.85	1 G	OPP	3 months
Boiled squid	0.68	0.6 G	OPP	3 months
Boiled and dried squid	0.69	0.6 G	OPP/PE	2 months
Boiled and dried small fish	0.63	1 G	OPP	3 months

*RT = Room temperature.

(v/v) ethanol while most molds were inhibited by 4% ethanol (Table 12.13; Freund Technical Information, 1985).

Effect of Ethanol Vapour on Food Spoilage/ Food Poisoning Bacteria

While most studies to date have focused on the use of Ethicap as an antimycotic agent, few studies have evaluated its potential to control food spoilage and food poisoning bacteria. Ethanol, when incorporated into agar media, has shown to be effective against several spoilage and pathogenic bacteria (Freund Technical Information, 1985; Shapero et al., 1978). Seiler and Russell (1993) also reported that ethanol at a level of 1% by product weight delayed the onset of "rope" caused by the growth of *Bacillus subtilis*. They also reported that low concentrations of ethanol (0.5–1%) by product weight inhibited bacterial growth in both whipping cream and custard, two well known vectors of food poisoning bacteria in filled bakery products. These studies clearly illustrate the antibacterial properties of ethanol when incorporated directly into media or a food product. More recently, Morris et al. (1994) examined the effect of Ethicap on the growth of *L. monocytogenes*. They observed that a 4 g sachet of Ethicap could control the growth of *L. monocytogenes* (Scott A) on agar media at 5, 10 and 15°C, the latter storage conditions representing mild temperature abuse. Further studies are now under way to determine the volume of ethanol vapour generated at these storage temperatures and the effect of these concentrations on the microbiological and chemical shelf life of fresh packaged pork.

Advantages and Disadvantages of Ethanol Vapour Generators

According to Smith et al. (1987), the advantages of Ethicap are:

(1) Ethanol vapour can be generated from sachets without having to spray ethanol directly onto the product surface prior to packaging.
(2) Sachets can be conveniently removed from packages and discarded at the end of the storage period.
(3) It eliminates the need for preservatives such as benzoic acid or sorbic acid to control yeast spoilage.
(4) It can control mold spoilage and delay staling in bakery products.

A disadvantage of using ethanol vapour for shelf life extension is its absorption by the product from the package headspace. However, the concentration of ethanol found in apple turnovers (1.45 − 1.52%) was within the maximum level of 2% by product weight when ethanol was sprayed onto pizza crusts prior to final baking (Smith et al., 1987). In addition, the level of ethanol in apple turnovers can be reduced to <0.1% by heating the product at 375°F prior to consumption. Therefore, while a longer shelf life may be possible by packaging product with Ethicap, its use as a food preservative may be limited to "brown and serve" or microwaveable products. Another disadvantage is the cost, which limits its use to products with a higher profit margin. Nevertheless, Ethicap is a viable alternative or supplement to gas-packaging to extend the shelf life of baked bakery products from both mold and yeast spoilage and to prevent staling.

CONCLUSION

In conclusion, the use of gas absorbents and ethanol vapour generators is, without doubt, one of the most exciting concepts of atmosphere modification available to the food industry. While both O_2 absorbent technology and ethanol vapour generators are used extensively in Japan to extend the shelf life and keeping quality of a variety of products, their use to date in the North American market is limited due to the cost of the sachets, consumer resistance to the inclusion of sachets in packaged products and lack of regulatory approval for Ethicap. Nevertheless, the use of gas absorbents/ethanol vapour generator sachets or labels offers the food industry a viable alternative to vacuum/gas flushing for shelf life extension of its products. As more companies become aware of the economic advantages of using these sachets/generators, and consumers accept these sachets or labels in their food products, the sachet generation will slowly emerge as the preservation technology of the 1990s.

REFERENCES

Abe, Y. and Kondoh, Y. 1989. *Controlled/Modified Atmosphere/Vacuum Packaging of Foods.* Food and Nutrition Press Publ., Trumbell, CT, pp. 149 − 158.

Alarcon, B. and Hotchkiss, J. H. 1993. "The effect of FreshPax oxygen-absorbing packets on the shelf-life of goods," Technical Report, Dept. of Food Science, Cornell University, NY, pp. 1 − 7.

Anon. 1988. "Ener-Getic all year long," *Packaging Digest*, 8:70, 72, 75.

Black, R. G., Quail, K. J., Reyes, M., Kuzyk, M. and Ruddick, L. 1993. "Shelf-life extension of pita bread by modified atmosphere packaging," *Food Australia*, 45:387–391.

Dallyn, H. and Everton, J. R. 1969. "The xerophilic mold *Xeromyces bisporus* as a spoilage organism," *J. Food Technology*, 4:339–403.

Ellis, W. O., Smith, J. P., Simpson, B. K. and Doyon, G. 1994. "Effect of films of different gas transmission rates on aflatoxin production by *Aspergillus flavus* in peanuts packaged under modified atmosphere packaging (MAP) conditions," *Food Research International* 27:505–512.

Ellis, W. O., Smith, J. P., Simpson, B. K., Khanizadeh, S. and Oldham J. H. 1993. "Control of growth and aflatoxin production by *Aspergillus flavus* under modified atmosphere packaging (MAP) conditions," *Food Microbiology*, 10:9–21.

FreshPax Technical Pamphlet. 1994. "Protect and preserve your products and profits with FreshPax," Multiform Desiccants, Buffalo, NY.

Freund Technical Information. 1985. "No-mix-type mould inhibitor Ethicap," Freund Industrial Co., Ltd., Tokyo, Japan, pp. 1–14.

Goodburn, K. E. and Halligan, A. C. 1988. *Modified Atmosphere Packaging—A Technology Guide*. Publication of the British Food Manufacturing Association, Leatherhead, U.K., pp. 1–44.

Harima, Y. 1990. *Food Packaging*. Academic Press Publ., London, U.K., pp. 229–252.

Lambert, A. D., Smith, J. P. and Dodds, K. L. 1991a. "Combined effect of modified atmosphere packaging and low-dose irradiation on toxin production by *Clostridium botulinum* in fresh pork," *J. Food Protection*, 54(2):97–104.

Lambert, A. D., Smith, J. P. and Dodds, K. L. 1991b. "Effect of headspace CO_2 on toxin production by *Clostridium botulinum* in MAP, irradiated fresh pork," *J. Food Protection*, 54(8):588–592.

Lambert, A. D., Smith, J. P. and Dodds, K. L. 1991c. "Effect of initial O_2 and CO_2 and low-dose irradiation on toxin production by *Clostridium botulinum* in MAP fresh pork," *J. Food Protection*, 54(12):939–944.

Minakuchi, S. and Nakamura, H. 1990. *Food Packaging*. Academic Press Publ., London, U.K., pp. 357–378.

Morris, J., Smith, J. P., Tarte, I. and Farber, J. 1994. "Combined effect of chitosan and MAP on the growth of *Listeria monocytogenes*," *Food Microbiology* (submitted for publication).

Nakamura, H. and Hoshino, J. 1983. *Techniques for the Preservation of Food by the Employment of an Oxygen Absorber*. Mitsubishi Gas Chemical Co., Tokyo, Japan, pp. 1–45.

Ooraikul, B. ed., 1993. *Modified Atmosphere Packaging of Food*. Ellis Horwood Publ., New York, pp. 49–117.

Pafumi, J. and Durham, R. 1987. "Cake shelf life extension," *Food Technology in Australia*, 39:286–287.

Palumbo, S. A. 1986. "Is refrigeration enough to restrain food borne pathogens?" *J. Food Protection*, 49:1003–1009.

Plemons, R. F., Staff, C. H. and Cameron, F. R. 1976. "Process for retarding mold growth in partially baked pizza crusts and articles produced thereby," United States Patent, 3,979,525.

Powers, E. M. and Berkowitz, D. 1990. "Efficacy of an oxygen scavenger to modify the atmosphere and prevent mold growth in meal, ready-to-eat pouched bread," *J. Food Protection*, 53:767–771.

Seiler, D. A. L. 1978. "The microbiology of cake and its ingredients," *Food Trade Review*, 48:339–344.

Seiler, D. A. L. 1988. "Microbiological problems associated with cereal based foods," *Food Science and Technology Today*, 2(1):37–41.

Seiler, D. A. L. and Russell, N. J. 1993. *Food Preservatives*. Blackie Publ., Glasgow, U.K. pp. 153–171.

Shapero, M., Nelson, D. A. and Labuza, T. P. 1978. "Ethanol inhibition of *Staphylococcus aureus* at limited water activity," *J. Food Science*, 43:1467–1469.

Smith, J. P. 1992. *MAP Packaging of Food—Principles and Applications*. Academic and Professional Publ., London, U.K., pp. 134–169.

Smith, J. P., Lyver, A. and Morris, J. "Effect of ethanol vapor on the growth of common mold contaminants of bakery products," *Food Microbiology* (submitted for publication).

Smith, J. P., Ooraikul, B., Koersen, W. J., and Jackson, E. D. 1986. "Novel approach to oxygen control in modified atmosphere packaging of bakery products," *Food Microbiology*, 3:315–320.

Smith, J. P., Ooraikul, B., Koersen, W. J., van de Voort, F. R., Jackson, E. D. and Lawrence, R. A. 1987. "Shelf life extension of a bakery product using ethanol vapor," *Food Microbiology*, 4:329–337.

Smith, J. P., Ramaswamy, H. and Simpson, B. K. 1990. "Developments in Food Packaging Technology. Part 2: Storage aspects," *Trends in Food Science and Technology*, 1:112–119.

Tomkins, R. G. 1932. "The inhibition of the growth of meat attacking fungi by carbon dioxide," *J. Society Chemical Industry*, 51P:261–264.

Toppan Technical Information. 1989. *Freshness Keeping Agents*. Toppan Printing Co., Ltd., Tokyo, Japan, pp. 1–8.

CHAPTER 13

The Applications of HACCP for MAP and Sous Vide Products

O. PETER SNYDER, JR.—*Hospitality Institute of Technology and Management, U.S.A.*

INTRODUCTION

THE other chapters of this book have provided the control technology for MAP and sous vide processes. However, technology must be integrated into a systematic zero-defect management plan and program (a plan put into action) if products are to be produced that meet the three elements of quality.

(1) They are safe (zero liability costs).
(2) They meet customer value satisfaction 100 percent of the time.
(3) They are produced with high productivity and effectiveness (no waste and high-speed lines).

HACCP provides a highly disciplined foundation for systematic, continuous process analysis and zero-defect control for hazards (zero company liability costs) [Bryan (1990); CFA (1990); ICMSF (1988); Microbiology and Food Safety Committee of the National Food Processors Association (1992); NAS (1985); NACMCF (1990a); Pierson and Corlett (1992); Sperber (1991)]. At the same time, all aspects of the process will improve because it requires full process management. The basic concept of HACCP, as implemented by The Pillsbury Company in the early 1970s (The Pillsbury Company, 1973) was an integrated management and technology system. For example, it specified that process records from the statistical process control program would be used to drive various types of experiments to achieve continuous safety (and quality) improvement.

THE FOOD PROCESS SYSTEM

The following is a systems flow diagram for HACCP-based food safety management (Figure 13.1). It identifies management as the criti-

(TQM)

MANAGEMENT (AND GOVERNMENT)

INPUT → **RETAIL PROCESS: FOODSERVICE, MARKETS, VENDING, HOME** → **SAFETY-ASSURED OUTPUT**

INPUT

Supplies and Material

- Environment contamination
  - Soil, water, air
  - Vegetation, plants, grains
  - Wild animals, birds, fish, insects, pests
- Supplier contamination
  - Pesticides, insecticides
  - Mold growth in grains
  - Filth contamination of food
  - Microorganisms, toxins, poisons, extraneous material
  - Poor nutritional food profiles due to feed supplies, condition of soil where food is grown
  - Hazardous feed additives
  - Container contamination of food
  - Time, temperature abuse
  - Inadequate facilities, equipment, and management
- Distribution contamination
  - Increase in pathogens, toxins, and poisons through mishandling
  - Nutrient loss in shipping
  - Food spoilage
  - Time, temperature abuse
  - Inadequate facilities, equipment, and management
- Wholesale processor contamination
  - Fecal contamination during slaughter
  - Pathogenic environmental organism contamination
  - Spoilage waste
  - Extraneous material
  - Unsafe chemical addition
  - Food mislabeling
  - Underprocessing
  - Overprocessing waste and nutrient loss
  - Packaging; container poisons
  - Time, temperature abuse
  - Inadequate facilities, equipment, and management

RETAIL PROCESS: FOODSERVICE, MARKETS, VENDING, HOME

Environment

- Safe air
- Insect and rodent control
- Safe water
- No soil on shoes

Facilities that are clean and maintained

Equipment that controls hazards or warns when it is not functioning correctly; construction from safe materials

- Refrigeration that keeps food at less than 32°F (0°C) and cools to less than 40°F (4.4°C) in less than 11 h
- Ovens that cook food from 40°F (4.4°C) to above 130°F (54.4°C) in less than 6 h
- Hot holding devices that keep food above 130°F (54.4°C)

Personnel

- Fingertip and hand washing control of transient organisms

Products and services

- Thawing
- Recipe food time and temperature control
- Proper food temperature measurement
- Food contact surface cleaning and sanitizing
- Unsafe chemicals control; — additive
- Control of carcinogens in cooking, as in broiling and grilling
- Nutrient loss minimization
- Food thermal pasteurization
- Food acid pasteurization

SAFETY-ASSURED OUTPUT

Consumer

- Proper balance between pleasurable and safe food
- Safe levels of hazards for consumer, based on immune threshold
- Consumer abuse control information
- Food sensitivities consumer communication
- Nutrition profile and contamination control (i.e., food components) for a long, physically excellent quality of life
- Hurdles
  - Temperature
  - Time
  - Water activity (a_w)
  - Oxidation/reduction (Eh)
  - Chemical and biological additive
  - Competitive micoorganisms
  - Packaging

FIGURE 13.1 The system for HACCP-based food safety total quality management.

cal element of control. It then shows the structure of the system from growing and harvesting to consumption. It must include consumption because in MAP-sous vide, the product is potentially hazardous (due to the possible presence of spores of *Clostridium botulinum, Bacillus cereus*, and *Clostridium perfringens*) until consumed. The times and temperatures are technically correct for safety and are not necessarily government regulatory standards. For example, cold holding safety standard is 32°F (0°C) because of *Listeria monocytogenes*, hot holding is >130°F (54.4°C) because of *C. perfringens*, and cooling to 40°F (4.4°C) within 11 h is based again, on *C. perfringens*.

SECTION I—APPLICATION OF HACCP TO MAP-SOUS VIDE

Typical MAP-Sous Vide Products

The current applications of MAP-sous vide are to perishable refrigerated products. The objective is to produce products that are minimally processed to achieve a balance between freshness and shelf life. Processing may be as simple as washing vegetables, cutting, and vacuum packaging, which is a specialized form of MAP processing that achieves a few days of shelf life. The other extreme is a product such as pasteurized crab that can be stored in a refrigerator at or below 38°F (3.3°C) for years. If the product is to be consumed without further processing, or if extended shelf life is wanted, the process includes a step to assure reduction of hazards in the raw product to a safe level.

There is a very important relationship between the shelf life and storage temperature, and pasteurization time and temperature. For example, with pasteurized milk, the lower the storage temperature, the longer its shelf life (Table 13.1).

With ultra-pasteurized liquid whole eggs [U.S. Patent No. 4, 808, 425 (February 28, 1989)], the longer the pasteurization time, the longer the refrigerated storage (Table 13.2).

With pasteurized canned crab, the same is true. The longer the pasteurization time at 85°C (185.0°F), the longer the refrigerated storage life (Rippen and Hackney, 1992) (Table 13.3).

The current concept of MAP-sous vide limits it to refrigerated-frozen food. But this could be broadened because many shelf-stable foods such as coffee, milk powder, etc. are packaged under modified atmospheres

Table 13.1 Effect of storage temperature on the shelf life of pasteurized milk.

Pasteurized Milk Temperature	Time to Spoilage
70°F (21.1°C)	12 h
60°F (15.6°C)	1 day
50°F (10.0°C)	3 days
45°F (7.2°C)	5 days
40°F (4.4°C)	10 days
32°F (0.0°C)	32 days

Table 13.2. Effect of pasteurization time at 70°C (158.0°F) on the refrigerated shelf life of liquid whole eggs.

Pasteurization Time	Time to Spoilage
1.3 s	8 weeks
3.8 s	12 weeks
10.3 s	16 weeks
40 s	20 weeks
150 s	24 weeks

Table 13.3. Effect of pasteurization time at 85°C (185°F) on the refrigerated shelf life of pasteurized canned crab.

Pasteurization Time	Shelf Life
10 to 15 min	1.5 months
15 to 20 min	2 to 4 months
20 to 25 min	4 to 6 months
25 to 30 min	6 to 9 months
30 to 40 min	9 to 18 months
>40 min	12 to 36 months

or vacuum for extended shelf life. This discussion, however, will be limited to refrigerated foods.

Figure 13.2 is a MAP-sous vide process-product tree. It is evident that some of these items have been in the market for many years, such as canned ham that is vacuum packed, cooked to 155°F and held refrigerated. So, MAP-sous vide is *not* new technology.

What Is Sous Vide?

Sous vide means "under vacuum." The term is actually very misleading because it really only was intended by French chefs to refer to a range of single-portion foods that were cooked and cooled in plastic pouches. The reason for the vacuum in today's sous vide products is simply to avoid an air insulating gap between the package wall and the product, thereby obtaining the best possible heat transfer coefficient between the process environment and the product, both during processing and reheating for service.

The initial concept of sous vide was to place one to a few portions of garnished food in a plastic bag, vacuum seal, and then immerse the food in a moderate temperature water bath to achieve pasteurization. It was stored at 5°C (41.0°F) or less and used within about five days. It was used by the chef or sold to another restaurant in town where it was used in a few days. To reconstitute the product, it was placed in an 80°C (176.0°F) hot water bath for a few minutes. The original sous vide item was goose liver paté. The chef who first experimented with sous vide wanted a higher yield and longer shelf life of this product. When processed under normal conditions in an oven, much of the fat melted out of the product, resulting in a low yield. Under sous vide processing conditions, the chef obtained a much higher yield and better flavour, as well as a longer shelf life for the goose liver paté. The concept has been extended by chefs to also include cooking, then putting into a bag, sealing, and cooling, as well as cooking, cooling, garnishing, putting into a bag, and vacuum sealing. If one removes the stigma of "under vacuum," it is evident that the movement toward sous vide is one of gentle food processing to achieve extended, safe shelf life through pasteurization or other forms of gentle preservation. Sous vide can perform a critical safety role for society. It can reduce pathogens on the grower/processor contaminated meat, poultry, and fish to a safe level. If society will not allow ionizing irradiation pasteurization to be used to make food safe, sous vide is a perfect substitute, just as for the past 90 years, milk has been pasteurized to make it safe.

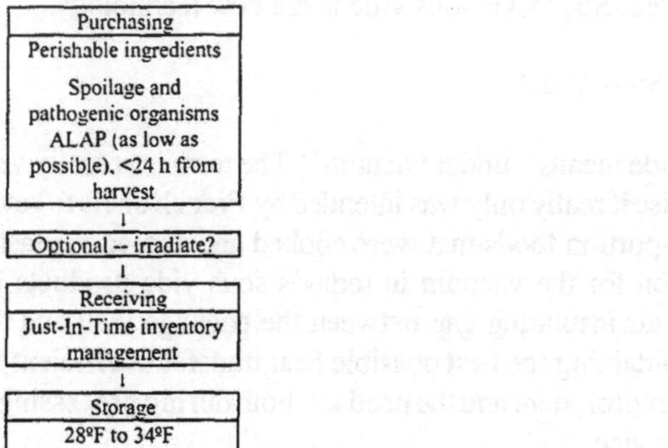

FIGURE 13.2 The MAP-sous vide process-product tree.

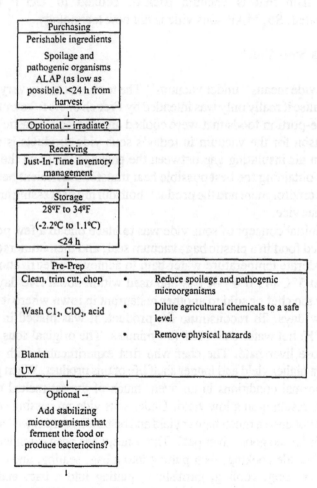

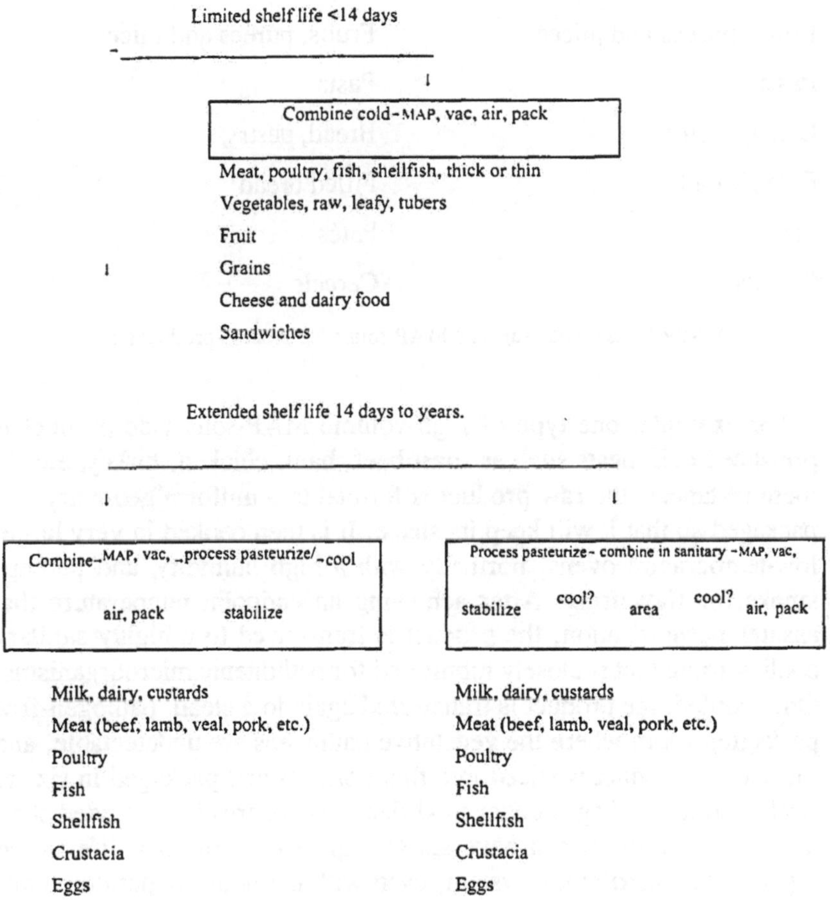

FIGURE 13.2 (continued) The MAP-sous vide process-product tree.

Combine–MAP, vac. air. pack – process pasteurize/ stabilize – cool	Process pasteurize/ stabilize – cool? – combine in sanitary area – cool? – MAP, vac. air. pack
Vegetables, purées and juices	Vegetables, purées and juices
Fruits, purées and juices	Fruits, purées and juices
Pasta	Pasta
Bread, pastry	Bread, pastry
Filled bread	Filled bread
Patés	Patés
Cereals	Cereals

FIGURE 13.2 (continued) The MAP-sous vide process-product tree.

For example, one type of high-volume MAP-sous vide product is pre-sliced deli meats such as roast beef, ham, chicken, turkey, etc. In these instances, the raw product is formed to a uniform geometry and packaged so that it will keep its shape. It is then cooked in very large, low-temperature ovens, normally with a high humidity, and perhaps smoke for flavouring. After achieving an endpoint temperature that assures pasteurization, the product is transferred to a highly sanitary cooling room that is closely monitored for pathogenic microorganisms. Once cooled, the product is transferred again to a clean, pathogen-free packaging room where the vegetative pathogens are undetectable, and the cooked product is sliced into thin portions and packaged in six- to twelve-ounce packages under modified atmospheres for extended shelf life. These products can have six to eight weeks of shelf life in the supermarket distribution system, even with the poor temperature controls associated with food markets.

It is necessary to be very careful of one point with MAP and sous vide products. They are both pre-processed before the consumer receives them. This pre-process at least washes and cuts a whole item into parts so that the finished product is more convenient for the consumer or the foodservice operation. The items could be pasteurized. The cutting and chopping process alone is very detrimental to shelf life in that it takes the spoilage and small numbers of pathogenic microorganisms on the surface of the product and moves them to the cut surfaces of the insides of the product. Because the amount of surface increases dramatically, the microorganisms will have more nutrients and more surface area on which to multiply, and the microbiological loads are much more difficult to control than they are on the original whole items. Because of the processing, these products will have a value-added price increase, which normally makes the product more expensive than whole, fresh product purchased from a local supplier.

SECTION II—THE SYSTEM: HAZARDS AND CONTROLS

Hazards in the MAP-Sous Vide Process

There are three categories of process-product defects that can become human hazards: biological, chemical, and physical (BCP). For all potential hazards there is a level at which a defect is non-hazardous, and a level at which the defect becomes a hazard. In microbiological terms, the latter is M (organisms per gram) in a 3 class sampling plan, or the probability of organisms per gram in a 2 class plan (ICMSF, 1986). In chemical terms it is commonly referred to in water quality as the health risk limit (HRL). For physical hazards it might be termed "food defect action level," except that most defects are simply filth and would not be health hazards from a physical safety viewpoint. A physical hazard probably should refer to objects that are larger than, perhaps, 1.6 mm. As each BCP defect is dealt with in a process, it will be realized that the process developer must consider all elements of process quality: product safety, customer satisfaction, and productivity. Therefore, it is necessary to define process deviations from a target value very carefully in terms of whether a deviation can allow a defect to become a hazard. If not, it is a quality or owner control problem.

A major problem exists in many countries' food regulations today in that critical limits for many hazards are not provided. This will be addressed later. There is also the problem that there are two populations being fed: immuno-competent and immuno-compromised people. The retail food industry assumes that it is feeding the immuno-competent person, and does only minimal processing because the customer can ask the owner directly how much the item is processed (i.e., how safe it is). Immuno-compromised people must ask for additional safety factors [e.g., eggs cooked above 160°F (71.1°C)] and avoid raw food as in salad bars, unless specifically designated for them. In a setting where there are only immuno-compromised people (e.g., hospital patients), food will be processed to a greater extent to reduce the level of the BCP in the products to a safe level for them.

There has been much discussion as to whether or not fraud defects such as under weight, species substitution, etc., should be included as hazards. They should not because this seriously dilutes the purpose of HACCP, which is to allow the operator to strive for zero liability costs. To keep this clear in doing a process analysis, it is appropriate to identify four different kinds of process controls. The government regulates the first two listed below, hazard controls and regulatory controls. Only hazard control is the true consumer safety control. The industry has complete responsibility for the last two, quality controls and owner controls.

(1) *Hazard controls:* Hazards are threats to the well-being of the consumer that can result in liability claims.
(2) *Regulatory controls:* These are normally fraudulent items such as incorrect labels, weight, amounts of ingredients, species substitution, etc.
(3) *Quality controls:* These refer to customer-desired product attributes which the company must achieve if the customer is to repeatedly purchase the product. The government must not set quality standards.
(4) *Owner controls:* These are the attributes upon which the owner insists in order to achieve desired productivity, operating costs, and profits so that the owner can remain in business and have resources for continuous quality improvement. Again, the government must not interfere with these.

Hazards and Controls

To begin the HACCP process, one must understand the hazards. Table 13.4 presents an overview of the hazards that can affect consumers. They are not necessarily based only on current regulatory thinking, but rather on liability to the food supplier and health cost to the consumer.

Let's examine each category of defects that can become hazards in order to identify historically common causes of problems. This then leads to the beginning of the hazard and control analysis of a MAP-sous vide process.

Table 13.4. Defects in food that can become hazards (goal—zero defects).

BIOLOGICAL	CHEMICAL (cont.)
Organisms and Their Toxins	Idiosyncratic reactions to food
Bacteria: vegetative cells and spores	Anaphylactoid reactions
Molds [mycotoxins (e.g., aflatoxin)]	**Nutrition**
Yeasts (*Candida albicans*)	Excessive addition of nutrients
Viruses and rickettsia parasites	Nutritional deficiencies and/or inaccurate formulation of synthesized formulas
Marine Animals as Sources of Toxic Compounds	Anti-nutritional factors
Fish	Destruction and unnecessary loss of nutrients during processing and storage
Shellfish	Inaccurate nutritional labeling
CHEMICAL	PHYSICAL
Poisonous Substances	**Extraneous Material**
Toxic plant material	Glass
Intentional (GRAS) food additives	Wood
Chemical created by the process	Stones, sand, and dirt
Agricultural chemicals	Metal
Antibiotic and other drug residues in meat, poultry, and dairy products	Packaging materials
Unintentional additives	Bones
Sabotage	Building materials
Equipment material leaching	Filth from insects, rodents, and any other unwanted animal parts or excreta
Packaging material leaching	Personal effects
Industrial pollutants	**Functional Hazards**
Heavy metals	Particle size deviation
Radioactive isotopes	Packaging defects
Adverse Food Reactions (food sensitivity)	Sabotage
Food allergies	
Food intolerances	
Metabolic disorder-based reactions	
Pharmacological food reactions	

Biological Defects

Organisms and Their Toxins

All raw ingredients for MAP-sous vide products are randomly contaminated with two forms of bacteria: vegetative cells and spores.

The vegetative cells can be controlled to some degree with modified atmospheres and can be destroyed by acids, radiation, heat, and disinfectants such as Cl_2, ClO_2, and H_2O_2. The spores, on the other hand, will always be a defect in the finished product, and a potential hazard.

Molds are a problem mostly in the supply system where nuts and grains, especially, are likely to have low levels of aflatoxin. Government safety levels have been established and the hazard appears to be controlled. In MAP-sous vide processing, special care must be taken to keep background mold and yeast levels in air low for shelf life extension. Yeasts are not a real hazard, merely a quality problem. Viruses and parasites are major problems, since very few are needed to cause illness. However, these cannot multiply in foods and viruses can be washed off to a safe level. The hazard is that city water plants cannot be trusted to remove them because many viruses and parasites are resistant to Cl_2, and only filtration or heat are adequate controls. Also, workers can contribute to the problem because the feces of ill workers can contain 10^9 microorganisms per gram, which is difficult to remove from fingertips.

Products are contaminated in a range of, perhaps, 1 to 100% with pathogenic organisms. Table 13.5 is an overview of some food contamination rates. In process development, the goal is to purchase ingredients from suppliers who have HACCP programs and produce products under conditions that minimize contamination, especially from low-temperature spoilage microorganisms.

From a practical standpoint, if microbiological data on ingredients are unavailable from a supplier, the following table (Table 13.6) provides an estimate of the level of pathogens on food that suppliers with HACCP programs can achieve.

Some day, it is hoped pathogens such as *Salmonella* spp. and pathogenic *Escherichia coli*, which are not normal contaminants of animals but get into animal food systems from the environment, can be eliminated from the food production environment by virtue of the way animals are grown. However, improving growing practices will happen only in the very distant future. Hence, for the time being, the processor, either a food manufacturing plant, a restaurant, or those who cook food at home, is responsible for the reduction of pathogenic vegetative cells on raw products to a safe level. This is where sous vide has a bright future.

Marine Animals as Sources of Toxic Compounds

Fish, shellfish, and other marine animals can be sources of toxic

*Table 13.5. Pathogen contamination of food products.**

Bacteria	Food	Percent Contaminated
Salmonella spp.	Raw poultry	40 to 100
	Raw pork	3 to 20
	Raw shellfish	16
Staphylococcus aureus	Raw chicken	73
	Raw pork	13 to 33
	Raw beef	16
Clostridium perfringens	Raw pork and chicken	39 to 45
Campylobacter jejuni	Raw chicken and turkey	45 to 64
Escherichia coli O157:H7	Raw beef, pork, and poultry	1.5 to 3.7
Bacillus cereus	Raw ground beef	43 to 63
	Raw rice	100
Listeria monocytogenes	Fresh potatoes	26
	Fresh radishes	30
Yersinia enterocolitica	Raw pork	49
	Raw milk	48
	Raw vegetables	46
Vibrio spp.	Raw seafood	33 to 46
Giardia lamblia, Norwalk virus	Water	30

*Smith, 1990.

compounds. For example, scrombotoxin may be present in members of the Scrombridae family which include tuna, red snapper, grouper, amberjack, and mahi mahi. Ciguatoxin may be present in fish taken from the Pacific Ocean and the Caribbean. Shellfish may also be sources of toxic compounds and can be potential sources of the following conditions: paralytic shellfish poisoning (PSP); diarrheic shellfish poisoning (DSP); neurotoxic shellfish poisoning (NSP); amnesiac shellfish poisoning (ASP). Potentially toxic fish should not be used without supplier testing and certification.

Chemical Defects

Poisonous chemical substances in foods can cause varying degrees of severity of illness and possible fatalities. Many substances are listed by the FDA in the Code of Federal Regulations (FDA, 1991). In each case, a person with expert training in the hazard should evaluate the process

*Table 13.6. Estimated pathogen contamination on raw food (number of pathogens per gram of food).**

Microorganism(s)	Meat and Poultry	Fish and Shellfish	Fruits and Vegetables	Starches
Salmonella spp., *Vibrio* spp., Hepatitis A; *Shigella* spp., *E. coli, Listeria monocytogenes*	10	10	10	10
Campylobacter jejuni	1,000	—	—	—
Clostridium botulinum	0.01	1.0	0.01	0.01
Clostridium perfringens	100	10	—	—
Bacillus cereus	100	—	100	100
Mold toxins		Below government tolerances		
Chemicals and poisons		Below government tolerances		

*Snyder 1993.

to ensure that these hazards are controlled. The following is a list of examples:

- Toxic plant material includes poisonous varieties of mushrooms and other plant material. Pyrrolizidine alkaloids are carcinogenic compounds which are present in some herbs and herbal teas.
- Intentional (GRAS) food additives include nitrites, sulfites; flavour enhancers (monosodium glutamate); colour additives (FD&C Yellow No. 5).
- Chemicals created by the process include higher amounts of nitrosamines, heterocyclic amines, and polycyclic aromatic hydrocarbons formed when meats are heated or cooked at high temperatures for prolonged periods of time.
- Agricultural chemicals include fertilizers, fungicides, pesticides and herbicides. The use of these compounds must be controlled and they must be removed as much as is possible in order to prevent toxigenic and carcinogenic effects that may result from ingesting foods containing their residues.
- Antibiotic and other drug residues in meat, poultry, and dairy

products include antibiotics found in milk, and growth hormones found in the muscle of meat.

- Unintentional additives are accidentally added during processing and service of food. Examples include machine oil or other lubricants and toxic cleaning compounds and sanitizing agents.
- Sabotage is the malicious, intentional addition of a toxic chemical to a food or food supply in order to cause harm.
- Equipment material leaching occurs when material of which equipment is composed, is dissolved by the food during processing. An example is leaching of lead or copper into acid foods when these products are stored in containers composed of material or parts containing either one of these metals.
- Packaging material leaching occurs when aromatic packaging compounds migrate to the packaged material. A few years ago, carbonless paper, which was initially processed with polychlorinated biphenyls (PCBs), was recycled and used to produce cartons for food packaging. The production of PCBs ceased in the U.S.in 1974. However, PCBs continue to cycle through the environment.
- Industrial pollutants are illegally dumped into the environment, and eventually become a part of plants and animals used for human consumption.
- Heavy metal poisoning occurs when lead, zinc, arsenic, mercury, and copper from the environment or equipment become a part of food and beverages being produced.
- Radioactive isotopes can occur in food naturally. For example, Potassium 40 is in all salt. With the development of radioactive materials for industrial application, there is a problem of leakage of these isotopes into the food system.

Adverse food reactions (food sensitivity) are a life-threatening problem for about 1 percent of the population. This hazard is partially controlled with correct, complete labeling of products, and/or providing individuals with a list of ingredients if this information is requested. Employees must be trained not to substitute ingredients without consulting food safety management personnel. The following are examples:

- Food allergies are adverse physiological and neurological reactions that have an immunological basis in some individuals after consumption of specific foods. For example, life-threatening reactions (e.g., anaphylactic shock) occur when sensitized individuals consume milk, eggs, nuts, corn, and/or fish.

- Food intolerances are illnesses that occur in some individuals due to metabolic disorder-based reactions (e.g., lactose intolerance, phenylketonuria). Pharmacological food reactions occur in individuals taking monoamine oxidase inhibitors. Gastrointestinal disturbances following ingestion of food also occur in people while they are taking some types of antibiotics.
- Idiosyncratic reactions to food are adverse reactions to food for which the mechanism is not always understood. Examples include: asthma induced in some people as a result of ingesting sulfites or FD&C Yellow No. 5, and "Chinese Restaurant Syndrome" due to consumption of excessive amounts of monosodium glutamate.
- Anaphylactoid reactions produce symptoms in individuals that are similar to true food allergies. However, the immune system is not involved. Histamines and other toxic compounds present in substantial amounts in food produce symptoms of illness.

Nutrition (as defined optimally) is the careful balancing of nutrients (carbohydrate, fat, protein, vitamins, and minerals) to provide growth (in the young), and maintain and sustain health in adults. Food producers must carefully analyze the nutritional value of their products so that it can be provided to consumers when necessary. The addition of nutrients to products must be accomplished accurately and precisely, utilizing both government and current nutritional guidelines. Hazardous conditions can develop in long-term care facilities such as hospitals and prisons if nutritional components of diet are not managed or controlled properly.

- *Excessive addition of nutrients:* Examples include excessive addition of fat soluble vitamins A and D.
- *Nutritional deficiencies and/or inaccurate formulation of synthesized formulas:* Examples of this hazard have occurred in the production of infant formulas, formulas for tube feeding hospital patients, and diet preparations for weight loss regimes.
- *Anti-nutritional factors:* These are found in plant material fruits and vegetables. These include oxalates, phytates, lathrogens, and thiamase.
- *Destruction and unnecessary loss of nutrients during processing and storage:* B-vitamins, ascorbic acid, and minerals are water soluble and are subject to loss during processing. Ascorbic acid is

quite subject to loss during prolonged heating and storage due to oxidation.

- *Inaccurate nutritional labeling:* This practice can be harmful to consumers if they are inaccurately informed. Inaccurate nutritional labeling is subject to litigation. Food products must accurately comply with government labeling regulations, if nutritional information is declared.

Physical Defects

Extraneous material is fragments of unwanted pieces of material in food which may be objectionable and/or hard enough to cause choking, broken teeth, laceration of the mouth and/or internal injury.

Extraneous material in foods include: glass; wood; stones, sand, and dirt; metal (bolts, screws, staples); packaging material (cardboard, plastic wrap, cellophane); bones; building materials (paint chips, insulation); filth from insects, rodents, and any other unwanted animal parts or excreta; personal effects [buttons, rings, earrings, contact lens, fingernails (false and real), hair].

Functional hazards include particle size deviation (e.g., whole allspice, peppercorns, and bay leaves); packaging defects (e.g., use of packaging materials that break or fragment easily); and sabotage, which is the malicious addition of any of the extraneous material listed previously in a direct attempt to cause injury to individuals and subsequent discreditation of the producer.

For control, specifications must be written for ingredients that specify particle size and defect level. Line employees must be trained to 1) recognize and identify hazards in the production environment, 2) notify management of potential hazard, and 3) take action to eliminate the hazard.

Some Biological Hazard Limits

Since "zero" is not a reality with biological hazards (or chemical, or physical), one must define the levels at which immuno-competent people and immuno-compromised people could be made ill. This will play a very important role because the objective for the government is to require only a minimum, safe process. If people can consume 1,000 *Staphylococcus aureus* cells per gram safely, then this bears on what limit should be set as the hazard vs. defect transition point. One of the best

Table 13.7. Foodborne illness hazards: threshold and quality levels.

Agent	Healthy Person (Estimated Illness Dose)	HITM* Suggested Purchaser Raw Food Quality Standards
Bacteria	(Number of Microorganisms)	(Number of Microorganisms)
Bacillus cereus	3.4×10^4 to 9.5×10^8/g(4)**	$<10^2$/g
Campylobacter jejuni	5×10^2 in 180 ml milk (17)	<1/g
Clostridium botulinum	3×10^3 (10) [a]	<1/g[b]
Clostridium perfringens	10^6 to 10^7/g(5)	$<10^2$/g
Escherichia coli	10^6 to $>10^7$ (dose)(2)	
Salmonella spp.		
S. anatum	10^5 to $>10^8$ (dose)(11)[c]	<10/g
S. bareilly	10^5 to $>10^6$ (dose)(12) [c]	<10/g
S. derby	10^7 (dose)(12)[c]	<10/g
S. meleagridus	10^7 (dose)(11) [c]	<10/g
S. newport	10^5 (dose)(12) [c]	<10/g
S. pullorum	10^9 to $>10^{10}$ (dose)(13)[c]	<10/g
S. typhi	10^4 to $>10^8$ (dose)(7)[c]	
Shigella spp.		
S. flexneri	10^2 to $>10^9$ (dose)(1,3,18)	<1/g
S. dysenteriae	10 to $>10^4$ (dose)(9)	<1/g
Staphylococcus aureus	10^5 to $>10^6$/g(6,16) [d]	$<10^2$/g
Vibrio cholerae	10^3 (dose)(8)	<1/g
Vibrio parahaemolyticus	10^6 to 10^9 (dose)(20)	<10/g
Yersinia enterocolitica	3.9×10^7 (dose)(15) [e]	$<10^2$/g
Listeria monocytogenes	Unknown (probably >1,000/g)	<10/g
Viruses		
Hepatitis A virus	?	<1/g
Norwalk virus	?	<1/g
Chemicals	(Amount in Food)	(Amount in Food)
Monosodium glutamate	0.5% (dose)(19)	<0.05%
Sodium nitrate	8 – 15 g(14)	<500 ppm
Sodium nitrite	?	<200 ppm
Sulfites	>0.7 mg/kg body weight/day(21)	<10 ppm
Extraneous Material	?	None

*HITM—Hospitality Institute of Technology and Management, St. Paul, Minnesota.

**Number in parentheses indicates references, given in the reference section of this chapter.

[a] Indicates the number of bacteria necessary to produce sufficient toxin for mouse LD_{50}.

[b] If a product is to be considered shelf stable above 50°F, then it should be heat processed to reduce a spore population of *C. botulinum* types A and B by 10^{12} or have a water activity (a_w) <0.86, or the pH of the product should be 4.1 or less, or a combination of processes should be used to control the growth of *C. botulinum* types A and B and *Salmonella* spp.

[c] Results from feeding studies. Data from outbreaks indicate lower values.

[d] Indicates number of pathogenic bacteria necessary to produce sufficient amount of illness-producing toxin.

[e] Probably lower.

compilations of microbiological standards from throughout the world is found in the book by Shapton and Shapton (1991).

As a general rule, it is assumed that immuno-compromised people can become ill with as few as one to ten infective vegetative pathogens such as *Salmonella* spp. or *E. coli* per gram of product. Fortunately, immuno-competent people can tolerate higher levels. Table 13.7 lists thresholds at which some pathogenic microorganisms cause illness in immuno-competent individuals. It also lists levels that should be set as targets in good MAP-sous vide processes for immuno-competent people. Note that it appears as though the more common an organism is to our environment, the higher the tolerance level for immuno-competent people. This includes *Salmonella* spp., *Vibrio parahaemolyticus*, *E. coli*, etc. This fits research data which demonstrate that when volunteers are fed increasing levels of pathogens they become increasingly resistant to infection and can tolerate higher doses (Levine et al., 1973; Shaughnessy et al., 1946; Tjoa et al., 1977). This has also been shown with people who travel throughout the world and choose to live in more highly contaminated environments. After a period, they become less susceptible to low levels of pathogens in their food. It may well be, since the environment will always be contaminated, that we should not strive for zero pathogens in all food.

Spore forms of microorganisms can be consumed by normally healthy individuals at levels of probably up to 1,000 per gram without ill effects. One exception appears to be the spores of *C. botulinum* types A and B by immuno-compromised people (Chia et al., 1986).

The fecal-oral microorganisms that are transferred to food on fingers from human feces and that are common to our diet such as Hepatitis A, Norwalk virus, and *Shigella* show low tolerance levels in healthy people. This is to be expected because in developed nations with good sewer and water systems, they are not normal to people's diets. This means that for people who handle food, hand washing is extremely important to reduce pathogens that are excreted by human beings to a level of less than ten on each finger. Fingertip washing, therefore, is a very important and sensitive process in terms of food safety. It must be assumed that every food worker sheds high levels of pathogens every day in his/her feces. Telling a person to stay home when he/she is ill is too late, if the employee has not been washing his/her fingertips with a fingernail brush.

Also included in this group are the uncommon organisms such as *Giardia lamblia*, *Cryptosporidium* spp. oocysts, and other parasitic organisms that could enter the food process through the water system or by unclean fingertips after a person uses the toilet. This puts forth the importance of assuring that a city's water is reprocessed when it enters the food facility, because no city water systems currently have HACCP programs. Water causes outbreaks throughout the U.S. each year, leading to thousands of people becoming ill.

Key Bacterial Growth and Death Characteristics

Table 13.8 provides an overview of the principal organisms that will govern the safety design, and analysis of MAP-sous vide processes (Snyder, 1993). There are thousands of pathogenic microorganisms, and each has different control characteristics. However, if the system is taken as a whole, it is possible to select key microorganisms on which to base process controls, such as shown in Table 13.8.

These microorganisms can be assumed to be on the raw food entering the MAP-sous vide process. Since *Yersinia enterocolitica* and *L. monocytogenes* both begin to multiply at 30 to 32°F (−1.1 to 0°C), food must be kept below a temperature of 32°F (0°C) if it is to be considered safe from multiplication of pathogens. Since *Salmonella* spp. will multiply at a pH as low as 4.1, it is essential that if food such as mayonnaise is made with raw eggs (notorious for being contaminated with *Salmonella*), the pH must be below 4.1 in order to assure that there is no *Salmonella* growth and, in fact, to assure its destruction. After about two days of room temperature at a pH of less than 4.1, *Salmonella* spp. and the other vegetative bacterial pathogens are no longer detectable in normal salad dressing manufacture (Smittle, 1977).

The data on *Campylobacter jejuni* point out that it grows very poorly outside the animal or human, and is quite easily destroyed. But, it currently contaminates, especially poultry, at an imminent illness level of thousands per gram. Therefore, the major problem with *C. jejuni* is cross-contamination in the process. It can be assumed that 10,000 *Campylobacter* spp. per cm^2 will be deposited on the food contact surface by raw food. To be safe, this must be reduced to undetectable levels. This is achieved when a surface containing a mixture of normal food microorganisms is cleaned to less than 2 APC/cm^2.

C. botulinum type E is a spore former that dies rapidly at 180°F (82.2°C). However, most sous vide foods are not cooked to this high a temperature. In fact, it probably is undesirable to heat to this temperature in most cases because spoilage spores that can multiply at 30°F (−1.1°C) are activated. It must be assumed, then, that all *C. botulinum*,

Table 13.8. Food pathogen control data summary.

Microorganisms	Temperature Range for Growth	pH Range and Minimal Water Activity (a_w) for Growth	G = Growth or Doubling Time, D = Death Rate for 10:1 Reduction Time, z = Temperature Increase for 10 × Faster Kill
Infective Microorganisms (inactivated by pasteurization)			
1. *Yersinia enterocolitica*	32 to 111°F (0 to 44 °C)	pH 4.6 to 9.0	*G* (32°F/0°C) = 2 d *G* (40°F/4.4°C) = 13 h *D* (145°F/62.8°C) = 0.24 to 0.96 min *z* = 9.2 to 10.4°F (5.1 to 5.8°C)
2. *Listeria monocytogenes*	32 to 112°F (0 to 44°C)	pH 4.5 to 9.5	*G* (32°F/0°C) = 5 d *G* (40°F/4.4°C) = 1 d *D* (140°F/60°C) = 2.85 min *z* = 10.4 to 11.3°F (5.8 to 6.3°C)
3. *Vibrio parahaemolyticus*	41 to 109.4°F (5 to 43°C)	pH 4.5 to 11.0 a_w 0.937	*D* (116°F/47°C) = 0.8 to 48 min
4. *Salmonella* spp.	41.5 to 114°F (5.5 to 45.6°C)	pH 4.1 to 9.0 a_w 0.95	*D* (140°F/60°C) = 1.7 min *z* = 10°F/5.6°C
5. *Campylobacter jejuni*	90 to 113°F (30 to 45°C)	pH 4.9 to 8.0	*D* (137°F/58.3°C) = 12 to 21 s *z* = 10.6 to 11.4°F (6.0 to 6.4°C)

(continued)

Table 13.8 (continued).

Microorganisms	Temperature Range for Growth	pH Range and Minimal Water Activity (a_w) for Growth	G = Growth or Doubling Time, *D* = Death Rate for 10:1 Reduction Time, *z* = Temperature Increase for 10 × Faster Kill
Toxin Producers and/or Spore-Formers (not inactivated by pasteurization)			
6. *Clostridium botulinum* (Type E and other non-proteolytic strains)	38 to 113°F (3.3 to 45°C)	pH 5.0 to 9.0 a_w 0.97	**Spores** *D* (180°F/82.2°C) = 0.49 to 0.74 min *z* = 9.9 to 19.3°F (5.6 to 10.7°C) **Toxin destruction (any botulinal toxin)** *D* (185°F/85°C) = 5 min *z* = 7.2 to 11.2°F (4.0 to 6.2°C)
7. *Staphylococcus aureus*	43.8 to 122°F (6.5 to 50°C)	pH 4.5 to 9.3 a_w 0.83	**Vegetative Cells** *D* (140°F/60°C) = 5.2 to 7.8 min *z* = 9.7 to 10.4°F (5.8 to 5.4°C)
	Toxin production 50 to 114.8°F (10 to 46°C)	pH 5.15 to 9.0 a_w 0.86	**Toxin destruction** *D* (210°F/98.9°C) = >2 h *z* = about 50°F (27.8°C)
8. *Bacillus cereus*	39.2 to 122°F (4.0 to 50°C)	pH 4.3 to 9.0 a_w 0.912	**Vegetative cells** *D* (140°F/60°C) = 1 min *z* = 12.4°F/ 6.9°C **Spores** *D* (212°F/100°C) = 2.7 to 3.1 min *z* = 11°F 6.1°C **Toxin destruction** Diarrheal: D (133°F/ 56.1°C) = 5 min. Emetic: stable at (249.8°F/121°C)

Table 13.8 (continued).

Microorganisms	Temperature Range for Growth	pH Range and Minimal Water Activity (a_w) for Growth	G = Growth or Doubling Time, D = Death Rate for 10:1 Reduction Time, z = Temperature Increase for 10 × Faster Kill
9. *Clostridium botulinum* (Type A and proteolytic B strains)	50 to 118°F (10 to 47.8°C)	pH 4.6 to 9.0 a_w 0.94	Spores D (250°F/121.1°C) = 0.3 to 0.23 min z = 18°F/10°C Toxin destruction (see above)
10. *Clostridium perfringens*	59 to 127.5°F (15 to 52.3°C)	pH 5.0 to 9.0 a_w 0.95	Vegetative cells G (105.8°F/41°C) = 7.2 min D (138°F/59°C) = 7.2 min z = 6.8°F/3.8°C Spores D (210°F/98.0°C) = 26 to 31 min z = 13°F/7.2°C

B. cereus, and *C. perfringens* spores will survive a sous vide pasteurization process. For absolute control, food must be stored below 38°F (3.3°C) in order to assure safety from the spores of non-proteolytic *C. botulinum* and *B. cereus*.

S. aureus begins to multiply at 43°F (6.1°C), but does not produce toxin until it reaches a temperature of 50°F (10.0°C). Since there is some likelihood of recontamination of food with *S. aureus* when people handle cooked, cooled food, as in making salads and sandwiches, if ingredients are pre-chilled to less than 50°F (10.0°C) and are kept below this temperature when processed, there will be no chance of *S. aureus* producing toxin. If the toxin is produced, the *D* value (decimal reduction time) of 2 h means that it is virtually impossible to destroy the toxin with heat. Therefore, reheating to 165°F (73.9°C), for example, should never be used as a critical control procedure with respect to staphylococcal enterotoxin. This is additionally important since many MAP-sous vide products are eaten cold. After food is cooked, the only zero-defect control is to prevent the production of toxin.

Proteolytic strains of *C. botulinum* types A and B do not begin to multiply and produce a toxin until they reach a temperature of 50°F (10.0°C). Again, if fruits and vegetables, which can be vacuum or MAP packaged and will be contaminated with low levels of *C. botulinum*, are kept at a temperature of less than 50°F (10.0°C), then *C. botulinum* types A and B will not cause illness. Since these foods have not been shown to have problems with non-proteolytic *C. botulinum*, 50°F (10°C) is an adequate control temperature.

Finally, there is *C. perfringens*. It will multiply up to 127.5°F (53.1°C). Therefore, the upper temperature limit for pathogenic microorganism control is 127.5°F (53.1°C) [rounded to 130°F (54.4°C)]. Because of its rapid doubling rate, as frequent as once every 7.2 min at 105.8°F (41.0°C), it dictates the heating and cooling rates for food (Shigahisa et al., 1985). Food such as sous vide roast must be heated from 40°F to 130°F (4.4°C to 54.4°C) in less than 6 h in order to assure no multiplication. Food must be cooled from 130°F to 40°F (54.4°C to 4.4°C) within 11 h in order to prevent the outgrowth of *C. perfringens* spores during cooling. (Note, the 2 h and 4 h cooling rates specified in many nations' codes are *not* based on correct *C. perfringens* cooling data.)

From these data, it is possible to identify growth control times (Table 13.9) and temperatures, and pasteurization times and temperatures (Table 13.10).

Since contamination of raw food by *L. monocytogenes* cannot be prevented, the process standard limits multiplication to five generations (1:32). Ten generations (1:1,024) would be considered very dangerous.

Two process standards are shown. Times are shown for 3D and 7D reductions. The 3D is a minimum for use in a retail operation such as a restaurant serving immuno-competent people. The 7D reduction would be for wholesale producers and situations where immuno-compromised people consume the food. Note that there must be a compromise. The more a food is processed, the greater the reduction in thermally sensitive nutrients and the poorer the sensory quality. When an immuno-competent person eats the food, as in a retail food operation, there is no need to process as if only immuno-compromised people were eating the food.

Facilities and Equipment Decontamination

The degree of decontamination required of the food processing area and food contact surfaces is a direct function of the following variables:

*Table 13.9. Suggested safe times and temperatures for holding foods.**

°F (°C)	Holding Time
<32 (<0.0)	No pathogen growth
32 (0.0)	12 d
35 (1.7)	9 d
40 (4.4)	5 d
45 (7.2)	2.5 d
50 (10.0)	2 d
55 (12.8)	30 h
60 (15.6)	20 h
65 (18.3)	15 h
70 (21.1)	10 h
75 (23.9)	6.5 h
80 (26.7)	5 h
85 (29.4)	3.75 h
90 to 125 (32.2 to 51.7)	2.5 h
>130 (>54.4)	No pathogen growth

*Based on five generations of multiplication of *L. monocytogenes*, because it begins to multiply at 32°F (0.0°C). The rates at various temperatures are derived from an Arrhenius plot (log growth vs. 1/*T*) of many *L. monocytogenes* research reports containing growth data (Snyder, 1993).

*Table 13.10. Suggested pasteurization times and temperatures for fully cooked, ready-to-eat foods.**

	Pasteurization Time	
°F (°C)	Immuno-Competent People; 10^3 Log Reduction	Immuno-Compromised People; 10^7 Log Reduction
130 (54.4)	51.9 min	121 min
135 (57.2)	16.4 min	38.3 min
140 (60.0)	5.19 min	12.1 min
145 (62.8)	1.64 min	3.8 min
150 (65.6)	31.1	72.6 s
155 (68.3)	9.84 s	23.0 s
160 (71.1)	3.1 s	7.3 s
165 (73.9)	0.984 s	2.3 s

*Inactivation times are based on *Salmonella* spp. and the USDA beef process standards (USDA, 1991d) which have been used to produce billions of pounds of deli-style roast beef (a sous vide product) without a problem when the rules are followed.

(1) Contamination of incoming ingredients
(2) Whether there is a food pasteurization step that follows
(3) Desired shelf life and expected temperatures in distribution and final use
(4) Whether the consumer is immuno-competent or immuno-compromised

The key to a clean, low microbiological count in a food processing area is frequency of cleaning and correct cleaning. For normal MAP-sous vide products, whether the processing area is refrigerated at, for example 10°C (50°F) or not, is mostly a question of shelf life of the food. If the food temperature can be rapidly brought back to less than 40°F (4.4°C) after assembly/processing as, for example, with a blast cooler or by spraying with CO_2 "snow," then a cold room is unnecessary. Cold rooms have some serious drawbacks. The humidity must be kept at less than 55 or 65%, or there can be mold and yeast growth, and bacteria will grow in wet areas. In a refrigerated room, it is much more difficult and expensive to maintain desired humidities. Also, a refrigerated process room can select for low-temperature pathogens such as *L. monocytogenes*, *Y. enterocolitica*, and *Aeromonas hydrophila*.

Some working standards for the number of spoilage microorganisms on pasteurized food contact surfaces are shown in Table 13.11.

The key to low biological counts is hot soap/detergent, water and physical action. For example, if a surface is to be power-cleaned, the water should be at 140 to 170°F (60.0 to 76.7°C), and at a pressure of about 150 psi. At higher pressures, there is splatter which spreads microorganisms, and at lower pressures, there is not enough physical force. If cleaning is done by hand, there must be a large volume of clean, hot [110 to 120°F (43.3 to 48.9°C)] water, as in the wash sink of a

Table 13.11. Suggested operating criteria for spoilage microorganisms on food contact surfaces.

Number of Spoilage Microorganisms	Rating
$<1/cm^2$ or $<1/ml$ of rinse solution	Excellent
2 to $10/cm^2$	Good
11 to $100/cm^2$	Clean-up time
101 to $>1{,}000/cm^2$	Out of control; shut down and find the problem

three-compartment sink, and a scrub brush must be used for mechanical force to remove the biofilm. The key to effective surface sanitizing is removal of the biofilm, not chemical sanitizers. Chemical sanitizers must come into direct contact with the microorganisms to be effective. Therefore, unless the biofilm is removed, the sanitizer is essentially ineffective. A properly cleaned surface can have microbiological counts of <5 CFU per cm^2.

At a minimum, surfaces being used for only one item should be cleaned every 4 h. If multiple items are processed, the surface should be cleaned before each new item. There should be no pathogens, for example, *L. monocytogenes*, in a 50 cm^2 swab of the floors, walls, and ceilings in a process area. It is critical to get a surface dry within 20 min after it is cleaned and sanitized. Otherwise, some residual organisms will begin to multiply.

SECTION III—IDENTIFY THE VARIABLES IN A FOOD SYSTEM

The Food System

When beginning the HACCP process, it is necessary to identify the food system and all major process variables in the system. Figure 13.3 provides an overview of the system components.

The system has inputs, processes, and outputs. The products produced by the company will have a specific process diagram/HACCP recipe procedure. The system process variables that must be BCP hazard controlled, in addition to the management and consumer processes, are personnel, environment, facilities, equipment, supplies, and product procedures.

Types of MAP-Sous Vide Processes

Table 13.12 lists the various MAP-sous vide processes. It provides examples of these products produced by these processes.

System Process Hazard and Control Analysis

The most complete information regarding process hazard and control analysis has been developed by the chemical process and nuclear process industries. The Center for Chemical Process Safety (1992) in its book,

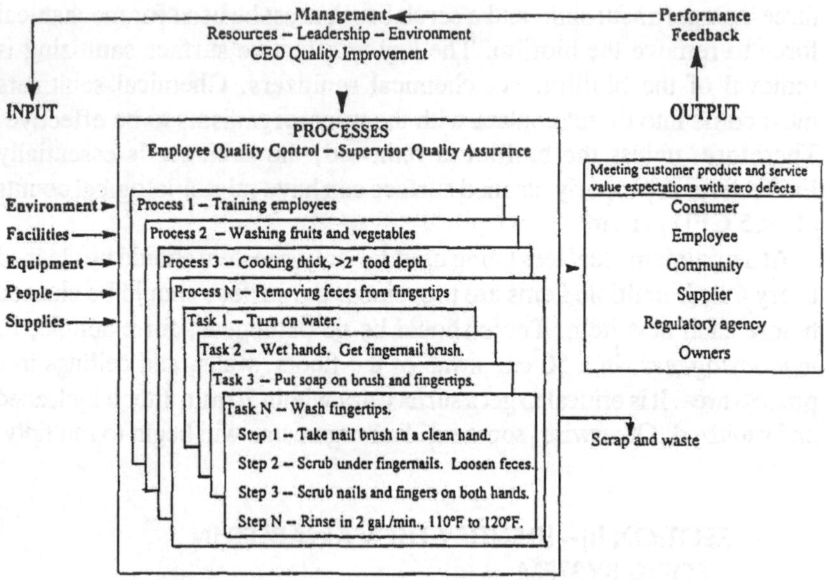

FIGURE 13.3 The HACCP-based TQM process.

Guidelines for Hazard Evaluation Procedures: Second Edition with Worked Examples, presents an extensive series of methodologies for hazard analysis at a process step. Included are

(1) Safety review
(2) Checklist analysis
(3) Relative ranking
(4) Preliminary hazard analysis
(5) "What if" analysis
(6) "What if" checklist analysis
(7) Hazard and operability analysis
(8) Failure mode and effect analysis (FMEA)
(9) Fault tree analysis
(10) Event tree analysis
(11) Cause-consequence analysis
(12) Human reliability analysis

Another excellent overview of HACCP in terms of occupational safety is provided by Johnson (1980). This and many other documents in the

*Table 13.12. Refrigerated food processes.**

Process	Examples of Foods
Cook-Package-Chill	
1. Assemble-Cook-Package Hot-Chill	
Pre-prep assemble → Cook pasteurize → Pump, package hot, MAP, vacuum, air → Chill	Stews, sauces, soups, sandwich spreads, fruit pie fillings, pudding
Package-Cook-Chill	
2. Assemble-Sear-Package-Cook-Chill	
a. Pre-prep assemble → Sear → Vacuum package → Cook pasteurize → Chill	Sous vide, rolls and roasts, canned crab and ham
b. Pre-prep assemble → Package (pan/tray), MAP, vacuum, air → Cook pasteurize → Chill	Casseroles, (meat, pasta, vegetable, sauce combination) pâtés, meat, fruit, and cream pies, quiches
3. Cook-Chill-Assemble-Package	
a. Pre-prep → Cook pasteurize → Chill → Assemble whole → Package, MAP, vacuum, air	Roast or fried chicken, other roasts, uncured sausage, bread
b. Pre-prep → Cook pasteurize → Chill → Slice/dice assemble → Package, MAP, vacuum, air	Uncured luncheon meat, diced meat
c. Pre-prep → Cook pasteurize → Chill → Assemble (pan/dish/tray) → Package, MAP, vacuum, air	Meat and pasta, dinners, meat and sauces, sandwiches and pizza, milk
d. Pre-prep → Cook pasteurize → Partial chill?? → Fill in dough → Chill → Package, MAP, vacuum, air	Meat pies, quiches, patties, pâtés, filled pastries
Assemble-Package	
4. Assemble with Cooked and Raw Ingredients-Package	
a. Pre-prep → Assemble cold raw and cold pasteurized ingredients → package, MAP, vacuum, air	Chef's salads, fruit salads, leafy salads, chicken salads/cold casseroles, sandwiches, pizza with raw ingredients
b. Pre-prep → Assemble cold raw/cold pasteurized and hot (liquid) ingredients → Package, MAP, vacuum, air → Chill/gel	Uncured jellied meats, gelled fruits

*Adapted from NACMCF, 1990b.

chemical processing, nuclear design, and occupational safety areas should be consulted if one intends to be a HACCP process authority.

The starting point in process hazard and control analysis is the development of a process flow diagram. Snyder (1993) has developed an extensive flow charting procedure. This procedure is based on the use of 3″ × 5″ cards (see Figure 13.4) to describe each process step. The card facilitates the flow diagramming process because very often, diagrams can be 100 to 200 or more steps, and developers have to inject new steps or rearrange steps. It carefully follows the logic and language of computer-structured programming, which is very simple and precise. The hazard analysis and control rules must be *very* simple and clearly understood if a worker on a process line is to be effective as a zero-defect manager. The only other control choice is to remove the worker and use the language and a computer to directly control a process step. This is often done in wholesale processing, and rarely done in retail MAP-sous vide processes. When the flow process diagrams are finally developed, HACCP'd, and adequate controls specified at a step, they are converted into written operating instructions for the employees. These detailed written instructions have the great advantage that they support employee-based continuous quality improvement.

Flow diagrams can become very long and complex. It is necessary to deal with each step of the process in order to find all of the possible BCP defects and control variables that can deviate and lead to a critical defect level. Remember, spores on incoming product are never eliminated in a MAP-sous vide system. Essentially, any step up to consumer consumption can become hazardous in terms of *C. botulinum*, *B. cereus*, or *C. perfringens* if the step is not controlled. While some additives such as acids, nitrates, and water activity lowering agents can minimize the problem, MAP-sous vide products normally have minimal additives. Suppliers want these products to be perceived by the consumer as having maximum freshness. This way the consumer will pay more for them, which is necessary for the added processing cost. The highly infective vegetative pathogens such as *Salmonella* spp. and *C. jejuni* should be lowered to a sufficiently low level of less than one per 25 g so that the consumer is not made ill. However, any break in container integrity or post-process contamination by a worker or food contact surface can make a product unsafe. Finally, a physical or a chemical agent getting into the process at any step can make the product unsafe. Therefore, it is truly necessary to deal with a process from growing and harvesting to consumption, step by step, in order to effectively strive for zero safety

Step No.	Procedures		
Process Type			
Food Start Temp		Food Item Name	Job Area Location
Food End Temp		Wt. Added Lost in Step	Employee Time
Temp around the Food		Wt. Output of Step	Preservative
Time to Perform Step		Equip. Item Used	Eh
Food Center Surf. Dist.		Equip. Use Time	
pH	a_w	Food Viscosity	

FIGURE 13.4 Process step description card (PSDC).

limit defects. Some of this analysis can be avoided if suppliers have HACCP programs and can identify BCP contamination levels.

System Analysis

As mentioned earlier, hazard analysis cannot be done in isolation from the basic development of a food process. However, the typical situation is that there is an existing process that has been developed and is known to meet customer expectations. The problem is that it has not been sized for volume production, nor is it quality assured. Nonetheless, one begins with a basic series of tasks and steps that are necessary to produce a MAP-sous vide item.

The first microbiological variable that must be controlled and to which a limit is set in any process in a food system, is volume to be produced. A given set of equipment in a building will have an upper limit for controlled (safe) time and temperature production. This needs to be set by the regulatory authority at the time a process system is being approved. If one is doing MAP-sous vide in a restaurant, the total production may be 100 pouches of a special breast of chicken, taking a total time of 1 h. If doing MAP-sous vide on a national production schedule, thousands of packages of chicken breast item would be produced. In reality, the processes themselves must be identical in time, temperature, and ingredients regardless of the volume, if the products are to have the same sensory characteristics. As a process is scaled up from tens per hour to thousands per hour, equipment changes dramatically in order to obtain high-volume production. Note, however, that the growth and death rates of the food-contaminating organisms do not change. Hence, the safe process times and temperature cannot change.

This is not the case with chemical and physical hazards, which can be very different, depending on the environment, facilities, and equipment in which the food is produced. When the equipment and facilities change, the chemical and physical hazards can change completely.

Another system process variable is incoming product. In a small operation such as a restaurant or food market, it is likely that the producer will have little control over the ingredients, and will have to accept whatever microbiological load is present on product coming into the city. With large-volume production, the manufacturing company may have its own farms and fields for major ingredients. In these cases, lower microbiological counts on raw ingredients can be achieved. Safety can be achieved with a much reduced process when the company has control over the pathogens on incoming product because the cleanliness of the growing, harvesting, and slaughtering processes can be controlled to company standards.

Process Charting and Analysis

The first step in process analysis is to precisely specify the output product in terms of all critical hazard control variables: temperature, atmosphere, pH, water activity, additives, and microbiological load. Then, each incoming ingredient must be described in terms of its microbiological load and other properties such as temperature and size/weight/thickness. When possible, only suppliers who have HACCP programs should be used. The goal of the process is to reduce the hazardous vegetative pathogens to an undetectable level, which, in practical terms, is less than one per 25 g.

The overall process flow charting methodology is shown in Figure 13.5. This technique is the same as for computer flow charting or industrial engineering process flow charting. Note that each step (using the PSDC) is identified in terms of one of five industrial engineering flow analysis terms: operation, inspect, transport, delay, and store. By assigning one of these five terms to each step, it becomes a simple matter to analyze one process vs. another to find out which one is more efficient and effective.

Overall, processes will be divided into the following tasks. They can take place in many different arrangements, as illustrated previously.

(1) Get ready
(2) Receiving, and storage
(3) Pre-preparation
(4) Preparation
(5) Package
(6) Chill
(7) Cold hold
(8) Transport, display, serve, consume
(9) Leftovers

This first chart is a general process chart used to decide the order in which various steps will be done. Each block has space for a step identification number and the symbol for the type of step. Then, it has space for a short narrative. At the bottom of each block is space for specifying the approximate time and temperature of the step. T_i is the temperature of the ingredients coming into the step, T_o is the temperature of ingredients out of the step, and t is the time for the step. By reading a food process flow chart, the process approving authority has a very clear idea of exactly what is taking place and where the potentially risky steps are. By applying the basic microbiological controls previously identified, critical process con-

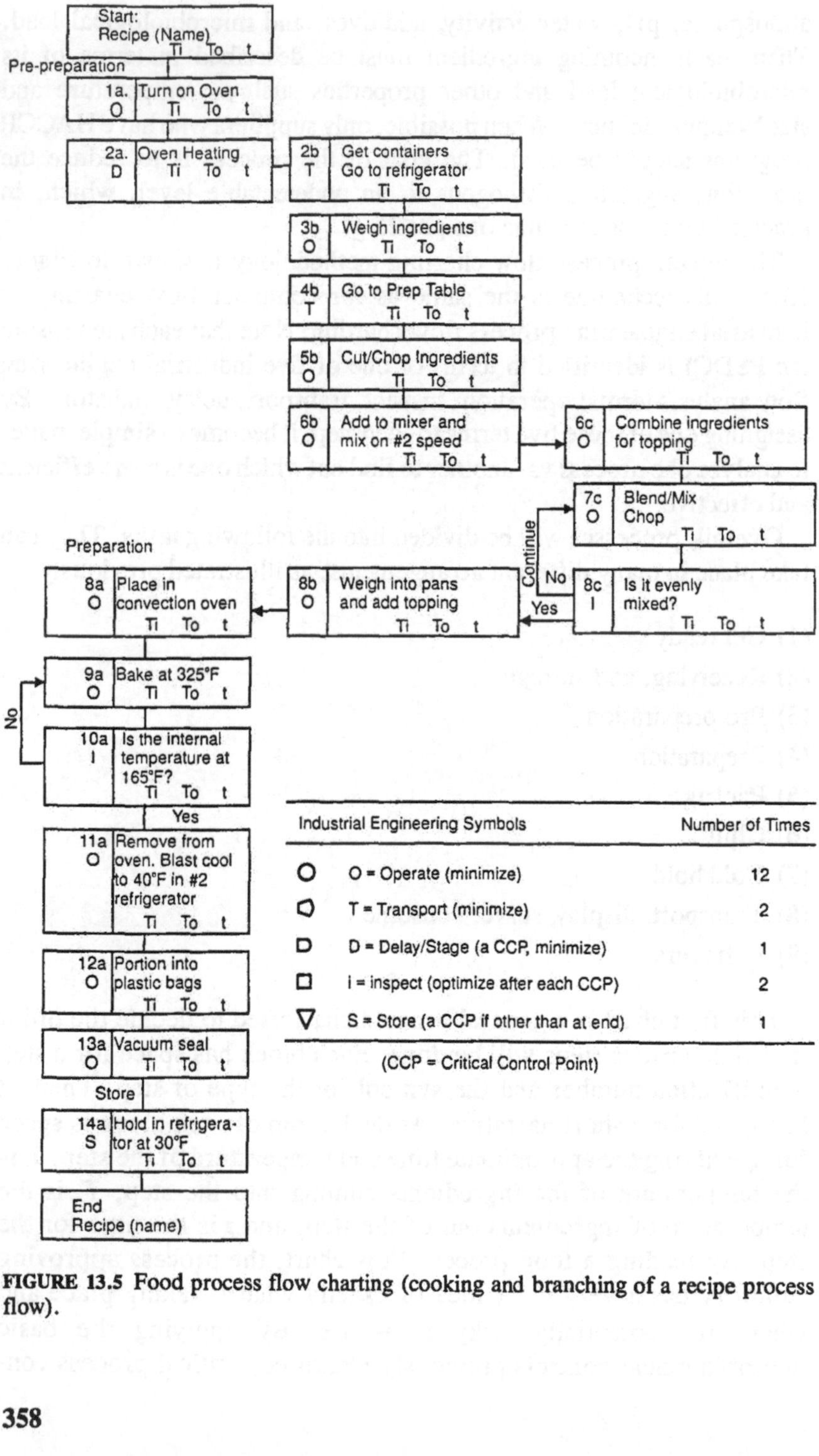

Industrial Engineering Symbols	Number of Times
○ O = Operate (minimize)	12
T = Transport (minimize)	2
D = Delay/Stage (a CCP, minimize)	1
□ i = inspect (optimize after each CCP)	2
▽ S = Store (a CCP if other than at end)	1

(CCP = Critical Control Point)

FIGURE 13.5 Food process flow charting (cooking and branching of a recipe process flow).

trol steps are easily identified. Note the use of standard decision yes/no blocks at inspection points where line employees make control decisions.

MAP-Sous Vide Processes and Controls

Figure 13.6 provides an overview of these nine tasks in terms of MAP-sous vide process controls and monitoring. These are general process controls and monitoring criteria. Note, this list presents only the obvious controls. Many accidental chemical and physical hazards cannot be identified until the operating process and process area are observed.

Pasteurized Food Process Standards

While there is a great deal of consumer interest in "fresh" food, the fresher a processed item is, the more it costs. It will never be as fresh as when the consumer performs the labour and consumes it immediately. For example, "fresh melon" is when the customer cuts the melon open, removes the seeds, and eats it immediately. Of course, if the melon just came out of the consumer's garden, this is even better for freshness. The same thinking can be extended to meat, poultry, fish, seafood, vegetables, etc. Fresh does not necessarily mean safer. It means mostly, lower spoilage microbiological counts and fewer chemical and enzymatic changes.

If there is to be more than a few days of shelf life, the food must be preserved/pasteurized to reduce the spoilage bacteria sufficiently to achieve the desired shelf life and reduce the vegetative pathogens to a safe level.

The canning industry in the 1920s developed good processing procedures to assure the safety of food. It was understood that pathogen destruction was an integrated lethality of time and temperature of the center of the food, when the center was in the zone of one z-value below the process end point temperature. Figure 13.7 illustrates this point. The lined area is the integrated lethality.

The U.S. government has not applied integrated lethality methodology to specifying lethality in pasteurized foods. The French government has. It is the scientifically correct way to control lethality, and anyone choosing to manage lethality should do the calculation. The current

Tasks and Controls	Monitoring
GET READY	
• Comply with GMPs where more strict than the following controls.	Observation; policies, procedures, and standards
• Food contact surface: subjected to a process that reduces APC/cm^2 >10^6 to <2 APC/cm^2	Swab
• Non-food contact surfaces: subjected to a process that reduces APC/cm^2 >10^6 to <10 APC/cm^2	Swab
• Air, if it must be environmentally controlled to clean room standards of <1,000 particles/ft.^{3} and be laminar flow wall to wall when process is pasteurize-cool-assemble, and storage is >5 days	Microbiological air sampling
• Hands and fingertips 10^6 APC reduction following double wash with fingernail brush	Glove juice test
• Ingredients, as low as possible APC counts, none >10^6, many <10^4	Surface swabbing, rinsing
• Filtered, treated water, physical and chemical control from the city	pH, micropore filter
RECEIVE AND STORE	
• <40°F (4.4°C) <1 day desired, <3 day maximum	Labels on containers
PRE-PREPARATION	
• Opening packages/cans to prevent contamination: clean the plastic bags of cold pasteurized meats to be sliced and not pasteurized	Observation; PP&S; swabs
• Control physical (extraneous material) defects. Sort out all rocks, metal, etc.	Observation; PP&S
• Double wash and clean vegetables, cold, clean water with surface active agents to help release contamination, filter, UV to get APC reduction $\geq 10^{-4}$	Microbiological sampling
• Cl_2 or ClO_2 to reduce APC by 10^{-2}	Oxidation-reduction, microbiological sampling

FIGURE 13.6 MAP-sous vide process tasks controls and monitoring.

Tasks and Controls	Monitoring
• Blanch in water or acid if needed	Temperature, pH, microbiological sampling
• Cut, chop, grind: keep temperature <50°F (10°C)	Temperature, microbiological sampling
• Preservatives (to modify pH, a_w, Eh, chemical, biological attributes of food)	Appropriate chemical test: accurate scale
PREPARATION	
• Heat pasteurize $F^{5.6°C}_{60°C}$ = 12.1 min. ($F^{10°F}_{140°F}$ - 12.1 min.)	Time, food center temperature, food thickness
10^3 cook, chill, hold, and serve <5 days to immuno-competent people	
10^7 cook, chill, hold, and serve >5 days to immuno-compromised people	
• Acid: 10^7 *Salmonella* spp. reduction	pH/titratable acidity, time, temperature
• Fermentation: 10^7 *Salmonella* spp. reduction	Microbiological counts, pH, time, temperature
PACKAGE	
• Pinholes, seals (heat or mechanical)	Visual, dyes, gas leak tests
• MAP	Gas analysis, microbiological sampling
• H_2O_2 to clean package and package filling equipment as it operates	Swabs, chemical tests
CHILL	
• <11 h to 40°F (4.4°C) for safety	Media, temperature, velocity
• Water bath 32°F (0.0°C), 5 ppm free Cl_2	Oxidation reduction potential
• Blast cooler <35°F (1.7°C)	
COLD HOLD	
• <5 d <40°F (<4.4°C)	Temperature and visual checks of labels
• 6 to 21 d <38°F (<3.3°C)	
• 22 to 45 d <32°F (<0.0°C)	
• >45 d <32°F (<0.0°C)	
TRANSPORT, DISPLAY, SERVE, CONSUME	
• Time and temperature (less than 5 generations multiplication *Listeria monocytogenes*) from production to consumption	Microbiological time-temperature histories
• Heating is not assumed necessary for safety	
• If heated for service, hot hold for <30 min	Time
• Warn customers not to abuse food	
• Ingredient labels to warn about allergenic ingredients	Visual
LEFTOVERS	
• Only re-serve once	Labeling of leftovers

FIGURE 13.6 (continued) MAP-sous vide process tasks controls and monitoring.

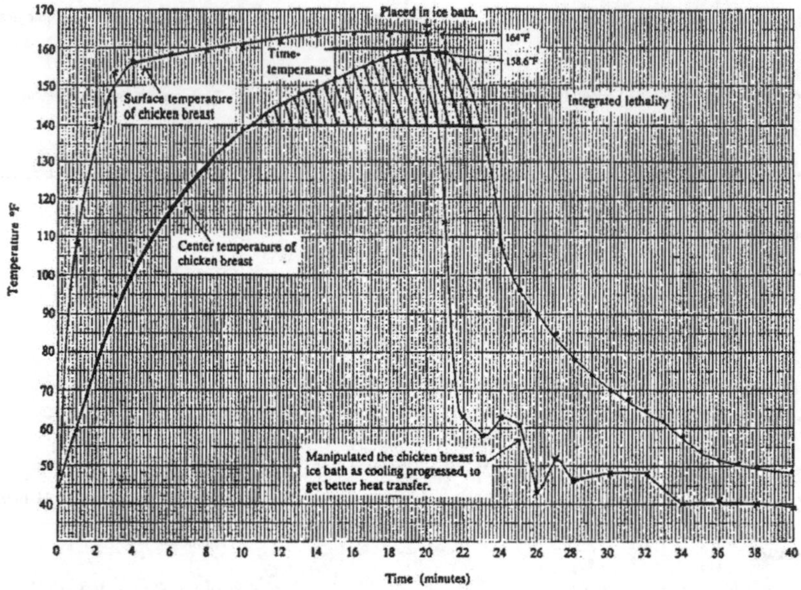

FIGURE 13.7 Heating and cooling curve for a vacuum-packed chicken breast cooked in a water bath.

pasteurization requirements are shown in Table 13.13. Observe the wide variation in requirements. A reason for most variation is that these standards have been based on the actual processes that were in existence when the government stepped in to set standards. Most are not based on sound microbiological logic and thorough research. They are "overkills." A long-term goal of the government working with industry is to set less stringent safety standards which then permit the food to be fresher.

Step Analysis

To do a process HACCP, the process must be specified to the step level. A HACCP team with expert knowledge and experience is assembled and the team studies the process and steps. Each step is described in sufficient detail so that all process variables are identified. If the process is basic, and the time and temperature are the only control variables, then items such as pH, a_w, and Eh can be ignored.

Recall that to do an analysis for all BCP hazards at a location, it is necessary to specifically identify the location in the processing area so that environment, facilities, and equipment variables can be checked. The equipment must be known in order for equipment variables to be

Table 13.13. Pasteurization standards.

Food	Microorganism	Reduction	*T*/*T*–*t* Int	Temp. [°F (°C)]	*D* (min)	*P*	z-Value [°F (°C)]
United States							
Cooked beef and roast beef (USDA, 1991d)	*Salmonella* spp.	10^7	*T*–*t*	140 (60.0)	1.73*	12.1	10 (5.6)
Baked meat loaf (USDA, 1991a)	NS	NS	*T*	160 (71.1)	—	—	—
Baked pork cut (USDA, 1991a)	NS	NS	*T*	170 (76.7)	—	—	—
Pork, to eliminate *Trichina* (USDA, 1991c)	*Trichina*	NS	*T*	140 (60.0)	—	1*	6.77 (3.76)*
Uncured poultry (USDA, 1991e)	NS	NS	*T*	160 (71.1)	—	—	—
Cooked salted duck (USDA-FSIS, 1991)	NS	NS	*T*	155 (68.3)	—	—	—
Jellied chicken loaf (USDA-FSIS 1991)	NS	NS	*T*	160 (71.1)	—	—	—

(continued)

Table 13.13 (continued).

Food	Microorganism	Reduction	T/T − t Int	Temp. [°F (°C)]	D (min)	P	z-Value [°F (°C)]
Partially cooked comminuted product (USDA-FSIS, 1985)	NS	NS	T − t	151 (66.1) 140 (60.0)*	— —	1 12.6*	— 10 (5.6)*
Hamburger (FDA, 1993)	*E. coli*	10^5	T − t	140 (60.0)	1.67	8.34	8.3 (4.61)*
France							
Ready meals hold <21 d at ~3°C (~37.4°F)	*Enterococcus faecalis*	10^{100}	Int	158 (70.0)	1.0	100	18 (10)
Ready meals hold <42 d at ~3°C (~37.4°F) (Veterinary Service for Food Hygiene, 1988)	*Enterococcus faecalis*	$10^{1,000}$	Int	158 (70.0)	1.0	1,000	18 (10)
Eggs, whole and pH 7 whites (USDA-ARS, 1969)	*Salmonella* spp.	$10^{8.75}$	T − t	140 (60.0)	0.4	3.5	8 (4.4)

*Calculated from given time-temperature data.
P = The pasteurization value or total time at temperature for the reduction specified.
NS = Not specified.
T − t = Temperature at a fixed time.
T = Temperature only; time is assumed instant.
Int = Integrated lethality.

checked. The source of the ingredients and the person doing the task must also be evaluated. A general scheme for step analysis applying the hazard and control logic previously presented for water bath pasteurization of vacuum packages of chicken breast can be seen in Figure 13.8.

SECTION IV— IMPLEMENTING HACCP

The Government Approach to HACCP

Up to this point, HACCP has been wrongly viewed as a government program rather than a company zero-liability-cost food safety program. The conventional government approach does not include a number of essential components in terms of what is necessary for an effective, continuous safety performance improvement process (risk reduction). For example, there really is no requirement for the management component in government HACCP. The government approach focuses only on process analysis, establishment of fixed control procedures, and records for government inspection. Records are really for the operator to use for continuous quality improvement. The current seven steps as listed in the latest publication of The National Advisory Committee on Microbiological Criteria for Foods (1992) are as follows.

(1) Conduct a hazard analysis.

(2) Identify the CCPs in the process.

(3) Establish critical limits for preventive measures associated with each identified CCP.

(4) Establish CCP monitoring requirements.

(5) Establish corrective action to be taken when monitoring indicates that there is a deviation from an established critical limit.

(6) Establish effective record-keeping procedures that document the HACCP system.

(7) Establish procedures for verification that the HACCP system is working correctly.

While the document provides many good ideas, it simply will not work as prescribed because it is incomplete and all food processes are dynamic. For instance, while there may be specifications for ingredients, there must be adaptability because the quality and safety of fresh ingredients that are key to MAP-sous vide are highly subject to

Prior steps

↓

Step No.	Procedures
55A	Heat the bags of chicken in the cooking rack for 20 min.
Process Type O	

Food Start Temp (7.2ºC) 45ºF	Food Item Name Chicken breast	Job Area Location Tank
Food End Temp (70.0ºC) 158ºF	Wt. Added/ Lost in Step ---	Employee Time 30 min.
Temp around the Food (71.1ºC) 160ºF	Wt. Ouput of Step 5 oz.	Preservative ---
Time to Perform Step 20 min.	Equip. Item Used Tank	Eh ---
Food Center Surf. Dist. inch <0.75	Equip. Use Time 20 min.	

pH	aw	Food Viscosity	
---	---	---	

↓

1. Conduct a FMEA of the environment, facilities, equipment, supplies, process, personnel, and customers to look for problems. As is immediately obvious, this analysis will also find opportunities to improve the quality and productivity of the process. This is why HACCP is merely the safety component of total process control. Use guide words such as NONE, MORE, LESS, OTHER THAN to test for failure of each control.
2. Assume that this step is the pasteurization step for a $7 \log_{10}$ *Salmonella* spp. reduction in a chicken breast vacuum sealed in a plastic pouch and immersed in water at 160°F (71.7°C). Document the following. (This has been abbreviated for this example.)

FIGURE 13.8 Process step description card example.

Deviation	Consequence	ROOT Cause	Removal	Prevention and Action to Improve
Environment				
Power goes off				
Facilities				
Room is 15°F (8.3°C) above normal				
Equipment				
Thermostat is out of calibration				
Chart recorder is not working				
Water circulating pump failed				
Supplies				
Chicken has 10^3 *Salmonella* spp. per gram, not 10^1				
Plastic pouch has a pinhole				
Pouch seal has a defect				
Personnel				
Not trained				
Has diarrhea and a cold				
Process				
Temperature set incorrectly				
Pouches touching each other				

3. Then the consequence is evaluated. If it has an insignificant consequence, stop and go to the next item.
4. To find the ROOT cause of problems, it is usually sufficient to ask a series of why-why questions of the HACCP expert team until the ROOT cause is identified. Any other analytical means that is appropriate can also be used. The WHY-WHY METHOD is approximately the same process as doing a cause-effect analysis.

FIGURE 13.8 (continued) Process step description card example.

Effect	Why	Why	Etc.	Removal	Prevention
Power goes off →	→Fuse blows on equipment	→ Fuse undersized	. . .		
→	→City power fails	→ Snow storm	. . . →		

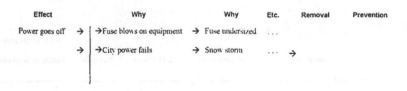

5. Next, for every significant problem, the HACCP team decides how to remove the problem. Experiments are conducted to find the best solution. The action plan is documented in terms of who will do what, where, when, how, to achieve what goal.
6. Prevention is the final step. The prevention is documented in the operations policies, procedures, and standards manual. Prevention, as is now obvious to the reader, will be founded in educating all line employees to follow instructions with zero defects (creating zero errors). To have an effective HACCP program, it is essential that there be a fully effective employee-centered Total Quality Management program.

FIGURE 13.8 (continued) Process step description card example.

weather conditions, capabilities of the grower/harvester, transporter, distributor, etc., and vary greatly. MAP and sous vide processes normally use fresh ingredients supplied just in time to minimize levels of spoilage microorganisms. As with raw milk that is pasteurized, the process might need to be adjusted with every new batch of ingredients, to compensate for changes in numbers of spoilage bacteria or thickness of the package in a sous vide process.

Why a Goal of Zero Defects?

The goal of any food processor (e.g., wholesale, retail, home operation) is customer satisfaction. If the "company" does not create repeat customer sales, it goes out of business. One attribute of product quality is that the food will not make the consumer ill. However, many times the consumer is actually selecting products based on value attributes such as least cost to provide healthy nutrition, and food safety is assumed. If zero safety defects is not the company's objective, what other stated food safety goal by the company president is possible? Can the food provider say, "I will make only one in 100,000 people who eat my food ill, and only one in ten of them will die?" Of course not. Anyone who prepares food must strive for zero defects.

Process Quality Goals—Safety, Customer Satisfaction, and Process Productivity

To do a HACCP plan, one must first at least have in mind a process in a facility with employees, facilities, equipment, supplies, and processes, etc., to analyze. In order to develop a zero-defect hazard control program, after all of the hazards are known and first-generation controls are specified, it is necessary to look at every possible mode of process variation and control failure at each step in the process. An example would be paint chips falling from the ceiling and into a cooking kettle because of a lack of facility maintenance. Initially in process development, the process developer simply works on designing a process which will produce a product that will meet customer expectations, yet stay within management cost constraints. Regulations are reviewed by the developer to assume minimum government standards are met. However, many potential hazards are not evident until a specific process is analyzed in operation.

Sometimes the government functions in a consultative and supportive role in terms of process development and improvement. In this case, regulators can have access to all steps in the process flow diagram. Yet sometimes, the government operates in a punitive and purely regulatory role. In this case, regulators will not be providing helpful information, and probably should only be given information related to the control requirements in their regulations. This is done by simply giving the government a generalized flow diagram, identifying the government-required control steps, and listing them individually on a form such as shown in Figure 13.9.

SECTION V—AN INDUSTRY-BASED HACCP FOOD SAFETY SYSTEM

Technology + Management = Total Quality Management

In any Total Quality Management (TQM) process, a fundamental rule is that everyone will communicate problems by using data, never guesses. Dr. Deming, the statistical "guru" of our times, has stated very properly, that judgment and visual assessments are almost always incorrect. Therefore, at the heart of an effective zero-defect HACCP-TQM program is a sound statistical process control program that collects factual data about the variables in the production processes. Causes of

Process Step (Describe)	Hazard or Regulatory Control	Biological, Chemical, Physical Hazards, or Regulatory Requirement	Critical Limits	Monitoring Procedures; Frequency; Person(s) Responsible	Corrective Action(s); Person(s) Responsible	Process Record	Verification Procedure; Person(s) Responsible

FIGURE 13.9 Process step HACCP listing.

problems can be separated into two categories: common causes and assignable causes. Deming has pointed out that almost 85% of causes of problems are common causes. They are system causes created by management, and the worker is powerless to do anything about them. Approximately 15% of production problems are assignable causes created by the unit operation: the machine, the worker, the supplies, the facilities, or the environment.

Figure 13.10 is a flow diagram depicting the four components of a TQM program cycle.

Statistical process control is the technically correct way that the line operator keeps a process on target, anticipates the need for correction, makes corrections, and avoids customer dissatisfaction and government recall problems. One requirement of the current government-described HACCP programs is that an operator must keep records of the process for the government to review. The entire reason for records to be kept is not for the government to find fault with the operator, but so that the operator can perform continuous quality improvement, in keeping with the modern concepts of statistical process control.

The Management Components of a HACCP-Based TQM System

There are six important components in evaluating whether a TQM program is capable of striving for zero process safety defects:

(1) Owner commitment to zero defects in safety and customer satisfaction
(2) Hazard analysis and control
(3) Written program
(4) Communications and training so that line operators are capable of zero defects
(5) Continuous measurement and reaction to process deviations to keep the system in control
(6) Rewards and enforcement of operating policies, procedures, and standards

When the TQM system functions correctly, the following occur or are present:

(1) Teamwork and trust among all personnel, and regulatory-company trust.

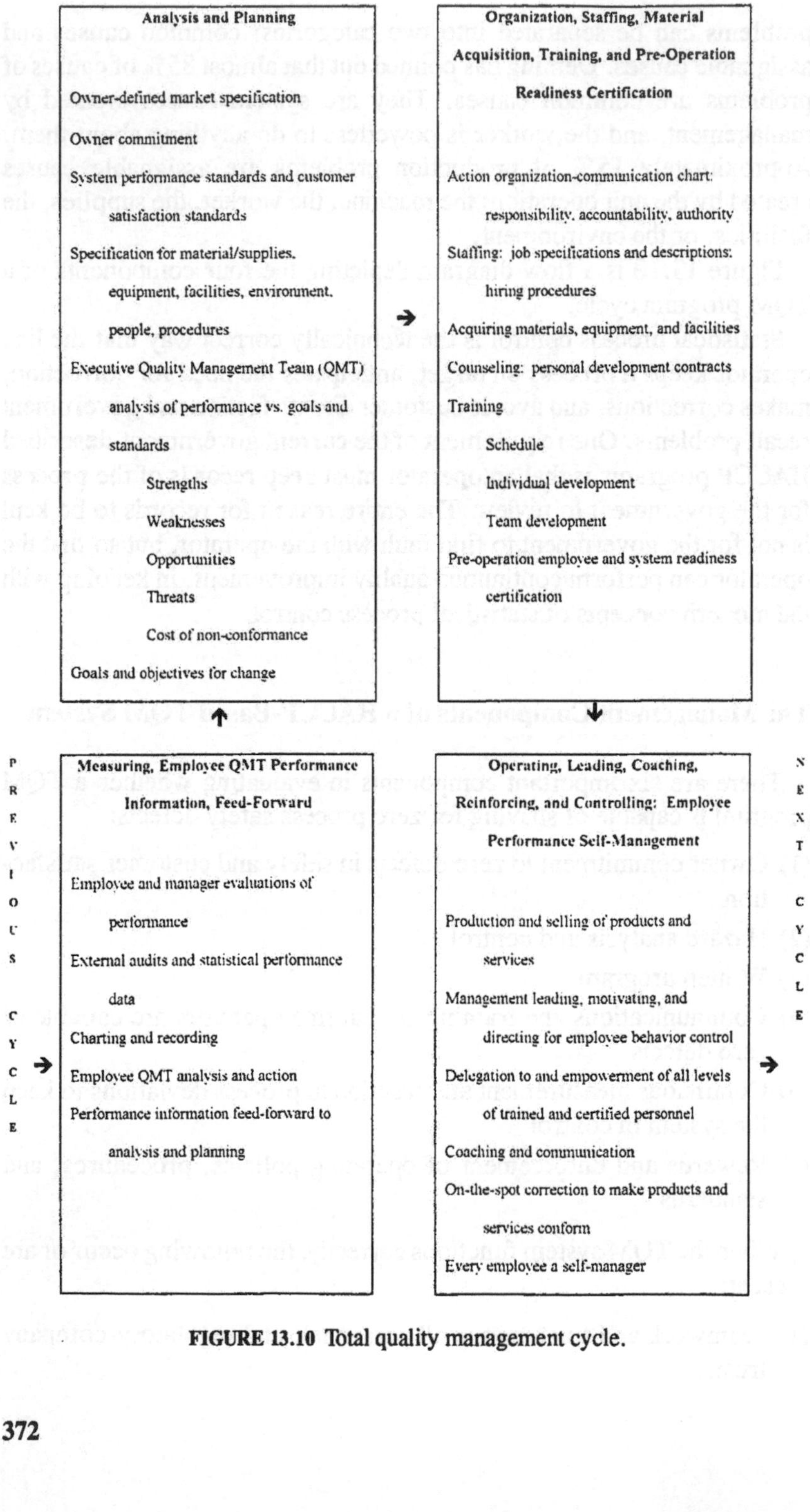
Analysis and Planning
Owner-defined market specification
Owner commitment
System performance standards and customer satisfaction standards
Specification for material/supplies, equipment, facilities, environment, people, procedures
Executive Quality Management Team (QMT) analysis of performance vs. goals and standards
Strengths
Weaknesses
Opportunities
Threats
Cost of non-conformance
Goals and objectives for change
Organization, Staffing, Material Acquisition, Training, and Pre-Operation Readiness Certification
Action organization-communication chart: responsibility, accountability, authority
Staffing: job specifications and descriptions: hiring procedures
Acquiring materials, equipment, and facilities
Counseling: personal development contracts
Training
Schedule
Individual development
Team development
Pre-operation employee and system readiness certification
Measuring, Employee QMT Performance Information, Feed-Forward
Employee and manager evaluations of performance
External audits and statistical performance data
Charting and recording
Employee QMT analysis and action
Performance information feed-forward to analysis and planning
Operating, Leading, Coaching, Reinforcing, and Controlling: Employee Performance Self-Management
Production and selling of products and services
Management leading, motivating, and directing for employee behavior control
Delegation to and empowerment of all levels of trained and certified personnel
Coaching and communication
On-the-spot correction to make products and services conform
Every employee a self-manager
PREVIOUS CYCLE
NEXT CYCLE

FIGURE 13.10 Total quality management cycle.

(2) Anticipation and prevention strategies, rather than inspection to find problems in finished products
(3) Striving for zero defects which drives the defects to such a low level that destructive product testing is totally unnecessary
(4) Virtually no risk in customer satisfaction or safety
(5) Precise measures to indicate to what degree the process is in control
(6) Full operator control of hazards

In a HACCP program that is capable of zero-defects, the following is true:

(1) There is management leadership and commitment, and adequate prevention resources are allocated.
(2) Hazard control is based on expert knowledge and data.
(3) There is realization that zero defects are achieved through employee behavioral management, and that antecedents and consequences are critically important. There is employee training and certification, and the employees are empowered to perform process control and strive for zero process deviations leading to defects.
(4) Supervisors are responsible for process pre-control, which means making it possible for employees and the process to perform with zero defects.
(5) There is measurement of the process using standard statistical process control methods, and there is immediate reaction to any deviation to keep the process in control.
(6) The system is simple.
(7) There is continuous improvement to reduce the probability of deviation beyond quality-owner control limits.
(8) All suppliers are partners with the producer, in striving for zero process defects.
(9) There is long-term commitment by management to employee education, continual improvement in quality of worklife, and empowerment.

An Integrated Industry Approach to HACCP

The following is an integrated industry approach to a zero safety defect program. It is adapted from the USDA TQC (USDA-FSIS, 1984), (USDA, 1991b) and NACMCF (1992) material. But it also includes the concepts of statistical process control for defect prevention as discussed in books such as *Defect Prevention: Use of Simple Statistical Tools* (Kane, 1989), *Statistical Quality Control Handbook* (AT&T Technologies, 1984), and *Quality Control Handbook* (Juran et al., 1979). The following is done by the operator. The government's role is to first and foremost, provide advice and then certification that a HACCP plan is adequate.

PLAN, DEVELOP, IMPLEMENT, AND VERIFY A HACCP PLAN

Assemble the HACCP Team

The team must be able to:

(1) Identify defects in the system processes that could lead to hazards. The system elements include: management, consumer, personnel, environment, facilities, equipment, supplies, and products.
(2) Specify limits where a defect becomes a hazard.
(3) Assign levels of risk (probability × consequences).
(4) Recommend controls (policies, procedures, and standards), procedures for monitoring (using statistical process control), and verification.
(5) Recommend appropriate corrective action if a deviation occurs.
(6) Recommend research if information is missing.
(7) Know the weak points in the HACCP plan and predict its risk of hazardous deviation.
(8) Describe the food products and methods of distribution and sales.
(9) Identify the intended use, consumer immune level, and potential abuse of the food.
(10) Develop a flow diagram that describes how ingredients are converted into finished product.
(11) Verify the flow diagram.

Conduct the Hazard and Control Analysis

(1) Identify the food system to be HACCP'd.
- Consider all steps from growing, harvesting, processing,

manufacturing, distributing, merchandising, to consumer consumption of the product.

- Describe the food products and methods of distribution and sales.
- Identify the intended use, consumer (immuno-competent or immuno-compromised), and potential abuse of the food.
- Develop a flow diagram that describes how ingredients are converted into finished product.
- Verify the flow diagram.

(2) Conduct a Failure Mode and Effect Analysis (FMEA) of each step in the flow process.
  - Identify the steps where defects can become hazards. Specify the limits at which defects become hazards. Conduct experiments as necessary to verify that limits, while minimal, are effective.
  - Do a FMEA and identify how defects may become hazards.

(3) If there is not adequate hazard prevention, find root causes and then control or eliminate variables that would allow a process defect to become a hazard.

(4) Revise the process written operating policies, procedures, and standards as necessary to prevent the defect or to prevent the loss of control that would allow the defect to become a hazard.

Establish Operational Control

(1) Establish employee process measurement and control procedures and train employees to keep deviations within safe limits and defects within specified tolerances.

(2) Establish effective statistical process control documents/charts that help the line operator to anticipate unacceptable deviation and provide data for continuous quality improvement. For example:
◑1▣ Employee control.
◑2▣ Employee control. Notify supervisor. Take action to improve the process.
◑3▣ Supervisor control. Take immediate action to improve the process.
Note that at ◑3▣ the deviation is still far below the point, perhaps another 3▣, at which the government would recall the product or shut down a line.

(3) Establish action to be taken when measurements indicate that there is a process deviation beyond a hazard limit.

Document the Plan

(1) Ingredients
- supplier certification documenting compliance with processor's specification
- processor audit records verifying supplier compliance
- storage temperature record for temperature-sensitive ingredients
- storage time records of limited shelf life ingredients

(2) Records relating to product safety
- sufficient data and records to establish the efficacy of barriers in maintaining product safety
- sufficient data and records establishing the safe shelf life of the product; identify if age of product can affect safety
- documentation of the adequacy of the processing procedures from a knowledgeable process authority

(3) Processing
- records from all monitored Critical Control Points (CCPs).
- records verifying the continued adequacy of the processes, that is, process stability

(4) Packaging
- records indicating compliance with specifications of packaging materials
- records indicating compliance with sealing specifications

(5) Storage and distribution
- temperature records
- records showing no product shipped after shelf life date on temperature-sensitive products

(6) Deviation and corrective action records

(7) Validation records and modification to the HACCP plan indicating approved revisions and changes in ingredients, formulations, processing, packaging and distribution control, as needed

(8) Employee training records

Establish Procedures for Verification by the Operator That the HACCP System Is Working Correctly

(1) Verification procedures may include:

- establishment of appropriate verification inspection schedules.
- review of the HACCP plan
- review of CCP records
- review of deviations and dispositions
- visual inspections of operations to observe if CCPs are under control
- random sample collection and analysis
- review of critical limits to verify that they are adequate to control hazards
- review of written record of verification inspections which certifies compliance with the HACCP plan or deviations from the plan and the corrective actions taken
- validation of HACCP plan, including on-site review and verification of flow diagrams and CCPs
- review of modifications of the HACCP plan

(2) Verification inspections should be conducted:

- routinely, or on an unannounced basis, to assure selected CCPs are under control
- when it is determined that intensive coverage of a specific commodity is needed because of new information concerning food safety
- when foods produced have been implicated as a vehicle of foodborne disease
- when requested on a consultative basis or established criteria have not been met
- to verify that changes have been implemented correctly after a HACCP plan has been modified

(3) Verification reports should include information about:

- existence of a HACCP plan and the person(s) responsible for administering and updating the HACCP plan
- the status of records associated with CCP monitoring
- direct monitoring data of the CCP while in operation
- certification that monitoring equipment is properly calibrated and in working order
- deviations and corrective actions
- any samples analyzed to verify that CCPs are under control, which may involve biological, chemical, and physical, or organoleptic methods
- modifications to the HACCP plan

- training and knowledge of individuals responsible for monitoring CCPs
- commitment by management to zero safety defects, adequate resources, and enforcement

ORGANIZE FOR ZERO DEFECTS

(1) Develop an action organization-communications chart and job descriptions showing who has what process task responsibilities, accountabilities, and authority.
(2) Staff to meet all needs. Ensure that people have the capacity to grow.
(3) Certify all suppliers.
(4) Acquire materials, equipment, and facilities, and improve existing processes.
(5) Train and improve the performance of all personnel to be able to master the improved system and process.
(6) Conduct a full pre-operation employee and system readiness certification.

OPERATE AND CONTROL FOR ZERO DEFECTS

(1) Produce and sell products to consumers who, in turn, consume the products.
(2) Management provides leadership, motivation, and coaching for employee behavior management.
(3) Employees exercise self-control based on their training.
(4) Empower the line employee, who is trained and performance-certified, to stop production when the process is out of "control."
(5) Supervisors coach and communicate.
(6) Every employee is a process-step manager.

MEASURE FOR ZERO DEFECTS

(1) Each employee collects performance data on the process control charts and data sheets regarding his/her process tasks and steps.
(2) The employee notes causes of process deviations.

(3) Managers evaluate process and system performance with checklists.
(4) External expert audits are conducted.
(5) Charts and recordings are made.
(6) Employee process Quality Management Teams (QMTs) meet and act.
(7) Information is forwarded to the executive TQM team for improvement of the processes and system.

SUMMARY

HACCP is not new. It is the part of total quality management that controls customer liability. (Note, the methods are just as applicable to employee injury prevention.) There are three classes of hazards that can hurt or kill customers: biological, chemical and physical. They are based on actual liability to the company. Because MAP-sous vide can be done by anyone in the food system from wholesale processors, through the retail food producers, to the home, it has been and will continue to be a major factor in bringing "fresher" and safer foods to the consumer. It cannot be expected that pathogen contamination of raw food will be reduced for many years. MAP-sous vide represents a major, immediate solution to the problem of contaminated food. The one condition is that it must be done correctly. The first result of lack of knowledge and management control is poor to very poor quality food. All experience today indicates that people tend to overprocess when they do not have controls. This results in poor quality and customer dissatisfaction, and operators return to cook-and-serve systems.

As is evident from this chapter, government personnel will need an in-depth knowledge of MAP-sous vide processes, since they are responsible for verifying that all hazards are adequately controlled. A key need, if the technology is to provide its maximum benefit, is for the regulatory agencies to begin an intensive program to develop minimum safe limits and provide requirements for management control, since these are for the most part missing.

REFERENCES

AT&T Technologies. 1984. *Statistical Quality Control Handbook.* AT&T Technologies, Indianapolis, IN.

Bryan, F. L. 1990. "Application of HACCP to ready-to-eat chilled foods," *Food Technol.*, 45(7):70–77.

Center for Chemical Process Safety. 1992. *Guidelines for Hazard Evaluation Procedures: Second Edition with Worked Examples.* American Institute of Chemical Engineers, New York.

Chia, J. K., Clark, J. B., Ryan, C. A. and Pollack, M. 1986. "Botulism in an adult associated with food-borne intestinal infection with *Clostridium botulinum*," *New Engl. J. Med.*, 315:289–241.

Chilled Foods Association (CFA). 1990. *Technical Handbook for the Chilled Foods Industry.* CFA, Atlanta, GA.

Food and Drug Administration (FDA) (HHS). 1991. 21CFR (Code of Federal Regulations). *189 Substances Prohibited from Use in Human Food.* Office of the Federal Register, National Archives and Records Administration, Washington, D.C., pp. 508–512.

International Commission on Microbiological Specifications for Foods (ICMSF). 1986. *Microorganisms in Foods: Vol. 2: Sampling for Microbiological Analysis: Principles and Specific Applications*, 2nd ed. University of Toronto Press, Toronto.

International Commission on Microbiological Specifications for Foods (ICMSF). 1988. *Microorganisms in foods. 4. Application of the Hazard Analysis Critical Control Point System to Ensure Microbiological Food Safety and Quality.* Blackwell Scientific Publications, London.

Johnson, W. G. 1980. *MORT Safety Assurance Systems.* Marcel Dekker, Inc., New York.

Juran, J. M., Gryna, F. M., and Bingham, R. S., eds. 1979. *Quality Control Handbook.* McGraw-Hill Book Company, New York.

Kane, V. E. 1989. *Defect Prevention: Use of Simple Statistical Tools.* Marcel Dekker, Inc., New York.

Levine, M. M., DuPont, H. L., Formal, S. B., Hornick, R. B., Takeuchi, A., Gangarosa, E. J., Snyder, M. J. and Libonati, J. P. 1973. "Pathogenesis of *Shigella dysenteriae* 1 (Shiga) dysentery," *J. Infect. Diseases*, 127(3):261–270.

Microbiology and Food Safety Committee of the National Food Processors Association. 1992. "HACCP and total quality management–winning concepts for the 90's: A review," *J. Food Protect.*, 55(6):459–462.

National Academy of Services (NAS). 1985. *An Evaluation of the Role of Microbiological Criteria for Foods and Food Ingredients.* NAS, National Research Council, National Academy Press, Washington, D.C.

National Advisory Committee on Microbiological Criteria for Foods (NACMCF). 1990a. *HACCP Principles for Food Production.* USDA-FSIS Information Office, Washington, D.C.

National Advisory Committee on Microbiological Criteria for Foods (NACMCF). 1990b. *Recommendations of the National Advisory Committee on Microbiological Criteria for Foods for Refrigerated Foods Containing Cooked, Uncured Meat or Poultry Products that are Packaged for Extended Refrigerated Shelf Life and That Are Ready-to-Eat or Prepared with Little or No Additional Heat Treatment* Washington, D.C.

National Advisory Committee on Microbiological Criteria for Foods (NACMCF). 1992. *Hazard Analysis and Critical Control Point System.* USDA-FSIS, Washington, D.C.

Pierson, M. D. and Corlett, D. A., eds. 1992. *HACCP Principles and Applications*. Van Nostrand Reinhold, New York.

Pillsbury Company. 1973. "Development of a food quality assurance program and the training of FDA personnel in hazard analysis techniques," Contract No. FDA 72-59. The Pillsbury Company. Minneapolis, MN.

Rippen, T. E. and Hackney, C. R. 1992. "Pasteurization of seafood: Potential for shelf-life extension and pathogen control," *Food. Technol.* 46(12):88–94.

Shapton, D. A. and Shapton, N. F. 1991. *Principles and Practices for the Safe Processing of Foods*. Butterworth-Heinemann Ltd., Oxford, England.

Shaughnessy, H. J., Olsson, R. C., Bass, K., Friewer, F. and Levinson, S. O. 1946. "Experimental human bacillary dysentery," *J. A. M. A.*, 132(7):362–368.

Shigahisa, T., Nakagami, T. and Taji, S. 1985. "Influence of heating and cooling rates on spore germination and growth of *Clostridium perfringens* in media and in roast beef," *Jpn. J. Vet. Sci.*, 47(2):259–267.

Smith, J. L. 1990. *Risks Associated with Foodborne Pathogens*. USDA-ARS Eastern Reg. Res. Ctr., Philadelphia, PA.

Smittle, R. B. 1977. "Microbiology of mayonnaise and salad dressing: A review," *J. Food Protect.*, 40(6):415–422.

Snyder, O. P. 1993. *HACCP-Based Safety- and Quality-Assured Pasteurized-Chilled Food Systems*. Hospitality Institute of Technology and Management, St. Paul, MN.

Sperber, W. H. 1991. "The modern HACCP system," *Food Technol.*, 45(6):116–120.

Tjoa, W. S., DuPont, H. L., Sullivan, P., Pickering, L. K., Holguin, A. H., Olarte, J., Evans, D. G. and Evans, D. J. 1977. "Location of food consumption and travelers' diarrhea," *Am. J. Epidemiol.*, 106(1):61–66.

United States Department of Agriculture (USDA). 1991a. 9CFR (Code of Federal Regulations) 317.8 False or misleading labeling or practices generally; specific prohibitions and requirements for labels and containers. Office of the Federal Register. National Archives and Records Administration. Washington, D.C., pp. 183–188.

United States Department of Agriculture (USDA). 1991b. 9CFR (Code of Federal Regulations) 318.4. *Preparation of Products to Be Officially Supervised; Responsibility of Official Establishments; Plant Operated Quality Control.* Office of the Federal Register, National Archives and Records Administration, Washington, D.C., pp. 196–200.

United States Department of Agriculture (USDA). 1991c. 9CFR (Code of Federal Regulations) 318.10. *Prescribed Treatment of Pork and Products Containing Pork to Destroy Trichinae.* Office of the Federal Register, National Archives and Records Administration, Washington, D.C., pp. 217–225.

United States Department of Agriculture (USDA). 1991d. 9CFR (Code of Federal Regulations) 318.17. *Requirements for the Production of Cooked Beef, Roast Beef, and Cooked Corned Beef.* Office of the Federal Register, National Archives and Records Administration, Washington, D.C., pp. 228–231.

United States Department of Agriculture (USDA). 1991e. 9CFR (Code of Federal Regulations) 381.150. *Requirements for the Production of Poultry Breakfast Strips, Poultry Rolls, and Certain Other Poultry Products.* Office of the Federal Register, National Archives and Records Administration, Washington, D.C., pp. 493–494.

United States Department of Agriculture Agricultural Research Service (USDA-ARS).

1969. *Egg Pasteurization Manual.* ARS 74-48, Poultry Laboratory of the Western Utilization Research and Development Division ARS, Albany, CA.

United States Department of Agriculture Food Safety and Inspection Service (USDA-FSIS). 1984. *Quality Control Guidebook* (Agriculture Handbook Number 612). U.S. Government Printing Office, Washington, D.C.

United States Department of Agriculture Food Safety and Inspection Service (USDA-FSIS). 1985. *Partially Cooked Comminuted Product.* FSIS Notice 92-1985, Washington, D.C.

United States Department of Agriculture Food Safety and Inspection Service (USDA-FSIS). 1991. *Standards and Labeling Policy Book.* USDA, Washington, D.C.

Veterinary Service for Food Hygiene. 1988. *General Guidance on Food Processing.* French Ministry of Agriculture, Paris.

References to Table 7

1. DuPont, H. L., Hornick, R. B., Dawkins, A. T., Snyder, M. J. and Formal, S. B. 1969. "The response of man to virulent *Shigella flexneri* 2a," *J. Infect. Dis.*, 119:296–299.
2. DuPont, H. L., Formal, S. B., Hornick, R. B., Snyder, M. J., Libonati, J. P., Sheahan, D. G., LaBrec, E. H. and Kalas, J. P. 1971. "Pathogenesis of *Escherichia coli* diarrhea," *N. Engl. J. Med.*, 285:1–9.
3. DuPont, H. L., Hornick, R. B., Snyder, M. J., Libonati, J. P., Formal, S. B. and Ganarosa, E. J. 1972. "Immunity in shigellosis. II. Protection induced by oral live vaccine or primary infection," *J. Infect. Dis.*, 125:12–16.
4. Goepfert, J. M., Spira, W. M. and Kim, H. U. 1972. "*Bacillus cereus:* Food poisoning organism. A review," *J. Milk Food Technol.* 35:213–227.
5. Hauschild, A. H. W. 1973. "Food poisoning by *Clostridium perfringens*," *Can. Inst. Food Sci. Technol. J.*, 6(2):106–110.
6. Hobbs, B. C. 1960. "Staphylococcal and *Clostridium welchi* food poisoning," *Roy. Soc. Health J.*, 80:267–271.
7. Hornick, R. B., Greisman, S. E., Woodward, T. E., DuPont, H. L., Dawkins, A. T. and Snyder, M. J. 1970. "Typhoid fever: Pathogenesis and immunologic control," *New Engl. J. Med.*, 283:686–691.
8. Hornick, R. B., Music, S. I., Wenzel, R., Cash, R., Libonati, J.P., Snyder, M. J. and Woodward, T. E. 1971. "The broad street pump revisited: Response of volunteer to ingested cholera vibrios," *Bull. N.Y. Acad. Med.*, [2]47:1181.
9. Levine, M. M., DuPont, H. L., Formal, S. B., Hornick, R. B., Takeuchi, A., Gangarosa, E. J., Snyder, M. J. and Libonati, J. P. 1973. "Pathogenesis of *Shigella dysenteriae* I (Shiga) dysentery," *J. Infect. Dis.*, 127:261–270.
10. Lubin, L. B., Morton, R. D. and Bernard, D. T. 1985. "Toxin production in hard-cooked eggs experimentally inoculated with *Clostridium botulinum*," *J. Food Sci.*, 50:969–970, 984.
11. McCullough, M. B. and Eisele, C. W. 1951. "Experimental human salmonellosis. I. Pathogenicity of strains of *Salmonella meleagridis* and *Salmonella anatum* obtained from spray-dried whole egg," *J. Infect. Dis.*, 88:278–279.
12. McCullough, M. B. and Eisele, C. W. 1951. "Experimental human salmonellosis. III. Pathogenicity of strains of *Salmonella newport*, *Salmonella derby*, and *Sal-*

monella bareilly obtained from spray-dried whole egg," *J. Infect. Dis.*, 89:209–213.

13. McCullough, M. B. and Eisele, C. W. 1951. "Experimental human salmonellosis. IV. Pathogenicity of strains of *Salmonella pullorum* obtained from spray-dried whole egg," *J. Infect. Dis.*, 89:259–265.
14. Magee, P. N. 1983. "Nitrate," In *Environmental Aspects of Cancer. The Role of Macro and Micro Components of Foods*, E. L. Wynder, G. A. Leveille, J. H. Weisburger, and G. E. Livingston, eds., Food and Nutrition Press, Westport, CT. pp. 198–210.
15. Moustafa, M. K., Ahmed, A. A-H. and Marth, E. H. 1983. "Behavior of virulent *Yersinia enterocolitica* during manufacture and storage of colby-like cheese," *J. Food Protect.*, 46:318–320.
16. Newsome, R. L. 1988. "*Staphylococcus aureus*," *Food Technol.*, 42(4):194–195.
17. Robinson, D. A. 1981. "Infective dose of *Campylobacter jejuni* in milk," *Brit. Med. J.*, 282:1584.
18. Shaughnessy, H. J., Olsson, R. C., Bass, K., Friewer, F. and Levinson, S. O. 1946. "Experimental human bacillary dysentery," *J.A.M.A.*, 132:362–368.
19. Snyder, O. P. 1985. Personal communication.
20. Stern, N. J. 1982. "Foodborne pathogens of lesser notoriety: Viruses, *Vibrio*, *Yersinia*, and *Campylobacter*," In ABMPS Report No.125, pp. 57–63, National Academy of Science Press, Washington, D.C.
21. Taylor, S. L. and Bush, R. K. 1986. "Sulfites as food ingredients," *Food Technol.*, 40(6):47–52.

CHAPTER 14

Potential Use of Time-Temperature Indicators as an Indicator of Temperature Abuse of MAP Products

BIN FU–*University of Minnesota, U.S.A.*
THEODORE P. LABUZA–*University of Minnesota, U.S.A.*

TEMPERATURE ABUSE GENERATING POTENTIAL HAZARD

PRODUCERS of chilled foods are usually careful about maintaining the temperature of their products during processing. However, distributors and retailers as well as consumers do not always maintain the temperatures recommended by the food processor. Temperature abuse is common throughout the distribution and retail markets. The Fresh Chef® line of salads, and the soups and sauces of Campbell Soup Co. were pulled out of the market in 1987 due to lack of temperature control and subsequent abuse. Culinary Brands used Federal Express to deliver its fresh products to try to overcome this problem but finally withdrew from the marketplace. Kraft General Foods Inc. and General Foods (Culinova Brand) installed their own cooler cases at retailer locations for their modified atmospheric packaging (MAP) line to minimize temperature abuse but they also withdrew because of quality problems (Rice, 1989). Maintaining proper refrigeration of MAP products purchased by consumers is even worse because consumers are generally unfamiliar with new products and how they should be handled. Van Garde and Woodburn (1987) reported that the temperature in 21% of household refrigerators was higher than 10°C. Recent data suggested that 33% of retail refrigerated foods were held in display cases above 7°C and 5% were held above 13°C (Anon., 1989). Temperatures were even higher in southern market regions. It has been estimated that the average temperature of a home refrigerator is 10–13°C. Obviously, the current distribution system and consumers' knowledge of handling practices concerning chilled foods cannot meet the requirements for the development of extended shelf life refrigerated (ESLR) products utilizing MAP technology. Temperature abuse may occur at any point in the perishable food distribution chain and jeopardize the product. Temperature abuse will not only cause economic loss, but more importantly, there is a potential hazard of foodborne illness.

The use of MAP technology to extend the shelf life of perishable products has been with us for a long time. The safety and quality of MAP foods depend on product characteristics, processing and packaging procedures, and environmental conditions during distribution and storage (Table 14.1). Temperature is a primary factor for maintenance of the safety and quality of refrigerated MAP foods. Temperature abuse or lack of control in distribution, resulting in potential pathogen growth or toxin production, has limited the popularity of MAP foods in the U.S. (Labuza et al., 1992).

The recommended storage temperature range for most nonrespiring refrigerated foods (e.g., meat, chicken, fish) is 0–2°C, and for most respiring foods (e.g., apple, strawberry, mushroom), 0–5°C (Lioutas, 1988; Smith et al., 1990). In the U.K., refrigerated products are grouped in three general categories according to their relative safety and spoilage characteristics (Anon., 1990). Category I includes fresh meat, poultry, fish and the like; the recommended temperature range is −1 to +2°C. Category II includes precooked refrigerated foods, sandwiches, fresh mayonnaise, etc.; the recommended temperature range is 0–5°C. Category III includes fermented meats, fruits and vegetables, and the like; the recommended temperature range is 0–8°C. With the limitations of the current refrigeration systems and the poor control in the distribution chain, temperature fluctuations are usually unavoidable. The questions to ask are how can we minimize these fluctuations or how much fluctuation is practically acceptable?

Table 14.1. Factors affecting the safety and quality of chilled MAP foods.

Food Composition	Processing	Packaging	Environmental Conditions
pH, acidity and buffering power a_w, water content and hysteresis Eh (Redox potential) Inhibitors (natural and additive) Microbial flora distribution and competition Nutrients (lipid, protein, etc.) Surface/volume ratio	Heat Irradiation Washing Dipping Grinding	Headspace Permeability Leakage Interactions Vacuuming Gas flushing	Temperature and fluctuation Relative humidity and fluctuation Light intensity

The potential risk in chilled MAP foods due to temperature abuse comes from several sources: 1) some pathogens, such as *Clostridium botulinum* type E, *Yersinia enterocolitica*, enterotoxigenic *Escherichia coli*, *Listeria monocytogenes* and *Aeromonas hydrophila*, are capable of growth at ≥5°C in foods; 2) some pathogens, such as *Campylobacter jejuni* and *Brucella* can survive for long periods at refrigerated temperatures; 3) some pathogens such as *Salmonella, Staphylococcus aureus, Vibrio parahaemolyticus*, and *Bacillus cereus* can grow slightly above 5°C up to 12°C (typical abuse temperature for refrigerated foods) (Palumbo, 1986; Farber, 1991); and (4) *Listeria* species may survive and multiply at temperatures down to 0°C, especially when various hurdles limit the growth of competitors (Gill and Reichel, 1989). The extended shelf life of many MAP products may allow extra time for these pathogens to reach dangerously high levels in a food (Conner et al., 1989), because the hurdles used to extend shelf life may limit the growth of spoilage organisms. Hence, temperature abuse could readily generate a hazard in a food.

A major question for MAP foods is whether organoleptic spoilage due to chemical or microbial action will occur before the pathogen numbers or toxin levels become a risk when a product undergoes cycling or is exposed to abuse temperatures. Post et al. (1985) reported that botulinal toxin was either present prior to, or occurred simultaneously with, sensory rejection of cod and whiting fillets for all vacuum or modified atmosphere treatments and temperature regiments. Some products appeared organoleptically acceptable for consumption even though botulinal toxin was found in them. This is extremely important from a regulatory viewpoint. Reddy et al. (1992) claimed that until this issue is addressed and resolved by the industry, MAP fish may not have a place in the retail market. The same problem exists for other chilled MAP foods, as discussed in the previous chapters.

There are several possible ways to solve or improve this problem: 1) minimize the initial load of microorganisms with the application of the Hazard Analysis and Critical Control Points (HACCP) system during handling, processing and packaging (Corlett, 1989); 2) optimize modified atmospheric (MA) conditions to eliminate pathogens or incorporate other processing steps, such as washing and low dose ionizing irradiation (Lambert et al., 1991a) so that the overall hurdles against pathogen growth or survival are increased; and 3) co-inoculate with lactic acid bacteria or other competitors to control *C. botulinum* outgrowth through biological competition in refrigerated foods or minimally-processed products such as sous vide foods (Gombas, 1989). All of these means warrant further investigation to ensure the safety of fresh foods packaged under modified atmospheres, especially when exposed to temperature-abuse storage conditions. Perhaps the most practical approach is to develop an effective cold distribution chain for chilled MAP foods, which requires the maintenance of low storage temperatures and minimizes temperature abuse in order to prevent pathogen growth, and to ship, sell and consume the product within a short period of time, just like the practice in Europe and Japan. A new important advance to further ensure organoleptic quality and safety at the point of consumption is the use of time-temperature integrators/indicators or electronic temperature recorders (TTIs) (Taoukis et al., 1991).

The food industry has for many years used some type of temperature recording devices for monitoring distribution, especially for frozen or refrigerated foods (Anon., 1992a). Originally these devices were windup motors with a bimetallic sensor that gave an ink pen trace on a chart. One had to scan the chart visually at the end of distribution to determine if the trace showed potential damage, e.g., going below freezing for fresh produce. The next generation of recorders used thermocouples or platinum-iridium thermistors and battery driven charts. Finally the computer/space age resulted in very small programmable units with a computer memory chip, which recorded the time/temperature exposure for subsequent downloading of the data to a computer spread sheet. There are over twenty companies marketing these latter products worldwide but they still require someone to look at the data to see if a problem occurred. One company erroneously developed a program that assumed the area under the curve of temperature versus time was the key. This will be subsequently discussed. It should be noted that all of these recorder devices are relatively expensive ($20 to >$500) and monitor the environment of a case, pallet load, truck, train, etc., and not necessarily the individual food pack. The next major technology step was the integration of the time-temperature data collection device with the kinetics of the food it was monitoring. Such devices were built for evaluation of fish quality as a function of storage temperature (Nixon, 1977). The idea was to determine if a delivered lot was acceptable to sell. These devices have not been commercialized. Since none of these devices give useful information to the consumer, the idea for an individual package time/temperature monitoring device that was consumer friendly, was born.

A time-temperature indicator or integrator is a small self-adhesive tag or label that can indicate temperature abuse or keep track of an accumulated time-temperature distribution function to which a perishable product is subjected from the point of manufacture to the display shelf of the retail outlet if on the case, or eventually to the consumer if on the individual food package. Generally, TTIs can be used for monitoring the cold distribution system, improving inventory management, assisting open dating of perishable foods, and ensuring food safety (Taoukis et al., 1991). Reviews and studies on TTI applications for a variety of food products have been done by several researchers (Fields and Prusik, 1986; Wells and Singh, 1988a), yet current use of TTIs in the food industry is limited to a few firms in the U.S. and in Europe. The use of TTIs for MAP foods is even less common. The major commercial use of TTIs is not by food companies but rather by the World Health Organization (WHO) for monitoring the cold distribution chain of vaccines (Taoukis et al., 1991). Other uses are by drug companies and distributors of sensitive chemicals. This chapter mainly presents the potential applications of TTIs for monitoring temperature abuse in the distribution chain and for use in predicting the safe shelf life of MAP foods. The principles of several types of TTIs and kinetic correlation approaches will be briefly presented.

HOW DO TTIs WORK?

The principles of TTI operation are either a mechanical, chemical or enzymatic irreversible change usually expressed as a visible response in the form of a mechanical deformation, colour development or colour movement. The visible response gives information on the storage conditions to which the tag has been exposed. Recently, Taoukis et al. (1991) gave a complete overview of time-temperature indicators. Only three companies have made successful commercial time-temperature indicators/integrators for food distribution, while two other companies have made such integrators for use in monitoring sterilization of medical instruments etc., in hospitals or clinics.

The LifeLines TTI System

The LifeLines Fresh-Scan® (previously LifeLines Freshness Monitor®, LifeLines Technology, Inc., Morris Plains, N.J.) (Fields and Prusik, 1986) is based on the solid state polymerization of a thinly-coated colourless acetylenic monomer that changes to a highly-coloured opaque polymer. The measurable change is the reflectance which is measured with a laser optic wand, with the data being stored in a handheld device. The indicator has two bar codes, one for identification of the product and the other for identification of the indicator model next to the colour response region, as shown in Figure 14.1a. The indicators are active from the time of production and prior to use have to be stored in the freezer, where they are most stable.

The LifeLines Fresh-Check® label is based on the same operation principle as the Fresh-Scan®. Its goal is to give distributors, retailers and consumers a better guide to the thermal abuse or shelf history of the product. With varied composition, the label can have a compatible shelf life to many foods. The current design looks like a "bulls-eye," as shown in Figure 14.1b. When the inner ring becomes as dark as the outer circle, the end of shelf life has been reached. The shape and design can be customized to user specifications. They can be combined with the open date in such wording as "Use by ________ unless center of ring is darker than outer ring." This eliminates the purchase of food which has not passed the expiration date but which is out of compliance with quality because of temperature abuse.

The 3M TTI System

The 3M Company Specialty Packaging Department (3M Co., St. Paul, Minnesota) offers a TTI system, called MonitorMark™ (Manske, 1983). This is based on a time-temperature dependent diffusion of a dyed fatty acid ester along a porous wick made of a high quality blotting paper as shown in Figure 14.2. The measurable response is the distance of the advancing diffusion front from the origin. Before use, the blue dye/ester mixture is separated from the wick by a barrier film so that no diffusion occurs. To activate the indicator, the barrier is pulled off and diffusion

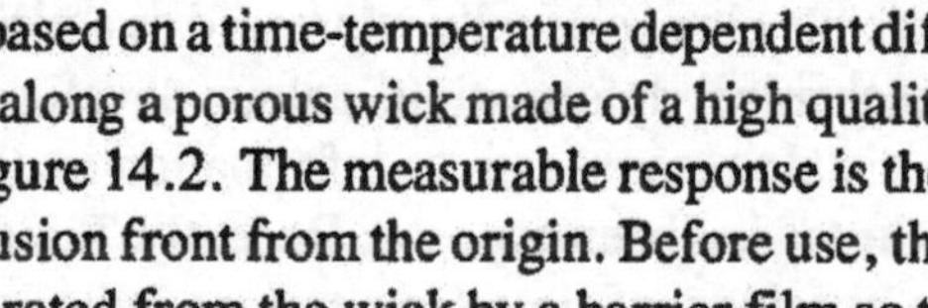

FIGURE 14.1a Lifelines Fresh-Scan® polymeric-based indicator.

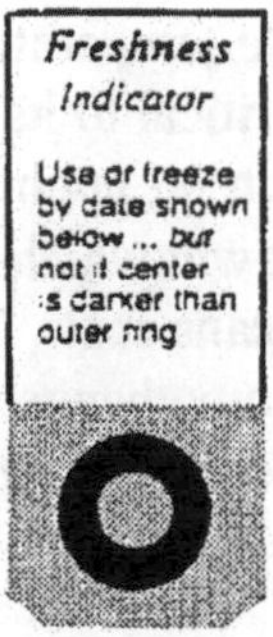

FIGURE 14.1b Lifelines Fresh-Check® consumer-readable indicator.

starts if the temperature is above the melting point of the ester, which determines the response temperature of a particular tag. Once activated, migration will only occur when the temperature is above the tag's response temperature. Thus, this type of TTI can be called a critical temperature/time integrator (CTTI). MonitorMark™ tags are available with response temperatures ranging from −17°C to 48°C (Manske, 1983). The MonitorMark™ tag can also be used as a consumer tag by manipulating the tag length or the marking symbols, i.e., "don't use product if hole shows a blue colour."

Another type of TTI made by 3M is the dual temperature indicator (DTI), which has a second, higher temperature set point on a CTTI, e.g., at 34°C for 29 hours. These dual temperature tags provide users with a greater degree of information, recording exposure and duration at one temperature level, as well as indication of temperature abuse at a second higher level.

The 3M Company also makes critical temperature indicators (CTIs) that show exposure above a critical temperature (T_c) after a certain period (a few minutes up to a few hours), but do not show the history of exposure above or below the temperature. They merely indicate that the product

FIGURE 14.2 3M MonitorMark™ diffusion-based indicator.

was exposed to an undesirable temperature for a short period of time sufficient to cause a change critical to its safety or quality (e.g., 5 min at 37°C). An example of such an application would be a tag for the indication of *C. botulinum* growth which occurs only above the critical temperature of 3.3°C (Notermans et al., 1990). Another tag, originally developed by 3M, shows a colour change when exposed to temperatures below freezing. This can be used to indicate freezing abuse of fresh fruits and produce.

The I-Point TTI System

The I-point® TTI (I-point Biotechnologies A.B., Malmo, Sweden) (Blixt, 1983) is based on a colour change caused by a pH decrease due to a controlled enzymatic hydrolysis of a lipid substrate. Before activation, the pancreatic lipase and the lipid substrate (tricaproin) are in two separate compartments. The TTI is triggered with a special activating device and can be applied manually or mechanically depending on the packaging line. At activation, the barrier that separated the compartments is broken, the enzyme and the substrate are mixed, the pH drops and the colour change starts, shown by the presence of an added pH indicator. The colour change can be visually recognized or instrumentally measured. The I-point® consumer tag is based on the same operation principle. The colour change is from dark green to yellow and can be compared to a printed reference ring as with the Fresh Check® device. Although the company suffered financial problems in the early 1990s, they are supposedly under a rebuilding process under new ownership.

MODELING TEMPERATURE SENSITIVITY OF FOODS

In order to specify an appropriate time-temperature indicator for a product, the most important information needed is the product's shelf life data. The end point of shelf life for MAP foods is generally determined by organoleptic quality, microbial spoilage and microbial safety. Sensory perceptions (e.g., meat colour, rancidity of fish, browning of lettuce), evidence of metabolic byproducts and types and levels of microorganisms are all valuable, and together give a full picture of food quality and safety. The change of quality index (A) with time (t) depends on reaction order. Most food quality deterioration processes follow either pseudo-first (e.g., microbial growth) or pseudo-zero kinetics (Labuza, 1984). There have been few kinetic studies on MAP food products stored at different temperatures, some of which will be discussed later. Although several models have been proposed to describe the temperature sensitivity of food quality deterioration (Labuza et al., 1992), only the Arrhenius and the log shelf life model have been employed to incorporate TTI responses.

Arrhenius Model

Temperature is the most important environmental factor affecting the growth and survival of microbes, and thus the quality and safety of MAP foods. Temperature affects the duration of the lag phase, the rate of growth, the final cell numbers, the nutrient requirements, and the enzymatic and chemical composition of the cells. The lag phase is shortest at the optimal temperature and prolonged when the temperature is lowered. Temperature also affects other quality deterioration processes in foods. Generally, the Arrhenius model can be used to describe the temperature sensitivity of microbial growth and other quality loss reactions where the rate of growth is exponentially proportional to the inverse of the absolute temperature by:

$$k_A = k_{oA}e^{-(E_{A\,(food)}/RT)} \quad (1)$$

In this equation, k_A is the rate constant at a given temperature (T), k_{oA} is the pre-exponential factor, $E_{A\,(food)}$ is the activation energy of the reaction or process that controls food quality loss, R is the universal gas constant and T is the temperature in degrees K. From growth kinetics, the increase in numbers of log cycles is directly related to the rate constant by:

$$\ln\frac{N}{N_o} = kt \quad (2)$$

Thus, a semilog plot of time to reach a certain number versus $1/T$ will also be a straight line if the Arrhenius relationship is followed.

The activation energy values of food-related chemical reactions usually fall between 10−30 kcal/mol. The value of E_A determines the temperature sensitivity, i.e., how much k_A changes as T is changed. Activation energy values can also be affected by other factors, such as a_w or CO_2 concentration. For example, from the data for shelf life of cut-up chicken by Ogilvy and Ayres (1951), the calculated E_A values for bacterial growth are 24.8 kcal/mol for 0% CO_2, 28.2 kcal/mol for 15% CO_2, and 31.9 kcal/mol for 25% CO_2. This implies that the microbial growth rate is more temperature sensitive at higher CO_2 levels.

Successful applications of the models for predictive microbiology are available in the literature for many different organisms (e.g., Reichardt and Morita, 1982), even though deviation at near the optimum temperature has been observed (Ratkowsky et al., 1982; Fu et al., 1991). Ogrydziak and Brown (1982) applied this model to describe temperature effects in MA storage of seafoods. The Arrhenius relationship can also be applied to model the temperature dependence of the lag phase, which would be critical for variable temperature predictions of foods starting out with low microbial loads, a necessity for ESLR foods under MAP conditions. The inverse of the lag time (i.e., lag rate) is used in constructing the Arrhenius plot (i.e., log t_{lag} vs $1/T$ gives a straight line).

For pathogenic organisms, the endpoint is essentially equal to the minimum detection level of the organism, which is based on the analytical method used. However, the presence of low numbers of certain pathogens (e.g., *S. aureus*, *Clostridium perfringens*, *B. cereus*) may not be hazardous and thus some action level must be set. In extended storage, toxin may be produced if there is a toxin-producing pathogen present; then the shortest (not the mean) lag time before the toxin can be detected at any growth condition should be used as the end of the shelf life (Baker and Genigeorgis, 1990).

Log Shelf Life Model

The stability information of a food is often collected by evaluating the sensory acceptance at several time intervals and under several constant storage temperatures. If the temperature range is small, then a simple plot of the shelf life on semilog paper vs. temperature (instead of inverse absolute temperature) is also a straight line. The log shelf life equation takes the form of:

$$t_s = t_o e^{-bT} \tag{3}$$

where t_s is the shelf life at temperature T in °C, t_0 is the shelf life at 0°C, b is the slope of a plot of ln t_s versus T. This plot can be used to establish the temperature sensitivity of the product without considering the true mechanism of shelf life loss.

Another term often used is Q_{10}, which is defined as the ratio of rate constants or shelf lives at temperatures differing by 10°C. The activation energy, b and Q_{10} are interrelated through the following equation:

$$\ln Q_{10} = 10_b = \frac{10\,E_A}{RT\,(T + 10)} \tag{4}$$

In essence, the Q_{10} value is a simplistic view of the temperature sensitivity for growth. Labuza and Fu (1992) have shown it to range from two to twelve times for the growth of organisms in the refrigerated temperature range. This large range reflects the growth characteristics of different organisms.

CORRELATION OF TTIs WITH FOOD

Direct Correlation

A number of experimental studies aimed at establishing correlations between the responses of specific TTIs and quality characteristics of specific products have been conducted (Taoukis et al., 1991). These studies offer useful information but do not involve any modeling of the TTI response as a function of time and temperature, and thus are applicable only for the specific foods and the exact storage conditions that were used. Extrapolation to other similar foods or quality loss reactions, or even use of the correlation equations for the same foods at other temperatures or for fluctuating conditions would be impossible.

Effective Temperature Approach

A kinetically based correlation scheme has been developed based on the Arrhenius model by Taoukis and Labuza (1989a). It relates the TTI response to food quality or shelf life through mathematical modeling. To calculate the cumulative effects of a variable time-temperature history, the concept of an effective temperature (T_{eff}) is introduced, which is defined as the constant temperature that results in the same quality change as the variable temperature distribution over the same period of time. Thus the effective rate constant (k_{eff}) at the T_{eff} can be expressed as:

$$k_{\text{eff (food)}} = k_{oA} e^{-(E_{A\,(\text{food})}/RT_{\text{eff (food)}})} \tag{5}$$

The change of a food quality function $f(A)_t$ for a known variable temperature exposure, $T(t)$, can be calculated through Equation (6):

$$f(A)_t = \int_{t_1}^{t_2} k_A dt = k_{oA} \int_{t_1}^{t_2} e^{-(E_{A\,(\text{food})}/RT(t))} dt = k_{\text{eff (food)}} t \tag{6}$$

The temperature sensitivity of TTIs is also described by the Arrhenius relationship. The $E_{A\,(\text{TTI})}$ values of the indicators cover the range of the most important deteriorative reactions in foods. Table 14.2 lists the activation energy and Q_{10} values of various types of TTIs. Based on the Arrhenius law, the effective temperature can be calculated from the TTI

Table 14.2. Values of activation energy and Q_{10} for several TTIs.

Producer	Model of TTI	E_A (kcal/mol)	Q_{10}
LifeLines	18[a]	27.0	5.8
	21[b]	21.3	4.0
	34[b]	17.8	3.2
	41[a]	20.5	3.8
	57[b]	21.3	4.0
	68[a]	19.7	3.6
	A20[c]	19.4	3.5
	A40[c]	19.5	3.6
3M	4Pa*	9.8	1.9
I-point	2090[d]	13.1	2.3
	2180[b]	14.3	2.5
	2220[b]	14.0	2.5
	3014[c]	11.4	2.1
	3270[d]	21.1	4.0
	4004[d]	40.0	13.5
	4007[a]	32.7	8.4
	4014[c]	24.3	4.9
	4021[a]	33.7	9.0

[a]From Taoukis and Labuza (1989a).
[b]From Wells and Singh (1988b).
[c]From Sherlock et al. (1991). Consumer-readable tags.
[d]Based on the data in McMeekin and Olley (1986).
*More models are available with similar kinetic parameters.

response, $f(x)_t$, for the same variable temperature distribution as the food product:

$$T_{eff\,(TTI)} = \frac{E_{A\,(TTI)}/R}{\ln\,[k_{o}t/f(x)_t]} \tag{7}$$

Thus, based on the assumption of $T_{eff\,(TTI)} = T_{eff\,(food)}$, the effective rate constant and the quality loss can be calculated from Equations (5) and (6), respectively.

Equivalent Time Approach

Similarly, one can introduce the concept of an equivalent time (t_{eq}), which is defined as the time at a reference temperature (k_{ref} as the rate constant) resulting in the same amount of quality changes as the variable time-temperature distribution. Then Equation (6) can be written as:

$$f(A)_t = \int_{t_1}^{t_2} k_A dt = K_{oA} \int_{t_1}^{t_2} e^{-(E_{A\,(food)}/RT(t))}\, dt = k_{ref\,(food)}\, t_{eq\,(food)} \tag{8}$$

Thus

$$t_{eq\,(food)} = \frac{f(A)_t}{k_{ref\,(food)}} = \int_{t_1}^{t_2} e^{-(E_{A\,(food)}/R)[1/T(t) - 1/T_{ref}]} dt \tag{9}$$

For a TTI exposed to the same time-temperature history and at the same reference temperature:

$$t_{eq\,(TTI)} = \frac{f(x)_t}{k_{ref\,(TTI)}} = \int_{t_1}^{t_2} e^{-(E_{A\,(TTI)}/R)[1/T(t) - 1/T_{ref}]} dt \tag{10}$$

Therefore the amount of food quality change for a certain time-temperature history can be calculated from Equation (8) based on the assumption that $t_{eq\,(food)} = t_{eq\,(TTI)}$ at the same reference temperature.

This approach will be able to provide direct information on lost or remaining shelf life of the food at a targeted storage temperature after exposure to a variable time-temperature history. For a continuously responding TTI, the difference between t_{eq} and t is the shelf life loss due to the temperature abuse. The remaining shelf life is equal to the open date minus t_{eq}. The open date is set based on the shelf life at the targeted storage temperature. The value of t_{eq} is the result of the temperature abuse in the case of using a CTI tag, thus the remaining shelf life is equal to the open date minus t_{eq} minus t, where t is the storage time.

Activation Energy Difference

In the effective temperature approach and the equivalent time approach discussed above, there is an important underlying assumption, i.e., $T_{eff\ (food)} = T_{eff\ (TTI)}$ or $t_{eq\ (food)} = t_{eq\ (TTI)}$, respectively, for a given temperature distribution. This is true when $E_{A\ (food)} = E_{A\ (TTI)}$ or when the temperature is constant throughout the distribution. Computer simulations with a large number of assumed variable temperatures have shown that if the difference between $E_{A\ (food)}$ and $E_{A\ (TTI)}$ is less than 10 kcal/mol, the error in the quality estimation is less than 15%, which in many cases is acceptable (Taoukis and Labuza, 1989a). Normally it is not difficult to find a TTI tag which has an activation energy value close to the food to be monitored. The smaller the difference between $E_{A\ (TTI)}$ and $E_{A\ (food)}$, the more accurately the TTI indicates actual shelf life. Besides the activation energy, the lifespan of a TTI itself should also be considered in choosing an appropriate tag.

USE OF TTIs

Distribution Monitor and Inventory Management Helper

A TTI or temperature recorder can be used to monitor the temperature exposure of food products during distribution and up to the time they are displayed at the supermarket. By attaching the monitor to individual cases or pallets or placing them in the immediate environment, they can give a measure of the preceding temperature conditions at each receiving place. These places would serve as information-gathering and decision-making centers. The information gathered from all these stations could be used for overall monitoring of the distribution system, thus allowing for recognition and possible correction of the weak links. The level of fluctuation for any distribution chain may be compared by using the difference of T_{eff} and T_{ref}, or the percentage shelf life loss due to the temperature abuse $[(t_{eq} - t)/t_s]$. Additionally, a TTI would allow targeting of responsibility and guarantee the producer and distributor that they can deliver a properly handled product to the retailer, thereby eliminating the possibility of unsubstantiated rejection claims by the latter. The presence of the TTI itself would probably improve handling, serving as an incentive and reminder to the distribution employees throughout the chain, of the importance of proper temperature storage (Fu and Labuza, 1992).

With the use of a TTI tag, an alternative to the traditional time-based first-in first-out (FIFO) issue policy would be to determine issue priority based on observed or estimated food quality change rather than elapsed time in storage. The advantage is to reduce food waste and to provide more consistent quality at time of issue for food items which have been exposed to various temperature conditions. LifeLines has developed the Inventory Management System, which is designed for checking product freshness throughout the distribution cycle (Anon., 1985). Using this system, one can determine which foods should be shipped to fill incoming orders or requisitions, based on freshness. An evaluation of the system in measuring the freshness of chilled fish was done by Tinker et al. (1985). Results showed that indicator labels could be used to satisfactorily monitor the freshness of irradiated fillets during distribution.

Safe Shelf Life Predictor

Shelf life predicting is an essential feature in marketing MAP products. In shelf life testing of MAP products, there are at least three quality parameters that need to be evaluated at various temperatures during the test period: microbial safety, microbial spoilage and overall organoleptic changes. When microbial hazards are minimal, tests for spoilage and organoleptic change take precedence in shelf life determination. This may be accomplished by monitoring microbial growth and changes in colour, odour, texture and overall acceptability of the product. Where a MAP product may pose a health risk after expiration, the indicator's shelf life prediction must be conservative enough to insure that the indicator will predict the end point well before a risk develops.

Predicting the shelf life of a food system with several different components becomes a problem when there are several modes of deterioration and each mode has its own temperature sensitivity (Labuza, 1984). Since different metabolic pathways and different organisms are affected differently by temperature changes (each has its own T, a_w, pH and gas dependence), this creates a real problem in terms of the ability to make microbial growth and shelf life predictions. In addition, as the a_w, O_2 and CO_2 levels change within the package, this will also affect shelf life and pathogen risk.

However, the shelf life of the whole product at any given temperature is determined by the mechanism of the component which proceeds fastest and thus results in the shortest life. When safety becomes a major consideration in marketing the product, every effort must be made to accurately determine its safe shelf life. For MAP foods, either spoilage or potential hazard because of pathogen presence, determines the end of shelf life. Figure 14.3 shows possible relationships using the log shelf life versus *T* function, between organoleptic spoilage (due to microbial spoilage, colour change, etc.) and potential hazard (due to pathogenic microbial growth and toxin production). Case (a) is safe since spoilage occurs prior to pathogenic microbial growth at all possible storage temperatures. The kinetic data for spoilage can be used to incorporate an appropriate TTI tag. Case (b) shows that the food is safe if stored below a critical temperature, where no pathogen will grow or the growth of pathogenic microorganisms to a hazardous level is slower than spoilage. A critical temperature indicator could be used to serve as a warning sign that the T_c has been exceeded for an unacceptable length of time. Case (c) shows the situation where the outcome of spoilage or a microbial hazard depends on the temperature of holding, i.e., above a

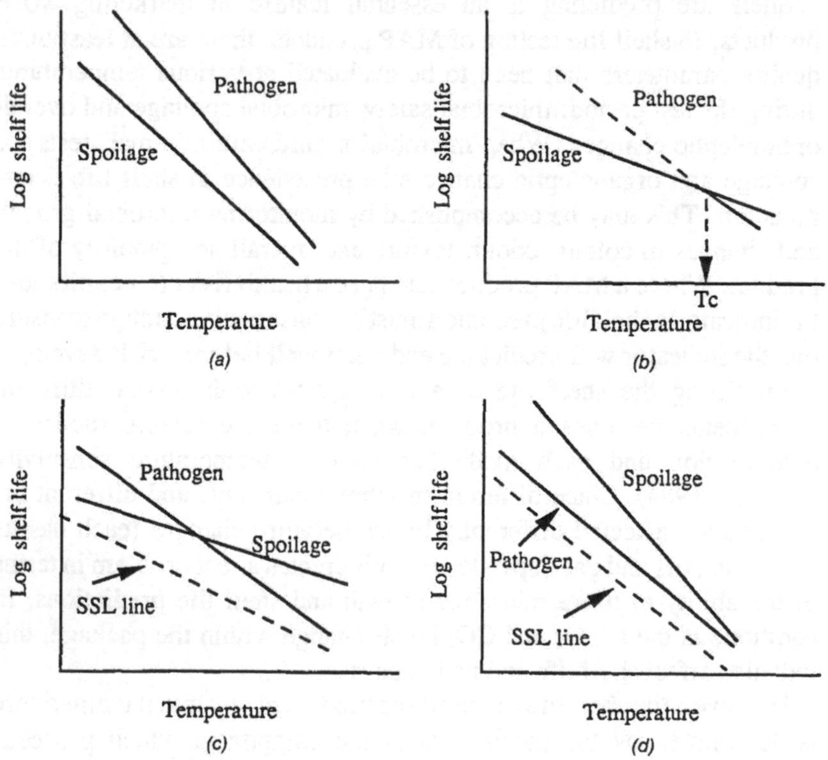

FIGURE 14.3 Possible relationships between organoleptic spoilage and potential hazard due to pathogen growth or toxin production.

critical temperature the pathogenic limit occurs before the food is organoleptically spoiled. The dashed line is an assumed safe shelf life (SSL) line, which could be used for TTI correlation. As noted, it is drawn at a shorter time for all temperatures. Case (d) indicates a product in which the pathogenic limit is reached before food spoilage, at all temperatures. Since setting the shelf life at this limit could lead to a potential safety problem if the data have typical biological variability, an assumed SSL line parallel to the pathogen line could be used for TTI correlation. In cases (c) and (d), the key problem is how to determine the safety margin, i.e., the SSL line. Kraft General Foods, Inc., used a safety margin of between one-third to one-fourth of the product's organoleptic shelf life as their printed shelf life date (Harris, 1989).

Fresh Meat and Poultry

For fresh red meat the two most important quality properties controlling shelf life are colour and microbial population. The colour of fresh meat depends on the relative amounts of three forms of myoglobin: reduced myoglobin, oxymyoglobin, and metmyoglobin (Young et al., 1988), which are controlled by the presence of O_2. TTIs generally cannot be used to monitor the colour change of meat, since it is much more O_2 dependent than temperature dependent. Thus the packaging method and packaging material used are the main factors which control the amount of O_2 in the package headspace and thus the colour of the meat. The microbial population of fresh meat is affected by many factors such as microbial species, health and handling of the live animal, slaughtering practices, chilling of the carcass, sanitation, type of packaging, and handling through distribution and storage (Young et al., 1988; Lambert et al., 1991b). During chilling of the carcasses, the relatively low temperatures give rise to predominately psychrotrophic bacteria such as *Pseudomonas* and *Achromobacter*. *Lactobacillus* organisms, which are facultative anaerobes, tend to thrive in vacuum-packaged meats. The dominant spoilage organisms differ depending on the storage temperature, the gas composition, and meat tissues (Jones, 1989; Gill and Molin, 1991).

In order to obtain the longest possible shelf life for meat, the storage temperature should be as close to 0°C as possible. The targeted shelf life of MAP fresh meat should be about 2–4 weeks under refrigeration (Powell and Cain, 1987). Since meat usually will be frozen after being purchased, a "freeze-by" date would be appropriate for open dating or "use-by" date for refrigeration, possibly coupled with a consumer tag.

Gill and Harrison (1989) studied the effects of vacuum and CO_2 on stability of pork cuts. Vacuum-packaged cuts were grossly spoiled by *Brochothrix thermosphacta* at both 3°C and −1.5°C, as were cuts packaged under CO_2 at 3°C. Growth of *B. thermosphacta* was suppressed in CO_2 packaged cuts at −1.5°C. At that temperature, the enterobacteria caused gross spoilage of the cuts. Figure 14.4 is the shelf life plot for pork, from which the activation energy and Q_{10} are calculated and listed in Table 14.3. These data can be used in choosing an appropriate tag. This situation is a good example of case (a) from Figure 14.3. Based on the data in Table 14.2, there are essentially only one or two tags with the same high temperature sensitivity.

Clark and Lentz (1969) reported that the growth of psychrotolerant, slime-producing bacteria (*Pseudomonas* and *Achromobacter*) on the surface of fresh beef was inhibited by CO_2. The degree of inhibition depended on the temperature, CO_2 concentration and the age of the culture at the time of CO_2 application. It was found that 20% CO_2 was the most practical concentration since it was effective at temperatures as high as 10°C, and since higher concentrations caused colour change in the meat but gave little additional inhibition of growth. Figure 14.5 is the shelf life plot for fresh beef packed under modified atmospheres. As seen, the level of CO_2 does not change the temperature sensitivity very much, rather only affecting out-growth time. The related kinetic

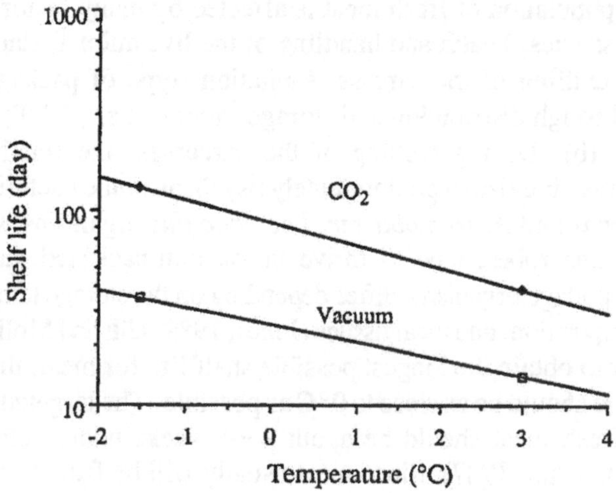

FIGURE 14.4 Shelf life plot for MAP pork cuts based on the results of Gill and Harrison (1989).

Table 14.3. Activation energy and Q_{10} values of various MAP food products.

Food System	MA Condition	E_A (kcal/mol)	Q_{10}	References
Raw pork	vacuum	31.3	7.7	Gill and Harrison (1989)
	CO_2	40.4	13.9	
Raw beef	air	15.8	2.8	Clark and Lentz (1969)
	10% CO_2	20.3	3.7	
	20% CO2	19.2	3.5	
	30% CO_2	19.9	3.6	
Raw chicken	air	15.9	2.8	Hart et al. (1991)
	30% CO_2 + air	23.5	4.6	
	30% CO_2 + N_2	23.5	4.6	
	100% CO_2	27.1	5.8	
Vegetable salad	air	29.1	6.6	Buick and Damoglou (1989)
	20% CO_2	20.9	3.9	

parameters are listed in Table 14.3. Since the CO_2 environment in a package can change with time, 20% CO_2 can be used to choose a tag. This situation is best represented by case (a) of Figure 14.3 in which spoilage occurs before outgrowth of pathogens. Based on Table 14.2, there are several tags with the same temperature sensitivity.

Hart et al. (1991) studied the effects of gaseous environment and temperature on the storage behavior of *L. monocytogenes* in chicken breast meat. The results of the study suggested that contamination with *L. monocytogenes* would not lead to any significant growth prior to spoilage under normal conditions of handling and storage. Holding the chilled meat under 100% CO_2 was the most effective means of extending the shelf life while ensuring that levels of *L. monocytogenes* were also suitably controlled. Figure 14.6 is the shelf life plot for chicken based on "off" odour development. The data for 100% CO_2 in Table 14.3 should be used for TTI correlation. This is similar to the case (a) or (b) of Figure 14.3.

In a conflicting study, Wimpfheimer et al. (1990) showed that MAP of raw chicken can substantially inhibit the aerobic spoilage microorganisms while allowing the levels of pathogenic *L. monocytogenes* to increase. Fortunately, the organism failed to grow at all on meat stored at 1–6°C under 100% CO_2, which is relevant to U.K. commercial practice for bulk storage of poultry meat under CO_2. But the beneficial effects of CO_2 were lost when the meat was exposed to an overt abuse

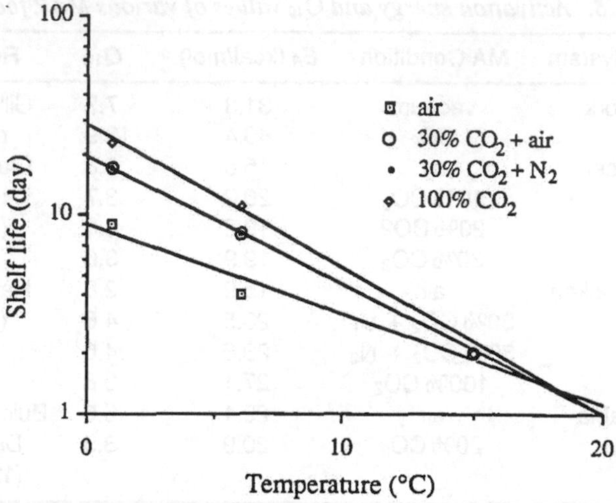

FIGURE 14.5 Shelf life plot for MAP fresh beef based on the results of Clark and Lentz (1969).

temperature of 15°C. This temperature may be used as the critical temperature for choosing a critical temperature indicator, as in case (b) of Figure 14.3. Thus a dual tag would be needed to indicate organoleptic shelf life and potential pathogen presence.

The temperature sensitivity of microorganisms is also affected by

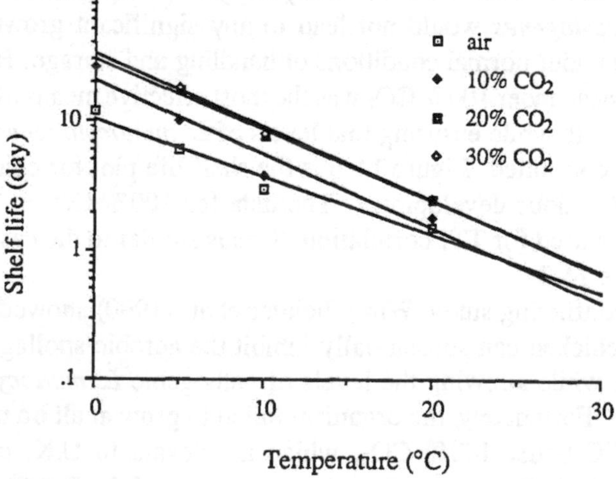

FIGURE 14.6 Shelf life plot for MAP fresh chicken based on the results of Hart et al. (1991).

MAP conditions. Based on the data reported by Wimpfheimer et al. (1990), the growth of aerobic microorganisms and *L. monocytogenes* on raw chicken under air had E_A values of 32.4 and 20.0 kcal/mol, and under MA (72.5:22.5:5, CO_2:N_2:O_2) had values of 23.2 and 21.0 kcal/mol, respectively. The data showed that MA did not affect the temperature sensitivity of *L. monocytogenes* while influencing the aerobic microorganisms significantly. When the temperature is lowered, the aerobic microorganisms are inhibited and *L. monocytogenes* may grow freely; thus a hazard issue arises. At this point, optimization of MA conditions is necessary. For example, use of 100% CO_2 can avoid this problem and a critical temperature tag set to 14°C (below 15°C) to be safe, would be valuable.

Fresh Fish

Pedrosa-Menabrito and Regenstein (1990) reviewed MA systems for fish preservation and pointed out that a temperature close to 0°C is critical for the inhibition of toxin production by *C. botulinum*. The use of TTI provides a method to resolve this by modeling the earliest time to toxin production at different temperatures. Baker and Genigeorgis (1990) predicted the safe storage time of fresh fish under modified atmospheres with respect to *C. botulinum* toxigenesis by modeling the length of the lag phase of growth. The earliest lag time was defined as the sampling period just prior to observation of toxicity. The margin of safety depends on the time interval for sampling. Figure 14.7 shows the lag time plot based on their data (the earliest lag times among all 927 experiments comprising 18,700 samples). The dashed line is also an assumed SSL line as in Figure 14.3, case (d). Based on this, the benefit of using MAP in extending shelf life is not significant since the predicted safe shelf life is only about 5 d at 0°C. More data are needed in the refrigeration temperature range of 0–10°C to improve on this correlation. Genigeorgis and coworkers have done a lot of work on predicting the probability of a microbial response under any given condition, which is appropriate for the concern of toxin production, but those models cannot be employed to correlate with the TTI response.

In the studies done by Post et al. (1985), naturally-contaminated fish were inoculated with *C. botulinum*, and then packaged in a modified atmosphere, as well as using fish packaged in air as a control. These products were then temperature abused, where 4°C was used as a control refrigeration temperature, 8°C a mild abuse temperature, 12°C a

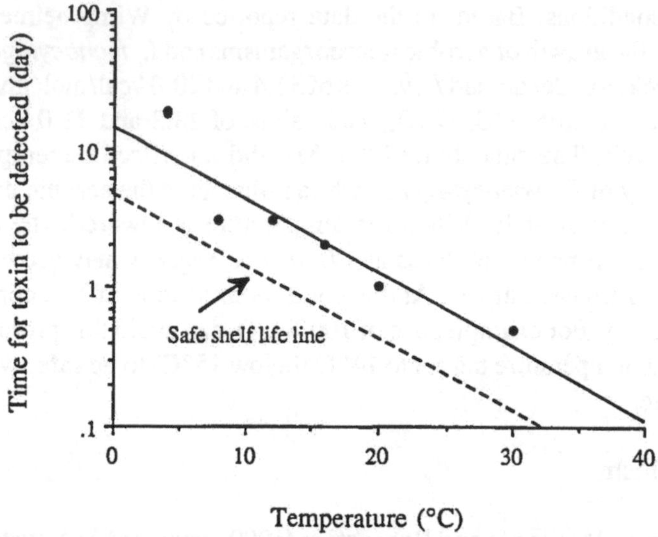

FIGURE 14.7 Lag time plot for *C. botulinum* toxin production in inoculated MAP fresh fish based on the results of Baker and Genigeorgis (1990).

moderate abuse temperature, and 26°C a severe abuse temperature. The samples were then analyzed at various times by splitting each sample. One portion was analyzed for toxin development, while the other sample was evaluated for signs of spoilage by a trained sensory panel. The time required for the product to develop detectable signs of spoilage was compared to the time for toxin detection. Figure 14.8 is the shelf life plot for cod fillets based on their results: (a) in vacuum, indicating spoilage preceded toxin formation and (b) in CO_2, indicating the reverse. For (a), the spoilage kinetic parameters can be used for choosing an appropriate tag. For (b), the assumed SSL line should be used. The problem is that not many temperatures were studied, especially in the refrigeration temperature range. Table 14.4 gives the kinetic parameters calculated for three MAP fish and several MA conditions, based on the results reported by Post et al. (1985). Again, using the data from Table 14.2, all three types of TTI could be used for some fish products.

Post et al. (1985) also studied the effects of cycling temperatures between 4 or 8 and 26°C on development of botulinum toxin and sensory deterioration during storage of vacuum and MAP fish fillets. In general, toxin was detected simultaneously or prior to normal organoleptic rejection. The shelf life model based on toxin detection time predicted a longer time than that required for actual appearance of toxin for all the cycling conditions, which emphasizes the need of a workable SSL line.

Stammen et al. (1990) and Reddy et al. (1992) reviewed shelf life extension and safety concerns about seafoods under MAP conditions and summarized that, with few exceptions, at temperatures above 20°C, organoleptic spoilage coincided with the appearance of toxin in many fresh fish products regardless of the MA used. However, at lower temperatures (4 – 12°C), organoleptic spoilage preceded toxin development in all fresh fish products except cod and whiting fillets held either in an air, vacuum, N_2 or CO_2 atmosphere. The time interval between toxin development and organoleptic spoilage of MA packaged fish products generally decreased as storage temperatures increased.

Garcia et al. (1987) reported that no toxin was detected in MAP fish inoculated with spores of different types of *C. botulinum*, stored at 8°C

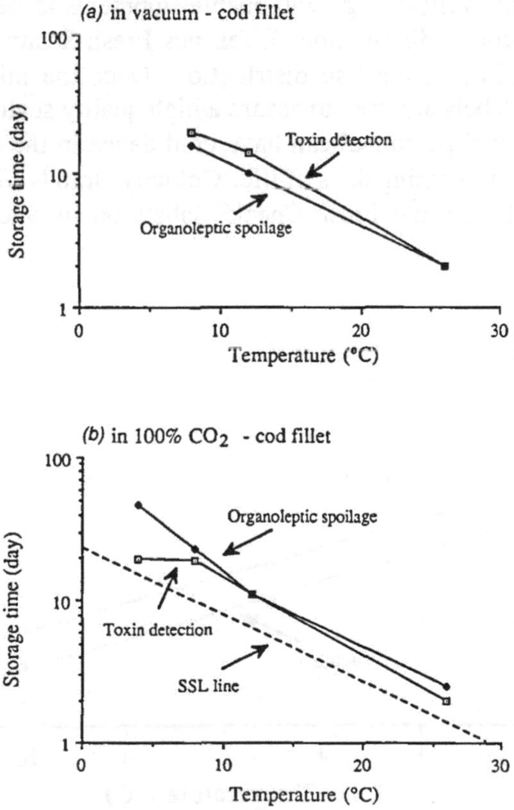

FIGURE 14.8 Shelf life plot for MAP cod fillet based on the results of Post et al. (1985).

over a 6 to 12 d period. Thus, if reliable data exist on the temperature behavior of different pathogens, a CTI tag could serve as a warning sign that a certain temperature has been exceeded for an unacceptable length of time.

Sensory quality may also be used for shelf life determination. A shelf life plot for packaged cod based on the time to reach an unacceptable cooked flavour score is shown in Figure 14.9, taken from the results of Gibson (1985). Kinetic parameters are listed in Table 14.4.

The shelf life of seafood under current icing and refrigerated storage conditions ranges from 2 to 14 d, depending on species, harvest location, and season. MAP technology has been shown to inhibit the normal spoilage microorganisms of seafood and double or triple shelf life (Stammen et al., 1990). A typical retailer of seafood would like at least 5 d of remaining shelf life when it gets to the store, with at least 2 d of life in the customer's hands (Prusik, 1990). Applications of TTIs to bulk products at the earliest stage will enable shipments to be conveniently monitored through distribution. LifeLines Fresh-Scan® labels can be used for monitoring the fish distribution. Once the initial quality is assessed and labels are used to assure a high quality seafood product at retail, store-level personnel can have confidence in the selling of fish with at least 2 remaining days of life. Culinary Brands (Sausalito, CA) used the bulls-eye-like Fresh-Check® labels on its vacuum-pouched

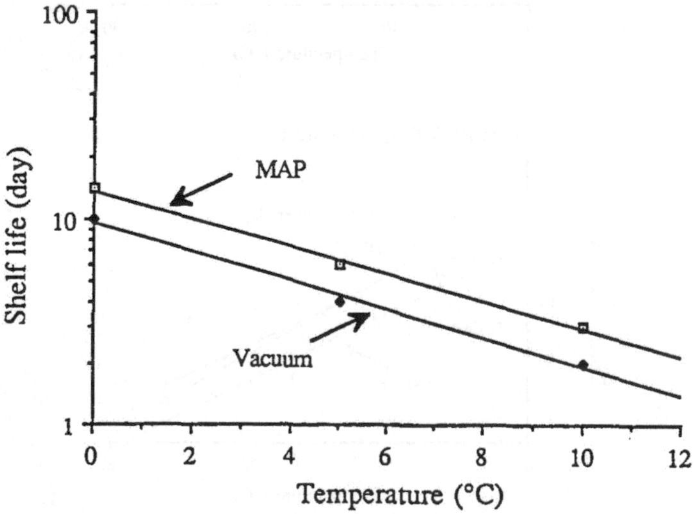

FIGURE 14.9 Shelf life plot for MAP cod fish based on the results of Gibson (1985).

Table 14.4. Activation energy and Q_{10} values of MAP fish products.

Food System	MA Condition	Mechanism	E_A (kcal/mol)	Q_{10}
Raw fish[a]	MAP	toxin	18.7	3.4
Cod fish[b]	MAP	spoilage	23.7	4.7
	vacuum	spoilage	24.7	5.0
Cod fillet[c]	air	spoilage	10.1	1.9
		toxin	14.5	2.6
	vacuum	spoilage	17.7	3.2
		toxin	20.1	3.7
	100% CO_2	spoilage	19.9	3.6
		toxin	16.9	3.0
	CO_2:N_2:O_2 (65:31:4)	spoilage	17.7	3.2
		toxin	18.8	3.4
Flounder fillet[c]	air	spoilage	11.9	2.2
		toxin	16.2	2.9
	vacuum	spoilage	20.5	3.8
		toxin	20.6	3.8
	100% CO_2	spoilage	11.6	2.1
		toxin	20.0	3.7
Whiting fillet[c]	air	spoilage	7.0	1.6
		toxin	11.6	2.1
	vacuum	spoilage	14.5	2.6
		toxin	18.6	3.4
	100% CO_2	spoilage	17.8	3.2
		toxin	19.7	3.6
	CO_2: N_2:O_2 (65:31:4)	spoilage	7.8	1.7
		toxin	10.7	2.0

[a]From the data of Baker and Genigeorgis (1990), where the MAP conditions included vacuum, 100% CO_2 or 70% CO_2 and 30% N_2.
[b]From the data of Gibson (1985), where the MAP condition was not specified.
[c]Calculated from the results reported by Post et al. (1985).

MAP fish products for the foodservice trade (Anon., 1989). The main benefit of the Fresh-Check® label was assuring the restaurant of freshness and food safety.

Fresh Fruits and Vegetables

The quality of fresh produce depends primarily upon the selection and careful handling of good products. Such measures as harvesting at optimal maturity, minimizing injury due to handling, reducing microbial infection through proper sanitation, and maintaining optimum temperature and humidity are important in maintaining postharvest quality. Once these primary requirements have been met, further maintenance of product quality can be achieved through modification of the atmosphere surrounding the product (Zagory and Kader, 1988). Temperature is the most important factor affecting MAP produce. Generally, fruits and vegetables will last longer at lower temperatures. However, many tropical fruits (e.g., avocados, mangoes, papayas) have a low temperature injury phenomenon and should not be stored below about 13°C. Non-chill-sensitive commodities (e.g., apples, broccoli, pears) can be stored near 0°C without adverse effects (Zagory and Kader, 1988). These critical temperatures can be used to choose an appropriate critical temperature indicator, e.g., a cold side indicator, to indicate that the produce has been exposed to lower detrimental temperatures. Since the quality change of produce is usually visible (e.g., colour change, mold growth), use of a TTI may not be beneficial for consumer purchasing, but can monitor distribution and help inventory management.

On the other hand, Berrang et al. (1989a, 1989b) demonstrated that extended shelf life could potentially increase the populations of pathogens in foods. They found that MAP could reduce spoilage and extend the shelf life of asparagus, broccoli and cauliflower by up to 33%. However, the growth of *L. monocytogenes* and *A. hydrophila* was not inhibited by the treatment. These bacteria continued to grow during the extended shelf life, and ultimately reached higher levels than would have been attained if the vegetables had deteriorated sooner. Solomon et al. (1990) reported that *C. botulinum* produced toxin before spoilage in shredded cabbage at room temperature under a modified atmosphere (70% CO_2, 30% N_2). Use of the purchased produce before the end of safe shelf life is critical. A consumer tag may be helpful to give the consumer a warning.

With the exception of mushrooms, fresh vegetables have not previously been implicated in botulism (Zagory and Kader, 1988) although shredded cabbage mixed with a coleslaw dressing (reduced pH) caused four cases of botulism in 1987 (Solomon et al., 1990). By choosing the right film (CO_2 permeability is generally three to five times greater than the O_2 permeability), or with the help of a CO_2 emitter and an ethylene scavenger, botulism may be avoided in MAP fruits and vegetables.

Sous Vide and Other Minimally Processed Prepared Foods

Cook-chill and sous vide are two additional processes which aid in the extension of refrigerated food shelf life. Both cook-chill and sous vide processes begin with fresh, raw products which are partially or fully cooked in a gas impermeable flexible bag or hot-filled into the bag (Snyder and Matthews, 1984). The sous vide processor first needs good quality ingredients and then must handle those ingredients properly to limit contamination. In terms of safety, the processor must check the preparation of sous vide at every stage by using the HACCP system (Baird, 1990). A fail-safe mechanism must be adapted where strict temperature control (0–2°C) is required. Either a TTI or a CTTI based on a SSL line can be employed. However, more research is needed to draw the SSL line, i.e., the information about the lag time for toxin or pathogen detection under refrigerated temperature and abuse temperatures. Culinary Brands tested the LifeLines TTI system for their sous vide refrigerated fish products (Fresh-Check® labels on individual packages, and Fresh-Scan® labels on master packs) (Rice, 1989). Labels were read throughout distribution to monitor performance. Recently, FDA urged the use of TTI for individual packages of sous vide products to indicate whether temperature abuse had occurred and whether a potential hazard might exist (Anon., 1992b).

Minimally-processed foods are not sterile, and thus have limited shelf life. Harris (1989) reported the shelf life data for several types of foods: entrees, 12–21 d; salads, 21–28 d; pasta and sauces, 90 d; and desserts, 21–40 d. To maintain quality and ensure product safety, there is a need to maintain refrigerated temperatures starting with raw material handling and moving right on through the production environment, storage and distribution of the products, display in the retail store, and use by the consumer. Notermans et al. (1990) pointed out that in order to ensure that the risk of botulism from refrigerated processed foods is controlled properly, the foods must be stored at a temperature of $<3.3°C$. If, however, this temperature cannot be guaranteed, the storage time has to be limited. Thus, at least a CTTI would be useful in this case. In one case, as pointed out by Solomon et al. (1990), the use of a pH 3.5 dressing and MAP for a shredded cabbage product was not enough to eliminate botulinum toxin production when the product was temperature abused by holding at room temperature. They showed a situation similar to Figure 14.3 case (b), where toxin production occurred before the product became organoleptically unacceptable, if held above some critical temperature.

Buick and Damoglou (1989) reported the effect of MAP on the microbial development and organoleptic shelf life of mayonnaise-based vegetable salad. The principal organisms causing spoilage of the salads were yeasts. Figure 14.10 is the shelf life plot for salad from which the kinetic parameters are calculated and given in Table 14.3. As seen from the graph, the extension of shelf life by MAP is about 18 d longer compared to storage in air at refrigeration temperature. Based on the Q_{10} of about 4 for the MAP product, several tag types listed in Table 14.2 would be applicable.

Use of MAP may greatly extend the shelf life of refrigerated cooked crab meat, but it cannot eliminate the risk of salmonellosis at abusive temperature (T_c = 7°C) (Ingham et al., 1990a). MAP was not effective at inhibiting growth of *L. monocytogenes* on cooked chicken loaf, especially at higher temperatures (7 and 11°C) (Ingham et al., 1990b). A CTI tag may be useful, but more data are needed to establish the criteria of time/temperature.

Ahvenainen (1989) found that the best gas composition for cooked meat products was a mixture of 20% CO_2 + 80% N_2, whereas for ready-to-eat products the optimal gas composition was very dependent on the product. Commercially produced sandwiches (including

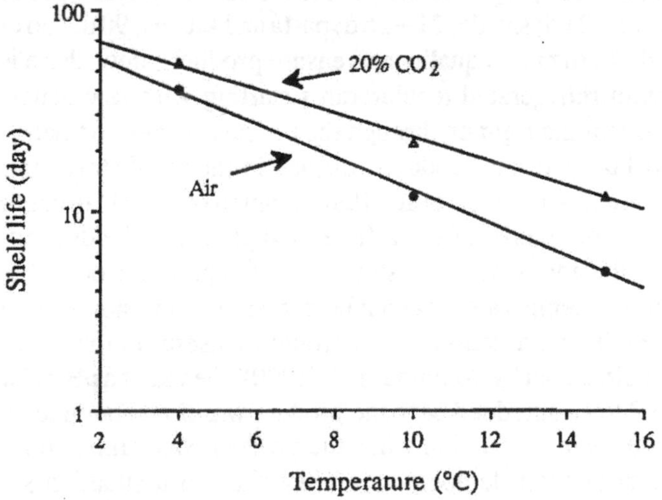

FIGURE 14.10 Shelf life plot for mayonnaise-based vegetable salad based on the results of Buick and Damoglou (1989).

processed meats, roast beef and hamburgers) packed in MA conditions have also been studied (McMullen and Stiles, 1989). There is a need not only for additional research on the storage life, but also on the safety of these products up to and beyond the manufacturer's stated storage life at different temperatures, so that the use of a TTI can be evaluated.

FURTHER CONSIDERATIONS

Complexity of Foods

The shelf life of a product based on the fixed final level of quality (A_s), is determined by its initial quality which, in most cases, is unknown. Thus, a safe estimate of the average initial quality is needed to do any evaluation of time-temperature histories. For most MAP foods, the initial quality is determined by the presence of naturally inherent and contaminating microorganisms. Determination of the initial microbial quality is even more difficult. One reason is that microbial distribution among food tissues is not uniform, so that local spoilage regions on the surface are possible. Different organisms have different temperature sensitivities and may compete with each other. In addition, some pathogens may not grow at the intended storage temperature, but may survive during storage. More importantly, the current routine microbial analytical techniques can require up to two days to get results for a specific pathogen, which is too long for quality assurance testing of chilled MAP foods. Hopefully, with better application of HACCP and more MAP practices, as well as the advancement of rapid microbial analysis techniques, food manufacturers should be confident in the initial quality of their products and be able to assess this quality quantitatively. With this added information, a reasonable value can be employed in any equation needed for prediction of shelf life.

Another major problem with MAP foods is that temperature changes also affect the permeability of the film. In general, film permeability increases as temperature increases, with CO_2 permeability responding more than O_2 permeability. This implies that a film that is appropriate for MAP at one temperature may not be appropriate at other temperatures (Zagory and Kader, 1988). This once again reinforces the importance of careful temperature management of MAP products. The change of partial pressure of O_2 and CO_2 at a certain temperature depends on their solubility in the formulation, the processing conditions, package permeability, headspace, surface to volume ratio, external pO_2 and pCO_2, aqueous pH, microbial load, biological oxidation rate, chemical oxidation rate, redox buffering capacity, redox equilibrium rate, temperature and storage time (Jones, 1989). A decrease in CO_2 in the headspace gases of MAP meats is common and has been attributed to diffusion through the package and absorption by the meat (McMullen and Stiles, 1991). Transfers between the product-package system and its environment can have significant consequences on the microbial activities necessarily associated with packaged foods (Bureau, 1986). The predominating microorganisms may change depending on different gas atmospheres and storage temperature (Jones, 1989; Gill and Molin, 1991). To overcome this in part, a new film which undergoes a glass transition to dramatically increase gas flow as temperature is increased, has been developed for MAP produce (Anon., 1992c).

Temperature abuse may also cause a history effect problem. That is, the previous time-temperature history has an effect on the microbial growth rate at the following storage conditions. For example, a warm storage temperature before storage at a lower temperature could have a much greater effect on quality and remaining shelf life than would be expected from rates measured during storage only at the lower temperatures. Bacterial enzymes formed at a higher temperature could well be active at the subsequent storage at lower temperature, whereas they would not be active if the product had never experienced the higher temperature regime (Bogh-Sorensen and Olsson, 1990; Fu et al., 1991). The relevant database in the literature for fluctuating temperature conditions is minimal compared to kinetic data and equations available to predict the destruction of microorganisms under variable heat processing conditions. Besides the history effect, an important factor that needs to be considered is the potentially different temperature-dependent behavior of different strains of the same species, as well as different competitive species, when exposed to a variable temperature sequence. Studies under cycling temperatures in the range of −5 to +5, +5 to +10, +10 to +15, and −5 to 25°C are recommended for different types of foods and distribution systems (George and Shaw, 1992), especially if a TTI is going to be employed.

Reliability, Applicability and Cost of TTIs

The reliability for TTI applications includes the variability in response of the device, the confidence of the determined kinetic parameters, the difference between the $E_{A\ (food)}$ and the $E_{A\ (TTI)}$, and the degree of temperature abuse. The applicability problem involves deviations from the Arrhenius relationship for both the TTI and food, the heat transfer problem (since a TTI tag is usually applied on a package surface and does not reflect the temperature response in the center of a pallet load), and the chemical and light sensitivity of TTIs. Taoukis and Labuza (1992) suggested that a realistic aim is an overall ΔT_{eff} between the food and TTI of less than 1° for the same time-temperature history. This corresponds to 12 and 8.3% error in the value of quality change estimation for a zero- and 1st-order reaction, respectively, of 20 kcal/mol E_A. This can be overcome by setting the tag response as shown in Figure 14.3, so that the tag always indicates the end of shelf life before the food deteriorates or pathogens or toxins appear. A significant amount of data on the food would be needed to accomplish this, however. The Campden Food and Drink Research Association in the U.K. has been in the forefront of setting reliability and performance standards for food use TTIs. Some of those standards are listed in Table 14.5. In the U.S., the only standard is in a proposal from the ASTM that will likely require the tag manufacturers to have an Arrhenius plot. It should be noted that LifeLines has published data which indicated that their products meet the U. K. standards, and the work of Taoukis and Labuza (1989a, 1989b) also indicated this situation for the 3M-type tag. Malcata (1990) estimated that an additional 15% of error was possible due to heat transfer lag when a TTI tag was applied on the package surface. In this case, of course, the tag expires before the food, which is safe but costly if there is no pathogen problem. The history effect of TTIs is usually negligible (Taoukis and Labuza, 1989b). The cost of a TTI, currently 0.5–50¢/unit, also depends on the quantity required. All of these aspects and their potential solutions have been discussed in detail by Taoukis et al. (1991).

Acceptance of TTIs by the Food Industry

The current situation is that many companies, at least in the U.S., are generally only concerned with whether their shipment is punctual or not. Late shipment translates to money lost. The major challenge for many companies is to deliver a quality product on time. However, the food industry has been reluctant to use a TTI tag on their products to help in this process. The industry has a number of concerns regarding TTI use: economic benefit, reliability, applicability, cost, recyclability of the package with a tag on and consumer acceptance (George and Shaw,

Table 14.5. Technical standards and procedures for the evaluation of time-temperature indicators.[a]

Test Procedure	Technical Standard
Temperature response test	
Frozen food	Reproducible end point
Temperatures: −25, −15, −10, −5 and +5°C	Maximum tolerance: ±6 days or 2.5% of the lifespan of the TTI
Chilled food	Reproducible end point
Temperatures: −5, +5, +10, +15, and +25 °C	Maximum tolerance: ±6 hours or 2.5% or the lifespan of the TTI
Evaluation of kinetic parameters of TTI	
Make an Arrhenius plot based on the data collected in the above temperature response test	Arrhenius plot and the values of activation energy and pre-exponential factor
Temperature cycling test	
Frozen food	The predicted rate constant at the reference temperature from the Arrhenius equation differs with the actually measured value by less than 10%.
Temperature range: −25 ↔ −15°C, −15 ↔ −10°C, −10 ↔ −5°C, −25 ↔ +5°C	
Cycling period: 10 times the intended lifespan of a TTI	
Chilled food	
Temperature range: −5 ↔ +5°C, +5 ↔ +10°C, +10 ↔ +15°C, −5 ↔ +25°C	Same as above
Cycling period: Same as above	
Accuracy of initial activation	
Statistical quality control	Depends on batch size, sample size and the acceptable quality level (e.g., 1.5%)
Simulated field test	
To simulate actual use	Applicable and reliable
Other tests	
Light (ultraviolet, visible and infrared)	TTI response should not be affected by these factors.
Vibration (including noise levels)	
Humidity	
Abuse tests (e.g., drop test)	

[a]From George and Shaw (1992).

1992). Several major manufacturers and retailers have expressed an interest in on-pack, time-temperature indicators but, before their application becomes widespread, it is essential that more information is obtained about potential customer understanding and use of such technology.

Consumer Education

Education is an important factor in assuring the safety of refrigerated foods. Results of a nationwide survey indicate the need for consumer education concerning proper food handling in the home. Education should be emphasized on practical foodborne disease information, temperature control, cross-contamination, and proper home food preparation practices (Williamson et al., 1992). Both food handlers and consumers need to be informed of the differences between many of the new refrigerated foods and the more traditionally processed foods with which they are familiar.

Another consumer survey showed that consumers clearly have a high concern for the freshness and quality of their purchased products and are confused by current open dating practices. They are very receptive to improved methods of freshness monitoring and will preferentially purchase products using TTIs (Prusik, 1990). The survey done by Sherlock and Labuza (1992) also showed consumer acceptance for TTI use for dairy food products. Use of TTIs could increase consumers' awareness of their own responsibility for keeping chilled foods at an appropriate temperature after purchase. They will quickly learn to store the product properly and consume their purchased product while it is still in a very high quality condition.

Regulation

Refrigerated products are currently the focus of intensive attention by manufacturers and regulatory agencies because of the concern that lowered O_2 levels and elimination of the typical spoilage organisms, as well as temperature abuse during distribution and storage, could lead to the growth of anaerobic pathogens. Based on the U.S. Food, Drug and Cosmetic Act, food manufacturers legally are considered to be responsible for the ultimate quality and product safety even when they have no control of the distribution chain. It is the responsibility of the manufacturer of the perishable product to determine the shelf life characteristics to be monitored by the indicator. Currently, no specific government standards on TTIs are available, except for the Campden publication previously mentioned. The delivered TTI should comply with all applicable country and local regulations relating to preparation, packaging, labeling, storage, distribution, and sale of the product in the commercial marketplace. The quality of the TTI must be of at least as high an order as the food product it seeks to monitor.

Consumers also have a responsibility for food safety. There is a concern that a consumer will not discard expensive products that still have good organoleptic quality (Brackett, 1992). Keeping refrigerated foods beyond their intended shelf life could increase the risk of the psychrotrophic pathogens present growing to hazardous levels. A leftover stored in a refrigerator may have an increased chance of becoming contaminated from other contaminated foods (Brackett, 1992). It was also reported that more than half the consumers questioned relied upon smell to assess freshness, a method which can be of risk with vacuum-sealed or MAP foods, since the contents may not display any of the usual signs of spoilage. This again emphasizes the importance of consumer education and the need for a TTI. Consumer instructions on the pack need to clearly indicate the action to be taken when there is a conflict between the end of product life indication as given by the date ("use by" or "best before") and the TTI. Regulations regarding the manufacture and sale of MAP and sous vide products will be discussed in detail in the last chapter of this book.

FUTURE LOOK

The application of TTI tags for monitoring the distribution chain and predicting the safe shelf life of MAP foods requires reliable microbial growth and shelf life data under different storage temperatures. More research will be expected in this area. Further information will also be required for new TTIs including their reliability, accuracy, precision, sensitivity to temperature change, linearity under isothermal storage, integrating performance under variable time-temperature conditions, stability prior to activation, and ease of reading. Additional TTI designs and varieties are expected to meet the requirements of different food products with different activation energies (Fu et al., 1992).

The combination of TTIs and computers in manufacturing and food handling offers a unique opportunity for the food industry. This particular method of protection is not the final answer, but shows significant progress. A consumer TTI tag plus a date code will give more information to the consumer about the stability or safety of a MAP product. Use of TTIs will influence buying patterns and will help to develop consumer confidence in the MAP food industry. The use of TTIs can help to change the negative image of refrigerated MAP foods to a positive one and help the overall industry advance to a growing mode. However, the successful application of a TTI tag depends on many factors, such as product stability analysis, reliability of a TTI, careful handling by the distributor and retailer and cooperation of consumers. Lack of control in any step may destroy the whole system.

ACKNOWLEDGEMENTS

The research was supported in part by the Minnesota-South Dakota Dairy Research Center and a 1992–1993 Graduate Dissertation Fellowship of University of Minnesota. It is published as paper no. 20201 of the contribution series of the Minnesota Agricultural Experiment Station.

REFERENCES

Ahvenainen, R. 1989. *Gas Packaging of Chilled Meat Products and Ready-to-Eat Foods*. Espoo: Technical Research Center of Finland, Publication 58.

Anon. 1985. "Computerized shelf life system tracks & grades product freshness," *Food Eng.*, (Aug.):125.

Anon. 1989. "Is it time for time/temperature indicators?" *Prepared Foods*, 158(11):219.

Anon. 1990. *Guidelines for the Handling of Chilled Foods*. 2nd ed. The Institute of Food Science and Technology, London. U.K.

Anon. 1992a. *Transit Temperature Recording–Origin and History*. Cox Recording Co., Long Beach, CA.

Anon. 1992b. "Time, temperature markers urged for sous vide packaging," *Food Chemical News*, (Sept. 21):22.

Anon. 1992c. "Temperature compensating films for produce," *Prepared Foods*, (9):95.

Baird, B. 1990. "Sous vide: What's all the excitement about?" *Food Technol.*, 44(11):92, 94, 98.

Baker, D. A. and Genigeorgis, C. 1990. "Predicting the safe storage of fresh fish under modified atmospheres with respect to *Clostridium botulinum* toxigenesis by modeling length of the lag phase of growth," *J. Food Prot.*, 53(2):131–140.

Berrang, M. E., Brackett, R. E. and Beuchat, L. R. 1989a. "Growth of *Aeromonas hydrophia* on fresh vegetables stored under controlled atmospheres," *Appl. Environ. Microbio.*, 55(9):2167–2171.

Berrang, M. E., Brackett, R. E. and Beuchat, L. R. 1989b. "Growth of *Listeria monocytogenes* on fresh vegetables stored under controlled atmosphere," *J. Food Prot.*, 52(10):702–705.

Blixt, K. G. 1983. "The I-point® TTM–A versatile biochemical time-temperature integrator," *IIR Commission C2 Preprints, Proceedings 16th Intl. Cong. Refrig.*, pp. 629–631.

Bogh-Sorensen, L. and Olsson, P. 1990. The chill chain, in *Chilled Foods: The State of the Art*, T. R. Gormley, ed., Elsevier Applied Science, London, pp. 245–267.

Brackett, R. E. 1992. "Microbiological safety of chilled foods: Current issues," *Trends Food Sci. Technol.*, 3(4):81–85.

Buick, R. K. and Damoglou, A. P. 1989. "Effect of modified atmosphere packaging on the microbial development and visible shelf life of a mayonnaise-based vegetable salad," *J. Sci. Food Agric.*, 46(3):339–347.

Bureau, G. 1986. Microbiological consequences of mass transfer, in *Food Packaging and Preservation: Theory and Practice*, M. Mathlouthi, ed., Elsevier Applied Science Publishers, London, pp. 93–114.

Clark, D. S. and Lentz, C. P. 1969. "The effect of carbon dioxide on the growth of slime producing bacteria on fresh beef," *Can. Inst. Food Technol. J.*, 2(2):72–75.

Conner, D. E., Scott, V. N. and Bernard, D. T. 1989. "Potential *Clostridium botulinum* hazards associated with extended shelf-life refrigerated foods: A review," *J. Food Safety*, 10(2):131–153.

Corlett, D. A. Jr. 1989. "Refrigerated foods and use of hazard analysis and critical control point principles," *Food Technol.*, 43(2):91–94.

Farber, J. M. 1991. "Microbiological aspects of modified-atmosphere packaging technology–A review," *J. Food Prot.*, 54(1):58–70.

Fields, S. C. and Prusik, T. 1986. Shelf life estimation of beverage and food products using bar coded time-temperature indicator labels, in *The Shelf Life of Foods and Beverages*, G. Charalambous, ed., Elsevier Science Publishers, Amsterdam, pp. 85–96.

Fu, B. and Labuza, T. P. 1992. "Considerations for the application of time-temperature integrators in food distribution," *J. Food Distr. Res.*, 23(1):9–17.

Fu, B., Taoukis, P. S. and Labuza, T. P. 1991. "Predictive microbiology for monitoring spoilage of dairy products with time-temperature integrators," *J. Food Sci.*, 56(5):1209–1215.

Fu, B., Taoukis, P. S. and Labuza, T. P. 1992. "Theoretical design of a variable activation energy time-temperature integrator for prediction of food or drug shelf life," *Drug Dev. Indu. Pharm.*, 18(8):829–850.

Garcia, G. W., Genigeorgis, C. and Lindroth, S. 1987. "Risk of growth and toxin production by *Clostridium botulinum* nonproteolytic types B, E & F in salmon fillets stored under modified atmospheres at low and abused temperatures," *J. Food Prot.*, 50(4):330–336.

George, R. M. and Shaw, R., eds. 1992. *A Food Industry Specification for Defining the Technical Standard and Procedures for the Evaluation of Temperature and Time-Temperature Indicators*. Campden Food & Drink Research Association, Technical Manual No. 35, London.

Gibson, D. M. 1985. "Predicting the shelf life of packaged fish from conductance measurements," *J. Appl. Bacteriol.*, 58(5):465–470.

Gill, C. O. and Molin, G. 1991. Modified atmospheres and vacuum packaging, in *Food Preservatives*, N. J. Russell and G. W. Gould, eds., Blackie and Son, Ltd., Glasgow, pp. 172–199.

Gill, C. O. and Harrison, J. C. L. 1989. "The storage life of chilled pork packaged under carbon dioxide," *Meat Sci.*, 26(4):313–324.

Gill, C. O. and Reichel, M. P. 1989. "Growth of cold-tolerant pathogens *Yersinia enterocolitica, Aeromonas hydrophila,* and *Listeria monocytogenes* on high-pH beef packaged under vacuum or carbon dioxide," *Food Microbiol.*, 6(4):223–230.

Gombas, D. E. 1989. "Biological competition as a preserving mechanism," *J. Food Safety*, 10(2):107–117.

Harris, R. D. 1989. "Kraft builds safety into next generation refrigerated foods," *Food Proc.*, 50(12):111–114.

Hart, C. D., Mead, G. C. and Norris, A. P. 1991. "Effects of gaseous environment and temperature on the storage behavior of *Listeria monocytogenes* on chicken breast meat," *J. Appl. Bacteriol.*, 70(1):42–46.

Ingham, S. C., Alford, R. A. and McCown, A. P. 1990a. "Comparative growth rates of *Salmonella typhimurium* and *Pseudomonas fragi* on cooked crab meat stored under air and modified atmosphere," *J. Food Prot.*, 53(7):566–567.

Ingham, S. C., Escude, J. M. and McCown, P. 1990b. "Comparative growth rates of *Listeria monocytogenes* and *Pseudomonas fragi* on cooked chicken loaf stored under air and two modified atmospheres," *J. Food Prot.*, 53(4):289–291.

Jones, M. V. 1989. Modified atmospheres, in *Mechanisms of Action of Food Preservation Procedures*, G. W. Gould, ed., Elsevier Applied Science, London, pp. 247–284.

Labuza, T. P. 1984. "Application of chemical kinetics to deterioration of foods," *J. Chem. Educ.*, 61(4):348.

Labuza, T. P. and Fu, B. 1992. Microbial growth kinetics for shelf life prediction: Theory and practice, in *Proceedings of the International Conference on the Application of Predictive Microbiology and Computer Modeling Techniques to the Food Industry*, R. L. Buchanan and S. Palumbo, eds., April 12–15, 1992, Tampa, FL.

Labuza, T. P., Fu, B. and Taoukis, P. S. 1992. "Prediction for shelf life and safety of minimally processed CAP/MAP chilled foods," *J. Food Prot.*, 55(10):741–750.

Lambert, A. D., Smith, J. P. and Dodds, K. L. 1991a. "Effect of headspace CO_2 concentration on toxin production by *Clostridium botulinum* in MAP, irradiated fresh pork," *J. Food Prot.*, 54(8):588–592.

Lambert, A. D., Smith, J. P. and Dodds, K. L. 1991b. "Shelf life extension and microbiological safety of fresh meat–A review," *Food Microbiol.*, 8(4):267–297.

Lioutas, T. S. 1988. "Challenges of controlled and modified atmosphere packaging: A food company's perspective," *Food Technol.*, 42(9):78–86.

Malcata, F. X. 1990. "The effect of internal thermal gradients on the reliability of surface mounted full-history time-temperature indicators," *J. Food Proc. Preserv.*, 14(6):481–497.

Manske, W. J. 1983. "The application of controlled fluid migration to temperature limit & time temperature integrators," *IIR Commission C2 Preprints, 16th Intl. Cong. Refrig.*, pp. 632–635.

McMeekin, T. A. and Olley, J. 1986. "Predictive microbiology," *Food Technol. Aust.*, 38(8):331–334.

McMullen, L. and Stiles, M. E. 1989. "Storage life of selected meat sandwiches at 4°C in modified gas atmospheres," *J. Food Prot.*, 52(11):792–798.

McMullen, L. M. and Stiles, M. E. 1991. "Changes in microbial parameters and gas composition during modified atmosphere storage of fresh pork loin cuts," *J. Food Prot.*, 54(10):778–783.

Nixon, P. A. 1977. "Temperature function integrator," U.S. Patent 41,061,033

Notermans, S., Dufrenne, J. and Lund, B. M. 1990. "Botulism risk of refrigerated, processed foods of extended durability," *J. Food Prot.*, 53(2):1020–1024.

Ogilvy, W. S. and Ayres, J. C. 1951. "Post-mortem changes in stored meats. II. The effect of atmospheres containing carbon dioxide in prolonging the storage life of cut-up chicken," *Food Technol.*, 5(3):97–102.

Ogrydziak, D. M. and Brown, W. D. 1982. "Temperature effects in modified-atmosphere storage of seafoods," *Food Technol.*, 36(5):86–96.

Palumbo, S. 1986. "Is refrigeration enough to restrain food-borne pathogens?" *J. Food Prot.*, 49(12):1003–1009.

Pedrosa-Menabrito, A. and Regenstein, J. M. 1990. "Shelf-life extension of fresh fish–A review. Part II–Preservation of fish," *J. Food Qual.*, 13(2):129–146.

Post, L. S., Lee, D. A., Solberg, M., Furgang, D., Specchio, J. and Graham, C. 1985. "Development of botulinal toxin and sensory deterioration during storage of vacuum and modified atmosphere packaged fish fillets," *J. Food Sci.*, 50(4):990–996.

Powell, V. H. and Cain, B. P. 1987. "The shelf life of meat during retail display," *Food Technol. Aust.*, 39(4):129–133.

Prusik, T. 1990. Freshness indicators: A tool for monitoring shelf-life and expanding distribution of fresh seafood products, in *Advances in Fisheries Technology for Increased Profitability*, M. N. Voigt and J. R. Botta, eds., Technomic Publishing Co., Lancaster, PA, pp. 225–233.

Ratkowsky, D. A., Olley, J., McMeekin, T. A. and Ball, A. 1982. "Relationship between temperature and growth rate of bacterial cultures," *J. Bacteriol.*, 149(1):1–5.

Reddy, N. R., Armstrong, D. J., Rhodehanel, E. J., and Kautter, D. A. 1992. "Shelf-life extension and safety concerns about fresh fishery products packaged under modified atmospheres: A review," *J. Food Safety*, 12(2):87–118.

Reichardt, W. and Morita , R. Y. 1982. "Temperature characteristics of psychrotrophic and psychrophilic bacteria," *J. Gen. Microbiol.*, 128(3):565–568.

Rice, J. 1989. "Keeping time/temp tabs on refrigerated foods," *Food Proc.*, 50(8):149–158.

Sherlock, M. and Labuza, T. P. 1992. "Consumer perceptions of consumer type time-temperature indicators for use on refrigerated dairy foods," *Dairy Food Environ. Sanit.*, 12(9):559–565.

Sherlock, M., Fu, B., Taoukis, P. S. and Labuza, T. P. 1991. "A systematic evaluation of time-temperature indicators for use as consumer tags," *J. Food Prot.*, 54(11):885–889.

Smith, J. P., Ramaswamy, H. S. and Simpson, B. K. 1990. "Developments in food packaging technology. Part II: Storage aspects," *Trends Food Sci. Technol.*, 1(5):111–118.

Snyder, O. P. and Matthews, M. E. 1984. "Microbiological quality of foodservice menu items produced and stored by cook/chill, cook/freezer, cook/hot hold and heat/serve methods," *J. Food Prot.*, 47(11):876–885.

Solomon, H. M., Kautter, D. A., Lilly, T. and Rhodehamel, E. J. 1990. "Outgrowth of *Clostridium botulinum* in shredded cabbage at room temperature under a modified atmosphere," *J. Food Prot.*, 53(10): 831–833.

Stammen, K., Gerdes, D. and Caporaso, F. 1990. "Modified atmosphere packaging of seafood," *Crit. Rev. Food Sci. Nutrit.*, 29(5):301–331.

Taoukis, P. S. and Labuza, T. P. 1989a. "Applicability of time-temperature indicators as food quality monitors under non-isothermal conditions," *J. Food Sci.*, 54(4):783–788.

Taoukis, P. S. and Labuza, T. P. 1989b. "Reliability of time-temperature indicators as food quality monitors under non-isothermal conditions," *J. Food Sci.*, 54(4):789–792.

Taoukis, P. S. and Labuza, T. P. 1992. "Assessing the food quality monitoring ability of a time-temperature indicator" (in preparation).

Taoukis, P. S., Fu., B. and Labuza, T. P. 1991. "Time-temperature indicators," *Food Technol.*, 45(10):70–82.

Tinker, J. H., Slavin, J. W., Learson, R. J. and Empola, V. G. 1985. "Evaluation of automated time-temperature monitoring system in measuring the freshness of chilled foods," *IJF-IIR Commissions C2, D3*, pp. 281–291.

Van Garde, S. J. and Woodburn, M. J. 1987. "Food discard practices of householders," *Am. Diet. Assoc. J.*, 87(4):322–329.

Wells, J. H. and Singh, R. P. 1988a. "Application of time-temperature indicators in monitoring changes in quality attributes of perishable and semiperishable foods," *J. Food Sci.*, 53(1):148–152.

Wells, J. H. and Singh, R. P. 1988b. "Response characteristic of full-history time-temperature indicators suitable for perishable food handling," *J. Food Proc. Preserv.*, 12(3):207–218.

Williamson, D. M., Gravani, R. B. and Lawless, H. T. 1992. "Correlating food safety knowledge with home food-preparation practices," *Food Technol.*, 46(5):94, 96, 98, 100.

Wimpfheimer, L., Altman, N. S. and Hotchkiss, J. H. 1990. "Growth of *Listeria monocytogenes* Scott A, serotype 4 and competitive spoilage organisms in raw chicken packaged under modified atmospheres and in air," *Internatl. J. Food Microbiol.*, 11(3):205–214.

Young, L. L., Reviere, R. D. and Cole, A. B. 1988. "Fresh red meats: A place to apply modified atmospheres," *Food Technol.*, 42(9):65–69.

Zagory, D. and Kader, A. A. 1988. "Modified atmosphere packaging of fresh produce," *Food Technol.*, 42(9):70–77.

CHAPTER 15

Regulations and Guidelines Regarding the Manufacture and Sale of MAP and Sous Vide Products

JEFFREY M. FARBER–*Health Canada, Ontario, Canada*

RATIONALE BEHIND THE REGULATORY APPROACH

MODIFIED-ATMOSPHERE packaged (MAP) and sous vide products have both evolved as a result of new processing and packaging technologies and changing consumer demands. These products have in some instances replaced home-prepared meals and packaged foods such as frozen and canned foods. These latter products are relatively safe from a microbiological viewpoint. However, a large percentage of both MAP and sous vide products fall under a class of foods designated as "potentially hazardous," and include low-acid (pH > 4.6), high-moisture (a_w > 0.85) products packaged in hermetically sealed packages, either under vacuum or with a modified atmosphere (MA). These "potentially hazardous" foods, besides not receiving a thermal treatment sufficient to achieve commercial sterility, have a shelf life of greater than 2 weeks. These perishable foods require refrigeration for microbiological safety and for preservation of quality.

The need for guidance becomes apparent when considering the following: 1) the potential microbiological hazards associated with the sale of MAP and sous vide products, especially at the retail level (the dangers associated with the growth of psychrotrophic pathogens in MAP and sous vide products have been dealt with in-depth in Chapter 4); 2) the long refrigerated shelf life of some of these products, e.g., MAP sandwiches with a 30-d shelf life are currently marketed in Canada; 3) for sous vide products refrigeration is, in many cases, the sole barrier to the growth of the non-proteolytic clostridia; 4) the average temperature of home refrigerators may well be in the range of 10–13°C, while 33% and 5% of retail refrigerated foods are being held in display cases above 7 and 13°C, respectively (see Chapter 14); and 5) MAP and sous vide foods are increasingly being produced on a small scale by individuals with limited experience and knowledge of food science and food microbiology.

It becomes evident from the above that there is a real need for either guidelines, recommended codes of practice or, ultimately, regulations for MAP and sous vide foods.

CURRENT CANADIAN GUIDELINES REGARDING MAP AND SOUS VIDE PRODUCTS

A document entitled "Guidelines for the production, distribution, retailing and use of refrigerated prepackaged foods with extended shelf life" has recently been published by Health Canada (Health and Welfare Canada, 1992a). The main subject of these guidelines are "refrigerated prepackaged foods with extended shelf-life" or REPFES. Products meeting the definition of REPFES include those that 1) rely largely on refrigeration for microbiological safety after processing, 2) are prepackaged using a packaging technology which does not adequately inhibit microbial growth, 3) are not commercially sterile and 4) have a shelf life equal to or greater than 10 d. Many MAP and sous vide products would fall under this definition of REPFES. Since REPFES are packaged in "hermetically sealed" packages and are not commercially sterile, they would fall under regulation B27.002 of the Canadian Food and Drugs Act. Subsection 2a states that low-acid foods packaged in a hermetically sealed container must either be kept under refrigeration and labelled "Keep Refrigerated," or frozen and labelled "Keep Frozen." Under section B.27.001, refrigeration is defined as a temperature of 4°C or less, but not frozen. The main purpose of the guidelines outlined in the document is to minimize the public health risk of REPFES. Specific guidelines are directed to a) processors, including those retailers processing their own REPFES; b) manufacturers of refrigeration cabinets, packaging equipment and materials; c) the retail sector; d) government agencies responsible for food safety; and e) the consumer. Salient points from each of the guidelines covered include:

Processors

It is recommended that ingredients used for REPFES should be of the highest quality with processors having a knowledge of the hazards associated with the ingredients. Processing should be conducted using the principles and practices of the hazard analysis and critical control point (HACCP) system, and products should be formulated and processed to significantly reduce microbial populations and limit their potential for growth, if exposed to abuse type temperatures. Record keeping is mentioned as an important part of the guidelines. Examples of areas where record keeping would be of great benefit include process parameters, changes in formulation, and tests carried out to determine the shelf life of a product. For those products having a shelf life > 10 d, it is recommended that upon request, the processor should make available information to regulatory authorities demonstrating that the food in question can safely be marketed for the intended shelf life. This information could include such things as microbiological challenge testing with foodborne pathogens such as *Listeria monocytogenes* and *Clostridium botulinum*, as well as shelf life testing to establish when spoilage occurs relative to the growth of foodborne pathogens.

It is also advised that labelling indicate that products be refrigerated at all times and that a "USE BY" date be prominently displayed (in addition to a "Best Before" date) on the package. In addition, clear preparation and cooking instructions, if needed, should be given. The use of time-temperature indicators to indicate that temperature abuse has taken place is also recommended. Processors are reminded to ensure that products are continuously maintained at appropriate refrigeration temperatures at all points along the food chain, including retail display. Also stressed is that the retailer should have mechanisms in place to properly dispose of out-dated or temperature-abused product.

Manufacturers of Refrigeration Cabinets, Packaging Equipment and Materials

The rationale behind these guidelines stems from practical observations on the misuse of refrigerated cabinets at the distribution/retail level. It is recommended in these guidelines that refrigeration cabinet manufacturers should issue specific instructions to the retail industry on the installation, operation and maintenance of equipment. As well, both the packaging equipment and material manufacturers should inform customers of the potential microbiological problems associated with the use of the packaging technology being advocated.

Retail Sector

As regards the retail sector, it is recommended that retailers follow the principles outlined in the *Canadian Code of Recommended Handling*

Practices for Chilled Food prepared by the Food Institute of Canada (FIOC, 1990).

The Code was established to assist in the production and distribution of refrigerated food, as well as to maintain product quality and safety up to the point of purchase. The Code itself is divided into different sections with the major ones covering the following areas: packers, manufacturers, and processors; public or distributor warehousing equipment; storage facilities for foodservice; retail handling practices and labelling criteria. Some of the more salient points covered include the following. 1) Product assembly should be carried out in an area kept at ≤10°C. 2) Products should be chilled as quickly as possible to between −1° and +4°C. 3) All refrigerated storage facilities, transportation vehicles and warehouses should be monitored with a suitable time/temperature recording device and maintained at temperatures between −1° and +4°C. Recommendations regarding transport and storage temperatures have been given for potentially hazardous foods within the various Provinces in Canada (Table 15.1). In addition, recommended transport and storage temperatures for various foods can be seen in Table 15.2. 4) All processed refrigerated products should have additional barriers or hurdles (see Chapter 11) built into the food for added safety. 5) Receivers of refrigerated food should check that products received are adequately

Table 15.1. Regulatory provisions for temperature control during storage and transportation of perishable or potentially hazardous foods across Canada.*

	Temperature Requirements	
Province	Storage	Transportation
Newfoundland	0–4°C	None
Prince Edward Island	<4°C	None
New Brunswick	voluntary 4°C	None
Nova Scotia	5°C	None
Quebec	<4°C	<4°C
Ontario	5°C	<4°C (milk only)
Manitoba	<5°C	<5°C
Saskatchewan	4°C	<4°C
Alberta	<5°C	<5°C
British Columbia	<4°C	<4°C
Yukon & N.W.T.	<4°C	None

*Usage of terms varies between provinces.

*Table 15.2. Recommended transport and storage temperatures for various food items.**

Product	Temperature (°C)
Fresh fish (fillets and round)	−30** to +2°
Live lobster	+2° to + 5°
Processed meats and meat products	−1° to +2°
Fresh meats	−1° to +4°
Chilled juices and punches	−1° to +4°
Prepared fruit and vegetables	< +5°
Refrigerated dairy products	+1° to +4°
Refrigerated bakery products	< + 5°

*Adapted from FIOC (1990).
**Lower end of range is too cold for some fish products, particularly eggs or roe-in fish, and will result in freezing of the eggs (−1°C is recommended).

maintained at −1 to 4°C. If the temperature of the product is >7°C or is frozen in error, the accountable manager should be notified immediately. 6) A "Keep Refrigerated" between −1 and 4°C should be prominently displayed on the shipping carton, and, on the inner carton or packing material a "Keep Refrigerated" and "Use by" date should be clearly visible. The Canadian Health Protection Branch recommends the use of a chilled/refrigerated symbol (thermometer) on the principal display panel and a time temperature indicator (see Chapter 14 on each retail package).

The major goal of the chilled food industry is to maintain *uniform* product temperatures of between −1° and +4°C as well as to ensure proper care and handling of chilled foods from the producer all the way to the consumer.

Another document dealing with REPFES is the *Canadian Code of Recommended Manufacturing Practices for Pasteurized/Modified Atmosphere Packaged/Refrigerated Food* (Agriculture Canada, 1990). The objective of this Code is to encourage the development of good manufacturing practices (GMPs) that lead to the safe processing, handling and distribution of products, and to maintain quality and safety up to the point of consumption. The major focus of the Code is on refrigerated products that are pasteurized either before or after being packaged in a hermetically sealed container under a MA. The whole approach of the document is based on the principles of HACCP. The

Code is divided into three major sections, with the first section dealing with product safety and identifying factors influencing microbiological contamination. Implicit in this is process development, i.e., product description and flow diagram, detailed specifications of manufacturing practices and monitoring at CCPs, identification of factors affecting microbiological contamination, etc. As well, a thorough risk analysis is recommended before the start of production to assess whether the developed process will actually maintain food safety. It is advised that risks be re-assessed and re-evaluated whenever necessary, especially when changes in product formulation occur.

The second section deals with the next level of control, i.e., the recommended manufacturing practices. It covers, in order, five major factors affecting microbiological contamination: raw ingredients, packaging, hygiene, time/temperature relationships and extra hurdles. Some of the more prominent points included in this section follow. 1) All raw materials arriving at the plant should be inspected in accordance with the principles of HACCP. 2) If possible, hot filling rather than cold filling is preferable to reduce chances of recontamination. 3) Supplier certification should be provided for pre-blended gases, or monitoring of gas blend ratios when a gas mixer is used. 4) The lethal potential of the pasteurization treatment (sous vide) should be based on process designs which show that the specific time/temperature combination used for each product is effective for producing a safe end product. Cooling should be done quickly so that food temperatures are reduced to 4°C or below in 2 h or less. It is recommended that product sampling and challenge studies be done and repeated regularly whenever the process, or product formulation is changed. In regards to target organisms and decimal reductions, in North America the target organism is *L. monocytogenes* (4 to 7D kill), while in France, the target organism chosen is *Enterococcus faecalis*.

The last section of the Code deals with the final level of control, namely monitoring of parameters at critical control points. Detailed examples of each type of critical control point are illustrated in an appendix at the back of the document. Microbiological criteria for final product are also discussed (Table 15.3). It is explained that these criteria should not be used as standards for acceptance/rejectance of lots or for routine testing, but rather for establishing a new process, or product formulation. This would ensure that a satisfactory microbiological standard is attained and maintained. Failure to meet criteria should not necessarily lead to lot failure, but rather to a critical review of all stages of the process.

Table 15.3. Microbiological criteria for pasteurized/MAP/ refrigerated foods.*

Organism	Criteria (cfu/g)
Salmonella	absent
coliforms	<100
Escherichia coli	absent
Staphylococcus aureus (coagulase-positive)	absent
Listeria monocytogenes	absent
Aerobic plate count**	<100

*These criteria are only recommendations and do not ensure product safety, especially under temperature-abuse conditions. Adapted from Agriculture Canada (1990).
**Plates incubated at 37°C/48 h.

A draft code of hygienic practice dealing specifically with MAP of fresh fish has also been written (Health and Welfare Canada, 1992b). The Code is mainly intended to manage the risks due to *C. botulinum* and other foodborne pathogens in MAP fish. The Code is presented as a HACCP document and, as such, the CCPs are specified. Discussion of the CCPs include:

(1) *Raw material specifications:* It is suggested that specifications for raw materials be set out with each batch being inspected upon arrival for proper temperature (≤5°C, 41°F), general quality and age (≤2 d from catch).

(2) *Temperature:* To obtain maximum benefit from the MAP process, it is recommended that product temperatures be maintained at −2 to +3°C (28−37°F) throughout production, distribution and retail, and at no time should fish temperatures exceed 10°C.

(3) *Washing:* It is recommended that fish be washed both before and after preparation with treated (chlorinated at 200 mg/L), refrigerated water.

(4) *MA composition:* It is realized that it is important to establish the optimal mix of gases and fish to gas ratio for each different type of fish. If the gas mixture differs from the intended one by more than 5%, the deviation should be corrected.

(5) *Shelf life determination:* It is advised that producers determine the shelf life of each product under the actual conditions prevailing throughout the distribution chain, as well as put a "Use Before" date on each package.

CURRENT U.S. GUIDELINES REGARDING MAP AND SOUS VIDE PRODUCTS

Although at present there are no regulations as such for MAP and sous vide products sold in the U.S., there are a number of documents dealing with either guidelines or recommendations for these products. The most comprehensive documents have been produced by the National Advisory Committee on Microbiological Criteria for Foods (NACMCF). The first paper written by this advisory group dealt with refrigerated ready-to-eat (RTE; may require small heat treatment) foods containing cooked, uncured meat or poultry products with an extended shelf life (NACMCF, 1990). Of particular interest were those meat products which rely on refrigeration as the primary barrier to controlling the growth of foodborne pathogens or those products which do not have an intrinsic second barrier such as a pH $\leq$ 4.6 or an $a_w \leq 0.93$. Products which were not a concern to this committee included some perishable products which have extensive commercial experience, well-known processes, products under regulatory control or products which have a good safety record, e.g., cooked beef and roast beef products. The document is divided into three major sections covering 1) epidemiological and microbiological considerations, 2) processing and packaging, and 3) recommendations.

The recommendations section includes discussions on HACCP, process controls, process categories, packaging, product distribution, equipment monitoring, regulatory inspection and review, production guidelines for nonfederally inspected establishments, distributors and retailers, imported and exported products, labelling, education and training and research. Some of the important and/or novel issues covered in these sections include in the process controls section, a requirement for a minimum 4D kill of *L. monocytogenes* which is recommended for validating the cooking process for cooked products. The current USDA cooking requirements for meat products can be seen in Table 13.13 of Chapter 13. In addition, it is recommended that challenge studies be done with *C. botulinum* to control the production of botulinal toxin from the time of production to consumption. Secondary barriers or hurdles formulated into the product are also recommended as an alternative to full thermal processing. Cooling parameters are also given and have been summarized in Table 15.4. It is stressed that processors must have and keep on file a verified HACCP plan *prior to* marketing and production. In terms of production guidelines for non-federally inspected establishments, it was also recommended that the size of an establishment

Table 15.4. USDA cooling requirements for refrigerated meat and meat products.

Product or Condition	Temperature (°C)
• Guidelines for refrigerated storage temperature and internal temperature control point	4.4
• Recommended for products requiring refrigeration periods exceeding 1 week	1.7
• For times not >1.5 h	Internal temperatures between 54.4 to 26.7°C
• For times not >5 h	Internal temperatures between 26.7 and 4.4°C
Intact muscle products	Chilling initiated within 90 min of cooking cycle • product should be chilled from 48 to 12.7°C in not >6 h • chilling shall continue and product should not be packaged for shipment until it attains a temperature of 4.4°C
Roast beef (for U.K. export)	• must be chilled to ≤20°C within 5 h after leaving the cooker and to ≤7°C within the following 3 h

Adapted from NACMCF (1990).

should not be a factor in allowing for the manufacture of these types of products, rather application of microbiological control practices that can assure product safety should be the sole deciding factor. In addition to a HACCP plan it was advised that the company have a documented control plan which includes some of the following: 1) product shelf life documentation; in the event that documentation is not available, the shelf life from manufacturer to retail sale should not exceed 10 d at 40°F; 2) establishment of ongoing employee training and education; and 3) establishments unable to comply with stated requirements would be limited to producing product of pH ≤ 4.6, an a_w ≤ 0.93, or products formulated to inhibit the outgrowth of *C. botulinum*. In the area of labelling it is recommended that because of the great temperature sensitivity of these products, a uniform label be written that reflects the potential hazards of temperature abuse, e.g., *IMPORTANT, MUST BE KEPT REFRIGERATED.

It is recommended that research in the following areas be done. 1) Determine the real incidence of disease resulting from these meat and poultry products. 2) Determine the incidence and survival characteristics of foodborne pathogens in MAs and other extended refrigeration products. The organisms considered most important to address (in decreasing order of priority) were *L. monocytogenes*, *E. coli* O157:H7, *Salmonella* spp., and *C. botulinum*. 3) Continue research on developing accurate and sensitive TTIs for use for consumer-sized packages. 4) Develop basic microbiological data and improved detection methods for those pathogens which may be able to exist in unique foods and environment.

The National Advisory Committee has also developed recommendations on the safety of vacuum packaging or MAP for refrigerated raw fishery products (NACMCF, 1992). Overall the committee found that temperature is the primary preventive measure against the possible hazard of toxin production by *C. botulinum*. The committee also recommended that the sale of VAC/MAP raw fishery products only be allowed when certain conditions are met. These include storage of the product at ≤3.3°C at all points from packaging onwards, the use of high quality raw fish, adequate product labelling respecting storage temperature, adequate shelf life and cooking requirements and the use of a HACCP plan. It was noted as well that at all times organoleptic spoilage and rejection by the consumer should come before the possibility of toxin production. Some of the general findings of the above report are also of interest. Firstly, it was discovered that the regulatory requirements for MAP/VAC raw fishery products are limited. Process approval is only required by the U.S. Department of Commerce (USDC), through a voluntary seafood inspection program. In some instances, as well, *C. botulinum* challenge studies are required. The FDA, although reviewing specific issues related to seafood safety of refrigerated MAP/VAC fishery products, has not officially addressed this issue at the retail level. However, the Association of Food and Drug Officials has advised against the use of MAP/VAC technology for fresh fish at the retail level (AFDO, 1990). Due to the findings of the committee that fish inoculated with high numbers of botulinal spores can become toxic after 6 to 8 d as temperatures approach 10°C, it was stated that fish storage temperatures from 4.4 to 10°C are a cause for concern. Some of the final recommendations of the committee (in addition to those mentioned above) include: 1) that the unrestricted use of MAP/VAC for fishery products should not be permitted; 2) the importance of having a HACCP plan, and, 3) that additional hurdles besides refrigeration should be used in order to increase product safety. Minimum conditions for VAC/MAP technology control are discussed under the following headings: raw fish quality, HACCP plan, hazard analysis/risk assessment, system design validation, labelling requirements and time and temperature records. With regard to the latter area, it was recommended that if, upon receipt, the product temperature of any shipment or carton exceeds 3.3°C, it should be rejected.

New proposed regulations covering the sale of refrigerated foods in reduced O_2 packages have recently been issued by the New York Department of Agriculture and Markets (NYDAM, 1992). The regulations would cover all segments of the food industry except for foodservice establishments and would cover the following categories of food: cook-chill, vacuum packaged, sous vide, MAP and CAP. It was recommended that a listing of products approved by the N.Y. State Commissioner must be posted in the processing area along with a warning against packaging unapproved foods.

Pertinent parts of the regulation dealing with reduced O_2 refrigerated foods include a requirement that all such foods be kept at 7°C or below, or frozen at −18°C and packaged with a bold prominent label stating "IMPORTANT–MUST BE KEPT REFRIGERATED AT 7°C" or "IMPORTANT–MUST BE KEPT FROZEN AT −18°C." Such foods must also have a "use-by" date which cannot exceed 14 d from retail repackaging, and cannot exceed the manufacturer's original "pull date" for that particular food. The regulations also stipulate which foods *cannot* be packaged in a reduced O_2 atmosphere. This would include foods such as raw or processed fishery products (unless frozen within 1 h after packaging and held frozen), soft cheeses such as ricotta, cottage cheese and cheese spreads (retail level), combination of cheese and other ingredients such as meat, fish or vegetables (retail level) and most meat or poultry products which are smoked or cured at retail. Also important is the requirement for a safety barrier verification in writing which must be updated every 12 months and which must be available at the processing site for regulatory review. To prevent against contamination at the retail level it is recommended that only unopened packages of good quality food be allowed for retail repackaging in a reduced O_2 environment and that if it is necessary to stop packaging for whatever reason for periods exceeding 30 min, that particular product should be diverted for another use in the retail store. There is also a section dealing with location in a store where packaging can take place. It is stated that an area specifically designed for this purpose should be established and that every means possible should be taken to prevent the risk of cross-contamination. In addition, only trained personnel should be allowed access to this special packaging area. Very importantly as well, food handling establishments involved in packaging low O_2 foods must be licensed and must obtain written permission from the State Commissioner before packaging such foods. Permission is granted on a product by product basis, i.e., not on a generic basis.

There are, however, departmental exemptions which can be granted. Exceptions, for example, could be granted for known non-hazardous foods as well as for those foods for which documentation is forwarded which demonstrates that the procedures used by the processor provide alternative safeguards to ensure that the food products will not represent an unacceptable public health risk. It should be noted that all or only part of the regulations can be waived for any particular food.

A code of recommended practices for chilled food handling and merchandising has also been published by a collection of different groups headed by the Food Marketing Institute and the American Frozen Food Institute (Farquhar and Symons, 1992). The code sets out "minimum precautions necessary to ensure that a wholesome, hygienic, safe product is supplied to the consumer." A series of do's and don'ts on retailer handling practices covering topics such as pre-distribution storage, general storage, and guidelines for the caterer are reviewed. Recommended temperatures for chilled foods are also discussed. For category I foods, which includes products such as fresh meat, poultry and fish and RTE open-pack cold meats and poultry, a temperature of -1 to $+1$°C is recommended, while for category II foods (all other foods, e.g., soft cheese, milk, coleslaw, bakery goods, etc.) a temperature of 0 to 4°C is recommended. It is also recommended that formal cook-chill catering products be kept at temperatures < 3°C (Farquhar and Symons, 1992).

CURRENT EUROPEAN GUIDELINES REGARDING MAP AND SOUS VIDE PRODUCTS

MAP Foods

Guidelines for the good manufacturing and handling of modified atmosphere packaged food products have recently been published by

The Campden Food and Drink Research Association (CFDRA) of the U.K., an industry association (Day, 1992). The major objective of the guidelines is to provide advice and recommendations on the safety of all MA packed foods to the food and related industries. The document initially provides background information on MAP, food spoilage, shelf life and food safety. It then delves into general recommendations for chilled manufacture and handling operations, and more specific guidance on chilled manufacture, storage and distribution, and retailing and catering. The last section is very useful in that it lists, in summary format, important information for a wide variety of MA packed food products. This includes such information as principal spoilage organisms associated with a particular MAP product, recommended storage temperatures, achievable shelf life, recommended gas mixtures, examples of typical gas mixtures, etc. (see Tables 15.5–15.9).

As with the other documents discussed in this chapter, the CFDRA recommends the HACCP system to identify specific hazards and control options for individual operational steps. Some of the more pertinent recommendations in the U.K. document which have not been previously discussed in this chapter include:

(1) Regarding gas mixtures, it is mentioned that to be fully effective, the gas mixture must completely surround the food product, and thus, in general, trays should be ridged on the bottom to allow for the free flow of gas completely around the product. A gas volume/product volume ratio ranging from 1:1 to 3:1 is recommended, with products requiring longer shelf life needing the higher gas volume/product volume ratios. In terms of gas mixtures for specific foods, ranges of gases rather than specific gas mixtures are recommended, and for individual food products, research is suggested to determine the optimal gas or gas mixture required.

(2) Regarding shelf life evaluation, the recommendation is made to study the effect of pack opening on subsequent shelf life, an important area which is not dealt with in other documents. It is further recommended that the shelf life of the MAP product be given, both for the unopened MAP product and for subsequent days after pack opening.

(3) MAP food manufacturers must comply with the U.K. Food Labelling Regulations of 1984, with its associated amendments. Generally, it is stated that MA packed chilled foods should be clearly and prominently labelled with a "use by" expiry date and a "Keep

Table 15.5. Information and recommendations for MAP red meats and offal.

Major types of red meat and offal	Beef, pork, lamb, veal, venison and offal (liver, kidneys, heart and brains)
Principal spoilage mechanisms	1. Colour change (red to brown) 2. Microbial, e.g., *Pseudomonas* species, *Acinetobacter/Moraxella* species, *Brochothrix* species, lactobacilli, micrococci, Enterobacteriaceae, yeasts and moulds
Possible food poisoning hazards	*Clostridium* species, *Salmonella* species, *Staphylococcus aureus*, *Bacillus* species, *Listeria monocytogenes* and *Escherichia coli*
Recommended storage temperature range	−1°C to +2°C
Achievable shelf lives	In air: 2 – 4 d In MA: 5 – 8 d
Recommended gas mixtures	Retail: red meats: 20 – 40% CO_2/60 – 80% O_2 Offal: 0 – 20% CO_2/60 – 80%O_2/0 – 10% N_2 Bulk: red meats: 20 – 40% CO_2/60 – 80% O_2
Typical MAP machines	Retail: TFFS and PTLF* Bulk: snorkle-type and vacuum chamber
Typical types of package	Retail (prepack): tray and lidding film Bulk: bag-in-box and master pack
Examples of typical MAP materials	Lidding film (top web): PET/PVDC/LDPE* PA/PVDC/LDPE* PC/EVOH/EVA Tray (base web): UPVC/LDPE HDPE EPS/EVOH/LDPE Bag-in-box and master pack: PA/LDPE PA/EVOH/LDPE

Reproduced from Day (1992) with permission of Campden Food and Drink Association.

*Acronyms as follows:

EPS—Expanded polystyrene foam; EVA—Ethylene-vinyl acetate; EVOH—Ethylene-vinyl alcohol; HDPE—High density polyethylene; HFFS—Horizontal form-fill-seal; LDPE—Low density polyethylene; MOPP—Metallised orientated polypropylene; MP—Microperforated film; MPET—Metallised polyester terephthalate; MPOR—Microporous film; OPP—Orientated polypropylene; PA—Polyamide (nylon); PC—Polycarbonate; PET—Polyester terephthalate; PVDC—Polyvinylidene chloride; PTLF—Preformed tray and lidding film; TFFS—Thermoform-fill-seal; UPVC—Unplasticised polyvinyl chloride; VFFS—Vertical form-fill-seal.

Table 15.6. Information and recommendations for MAP fish and seafood products.

Major types of fish and other seafood products	White fish: cod, haddock, plaice, Dover and lemon sole, coley, halibut, whiting, skate, swordfish, shark and hake Oily fish: herring, mackerel, salmon, trout, sardines, whitebait and Greenland halibut Crustaceans and shellfish: prawns, lobster, crab, squid, mussels and cockles
Principal spoilage mechanisms	1. Oxidative rancidity 2. Microbial, e.g., *Pseudomonas* species, *Acinetobacter/Moraxella* species, lactobacilli, streptococci, *Shewenella* species, *Flavobacterium* species and *Alteromonas* species
Possible food poisoning hazards	*Clostridium botulinum* (non-proteolytic E, B and F), *Vibrio parahaemolyticus*, scombrotoxin
Recommended storage temperature range	−1°C to +2°C
Achievable shelf lives	In air: 2–3 d In MA: 4–6 d
Recommended gas mixtures	Retail: whitefish 35–45% CO_2/25–35% O_2/25–35% N_2 Oily fish: 35–45% CO_2/55–65% N_2 Crustaceans and shellfish: 35–45% CO_2/25–35% O_2/25–35% N_2 Bulk: all types of fish and seafood: 70–80% CO_2/20–30% N_2
Typical MAP machines	Retail: TFFS and PTLF snorkle-type and vacuum chamber Bulk: snorkle-type and vacuum chamber
Typical types of package	Retail (prepack): tray and lidding film Bulk: bag-in-box and master pack

(continued)

Table 15.6. (continued).

Examples of typical MAP materials	Lidding film (top web): PET/PVDC/LDPE PA/PVDC/LDPE PC/EVOH/EVA Tray (base web): UPVC/LDPE HDPE EPS/EVOH/LDPE Bag-in-box and master pack: PA/LDPE PA/EVOH/LDPE

Reproduced from Day (1992) with permission of Campden Food and Drink Association. See Table 15.5 for definitions of acronyms used.

refrigerated below 5°C." In addition, as stated above, a label indicating the remaining shelf life after pack opening is preferred, i.e., "Use within 3 d of opening." Labelling recommendations for master packs are also given. Recommendations include that the words "Do not open before (a certain date)" appear on the package so that these master packs are not opened prematurely at the retail level, with the implications that such premature opening may shorten the overall shelf lives of these products. MA packed ambient stored foods such as bakery and dried products should be labelled with a "best before" and "Store in a dry place" and not with a "sell by" label.

General information and recommendations for all food categories are also given (Tables 15.5–15.9). It is recognized in this document that a major part of the success of extending the shelf life and quality of MAP foods while still maintaining safety, is via a combination of the type of MAs, using proper refrigerated temperatures, the use of additional hurdles, and the control of contamination and use of GMPs through the application of HACCP principles. Some of the more specific recommendations for individual food products include the following. 1) Offal tends to suffer from excessive drip, especially in a CO_2 environment, and so only low concentrations ($\leq 20\%$) of CO_2 are recommended for these products. 2) For MAP fish products, it is highly recommended that MA packs be stored within the recommended temperature range, especially for bulk MA master packs and retail MA packs of oily fish. For these latter products O_2 is usually not added because of potential oxidative rancidity problems, and therefore the anaerobic environments produced could allow for toxin production by non-proteolytic *C. botulinum* type E. The Sea Fish Industry in the U.K. (SFIA) has also published specific

Table 15.7. Information and recommendations for MAP poultry products.

Major types of poultry products	Chicken, turkey, duck, pheasant, quail, goose and poussin
Principal spoilage mechanisms	Microbial, e.g., *Pseudomonas* species, *Achromobacter* species, *Flavobacterium* species, *Acinetobacter/Moraxella* species, *Alcaligenes* species, *Aeromonas* species, *Alteromonas* species, *Brochothrix* species, lactobacilli and yeasts
Possible food poisoning hazards	*Salmonella* species, *Clostridium* species, *Campylobacter* species, *Staphylococcus aureus* and *Listeria monocytogenes*
Recommended storage temperature range	−1°C to + 2°C
Achievable shelf lives	In air: 4 – 7 d In MA: 10 – 21 d
Recommended gas mixtures	Retail: 25 – 35% CO_2/65 – 75% N_2 Bulk: 80 – 100% CO_2/0 – 20 N_2
Typical MAP machines	Retail: TFFS and PTLF Bulk: snorkle-type and vacuum chamber
Typical types of package	Retail (pre-pack): tray and lidding film Bulk: bag-in-box and master pack
Examples of typical MAP materials	Lidding film (top web): PET/PVDC/LDPE, PA/PVDC/LDPE, PC/EVOH/EVA Tray (base web): UPVC/LDPE, HDPE, EPS/EVOH/LDPE Bag-in-box and master pack: PA/LDPE, PA/EVOH/LDPE

Reproduced from Day (1992) with permission of Campden Food and Drink Association. See Table 15.5 for definitions of acronyms used.

Table 15.8. Information and recommendations for MAP dairy products.

Major types of dairy products	Hard and soft cheeses, butter, yoghurt, cream, aerosol cream, cream cakes and milk
Principal spoilage mechanisms	1. Microbial, e.g., *Pseudomonas species*, *Acinetobacter/Moraxella* species, lactobacilli, streptococci, micrococci, yeasts and moulds 2. Oxidative rancidity 3. Physical separation
Possible food poisoning hazards	*Listeria monocytogenes*, *Staphylococcus aureus*, *Escherichia coli*, *Salmonella* species, *Bacillus* species and *Clostridium* species
Recommended storage temperature range	0°C to +3°C
Achievable shelf lives	Hard cheeses: in air: 2 – 3 weeks in MA: 4 – 10 weeks Other dairy products: in air: 4 – 14 d in MA: 1 – 3 weeks
Recommended gas mixtures	Hard and soft cheeses: Retail: 10 – 40% CO_2/60 – 90% N_2 bulk: 30 – 100% CO_2/0 – 70% N_2 Other dairy products: retail and bulk: 100% N_2
Typical MAP machines	Retail: TFFS, HFFS, VFFS and PTLF Bulk: snorkle-type and vacuum chamber
Typical types of package	Retail (prepack): pillow-pack and tray and lidding film Bulk: bag-in-box and master pack
Examples of typical MAP materials	Lidding and/or pillow pack film: PET/PVDC/LDPE PA/PVDC/LDPE PC/EVOH/EVA MPET MOPP OPP/PVDC Tray (base web): UPVC/LDPE HDPE Bag-in-box and master pack: PA/LDPE PA/EVOH/LDPE

Reproduced from Day (1992) with permission of Campden Food and Drink Association. See Table 15.5 for definitions of acronyms used.

Table 15.9. Information and recommendations for MAP fruit and vegetables.

Examples of fruit and vegetables	Onion, cabbage, beetroot, celery, cucumber, tomato, lettuce, peppers, carrots, parsnips, potatoes, mango, pineapple, apples, pears, oranges, cauliflower, berries, asparagus, spinach, watercress, beans, sweetcorn, mushrooms, peas, broccoli, Brussels sprouts and bananas
Principal spoilage mechanisms	1. Microbial, e.g., *Erwinia* species, *Pseudomonas* species, lactobacilli, yeast and moulds 2. Enzymic browning of cut fruit and vegetables 3. Moisture loss
Possible food poisoning hazards	*Clostridium* species, *Bacillus* species, *Listeria monocytogenes*
Recommended storage temperature range	Most fruit and vegetables: 0°C to +3°C Whole tropical fruits, bananas, tomatoes and cucumbers: +10°C to +15°C
Achievable shelf lives	In air: 2 – 14 d In MA: 5 – 35 d
Recommended gas mixtures	Retail: air or 3 – 10% CO_2/3 – 10% O_2/80 – 94% N_2 Bulk: air, vacuum or 3 – 10% CO_2/3 – 10% O_2/80 – 94% N_2
Typical MAP machines	Retail: VFFS, HFFS, TFFS and PTLF Bulk: snorkle-type and vacuum chamber
Typical types of package	Retail (prepack): pillow-pack and tray and lidding film Bulk: bag-in-box
Examples of typical MAP materials	Pillow-pack and/or lidding film: OPP, OPP/LDPE, EVA, MP, MPOR Tray (base web): UPVC/LDPE, HDPE, EPS/EVOH/LDPE Bag-in-box (vacuum): PA/LDPE Bag-in-box (air or glass flush): OPP, OPP/LDPE, EVA, MP, MPOR

Reproduced from Day (1992) with permission of Campden Food and Drink Association. See Table 15.5 for definitions of acronyms used.

guidelines on fish handling (SFIA, 1985, 1987, 1989). 3) For bulk master packaging of retail poultry products, gas atmospheres containing 80 to 100% CO_2 are recommended. However, it is stated that for retail size MA packs, initial CO_2 levels should not exceed 35% due to problems associated with excessive drip and pack collapse. 4) Specific recommendations are also given for MAP dairy products. As examples, for retail MAP packs of soft or hard cheese, CO_2 levels should be no higher than 10 to 40% to avoid pack collapse which could result in excess strain on heat seals leading to a loss of seal integrity. For MAP of creams and cream-containing products, 100% N_2 is recommended because in the presence of CO_2 acidification of the products can occur, giving the product a sharp taste. MAP is *not* recommended for mould-ripened cheeses because the fungi would be inhibited by the CO_2/N_2 mixture. 5) For fruits and vegetables, it is recommended that a minimum level of 2 to 3% O_2 be used to ensure that potentially hazardous environments do not develop. However, it should be remembered that research has shown that *C. botulinum* can produce toxin even in high initial O_2 environments (Lambert et al., 1991). Besides the normal practice of matching a commodity's respiration characteristics with a particular film permeability, it is noted that only produce of the highest quality should be used for MAP. In addition, to ensure low microbial levels prior to chilled storage and distribution, it is recommended that produce be treated with chilled chlorinated water and then be rinsed and dryed before being placed in a MA.

Another document entitled "Proposed draft code of hygienic practice for fish and fishery products in controlled and modified atmosphere packaging" has been prepared by the Codex Committee on Fish and Fishery Products. Norway has taken responsibility for writing the original and subsequent drafts of the code. There will undoubtably be many changes to the code before the document is finalized. Nevertheless, the document contains eight sections dealing with subject areas such as raw materials and equipment, plant facilities, procedures for heat processing, packaging and pasteurization processes (Codex Alimentarius Commission, 1994).

Sous Vide

Similar to the document on guidelines for MAP foods put out by the Campden Food and Drink Research Association, the same association has prepared a paper on the microbiological safety of sous vide products

(Betts, 1992). The intent of the document is to serve as a general guide on safe sous vide processing. It is stated at the outset that at present, there are no specific UK recommendations on the maximum shelf life of sous vide foods, although for in-house sous vide operators, a maximum shelf life of 8 d has been recommended (SVAC, 1991). It was interesting to note that the document lists certain categories of food products based on their microbiological risk. The high risk category 1 products include foods such as infant formulae, commercially sterile shelf stable products with a pH > 4.6, raw foods of animal origin and products containing eggs, fish, dairy ingredients, etc., which are normally stored at 5 to 8°C. It is in Category 1 that sous vide products would fall. Because of this, it is stressed that sous vide products require excellent hygienic practices during production, a working, effective HACCP system and vigilant monitoring of critical process parameters to ensure product safety.

It is generally agreed upon in the document that because of the mild heat treatment used in sous vide processing along with the reduced O_2 levels in the packages, that from a process viewpoint, the target organism of concern should include the psychrotrophic *C. botulinum* and not only vegetative bacteria such as *L. monocytogenes*. In fact, the Richmond Committee (1990) recommended a "6D" kill of psychrotrophic strains of *C. botulinum* for extended shelf life foods. To achieve this 6D kill, a 7 min heat treatment at 90°C (or equivalent thermal process) would be required (Gaze and Brown, 1990), although for added safety 10 min at 90°C is recommended for sous vide products with a shelf life greater than 10 d.

As with other documents on REPFES, the use of a HACCP based approach to the microbiological safety of sous vide products is strongly recommended. Papers dealing specifically with both sous vide and HACCP have been published (Smith, 1990; Adams and Banks, 1992; Agriculture Canada, 1990). The document also gives an excellent review of the current literature on refrigerated ready meals and specifics on sous vide products. Recommended process parameters for sous vide products from a number of different countries covering parameters such as pasteurization, cooling parameters, storage temperature, distribution temperature, shelf life, reheating and proposed microbiological criteria can be seen in Table 15.10. For example, French legislation for "sous vide" foods is based on the thermal destruction of a strain of *Enterococcus faecalis* having a D-value at 70°C of 2.95 min and a z-value of 10°C (Anon., 1974, 1988). This equates to a recommended process of 100 min at 70°C (equivalent to 1 min at 90°C) for products with a 21 d shelf life and 1000 min at 70°C (equivalent to 10 min at 90°C) for products

Table 15.10. Recommended process parameters for sous vide processes.

References (Country)	Prepasteurization Storage °C	Pasteurization	Cooling	Storage Temp°C	Distribution Temp °C
DOH Guidelines (1989) (U.K.)	Prepared food $< +10°C$	Core temperature of 70°C/2 min should be achieved.	The temperature of the food should be reduced to 0 to +3°C within 90 min. Chilling should begin within 30 min of end of cooking.	0 to +3°C	0 to +3°C
Anon. (1974, 1988) French Regulations (France)	0 – +3°C	Dependent on proposed shelf life. The following temperature and timescale recommended. Temp (°C) Time (min) Shelf life 70 40 6 d Temp (°C) Time (min) Shelf life 70 100 21 d Temp (°C) Time (min) Shelf life 70 1000 42 d	The core temperature should be reduced to $< +10°C$ within 120 min	0 to + 3°C	0 to +3°C

Table 15.10. (continued).

References (Country)	Shelf Life	Reheating	Microbiological Criteria	
DOH Guidelines (1989) (U.K.)	5 d including day of cooking and consumption	A core temperature of 70° C/2 min should be achieved. Heated food should be served within 15 min and the temperature should not fall below 63°C. Reheating should begin with 15 min after food is removed from chiller.	These criteria apply to foods immediately before reheating.	
			Total aerobic count	$<10^5$/g
			E. coli	<10/g
			S. aureus (coagulase +ve)	$<10^2$/g
			C. perfringens	$<10^2$ g
			L. monocytogenes	absent in 25 g
			Salmonella spp.	absent in 25 g
Anon. (1974, 1988) French Regulations (France)	6 d up to 21 d or up to 42 d dependent on pasteurization process achieved	Product must achieve a minimum temperature of 65°C in 1 h. The food must not fall below this temperature before serving.	These criteria apply throughout the shelf life	
			Staphylococcus spp.	$<10^2$/g[1]
			Coliforms	$<10^3$/g
			E. coli	<10/g
			Sulphite reducing anaerobic count	<30/g
			Mesophilic anaerobic count	$<3 \times 10^5$/g

(continued)

Table 15.10. (continued).

References (Country)	Prepasteurization Storage °C	Pasteurization	Cooling	Storage Temp °C	Distribution Temp °C
SVAC (1991) (U.K.)	Perishable ingredients 0 to +3°C. Fresh fruit and vegetables 5 – 7°C. Prepared foods ≤10°C.	Aimed to achieve a six log reduction of non-proteolytic *C. botulinum*. The following equivalent temperatures and times must be achieved at the slowest heating point. Temp (°C) Time (min) 80 26 85 11 90 4.5 95 2	Core temperature of 0 to +3°C to be achieved within 90 min. Chilling to start within 30 min.	0 to +3°C	0 to +3°C

Table 15.10. (continued).

References (Country)	Shelf Life	Reheating	Microbiological Criteria
SVAC (1991) (U.K.)	≤8 d	Minimum core temperature of 75°C must be achieved. Reheating should begin within 30 min of leaving chiller. Reheated food should be served within 15 min and the temperature should not fall below 65°C.	Not stated

(continued)

Table 15.10. (continued).

References (Country)	Prepasteurization Storage °C	Pasteurization	Cooling	Storage Temp °C	Distribution Temp °C
SYNAFAP (1989) (France)	0 to +3°C is recommended	Recommended pasteurization process dependent on shelf life required. The following equivalent temperatures and times are recommended. Temp (°C) / Time (min) / Shelf life 70 / 40 / ≤6d 70 / 100 / ≤21 d 70 / 1000 / ≤42d	Core temperature of < +10°C within 120 min	0 to +3°C at product core	0 to +3°C
Sheppard (1988) (U.K.)	Perishable foods 0 to +3°C. Prepared foods < +10°C	Minimum core temperature of 70°C to be achieved.	Core temperature 0 to +3°C to be achieved within 90 min (cooling to begin within 30 min of completion of cooking).	0 to +3°C	0 to +3°C

Table 15.10. (continued)

References (Country)	Shelf Life	Reheating	Microbiological Criteria
SYNAFAP (1989) (France)	6 d, up to 21 d or up to 42 d dependent on the pasteurization value achieved	Minimum core temperature of 65°C must be achieved and maintained for 60 min. The food must be maintained at 65°C until served.	The following microbiological criteria apply to the end of the shelf life Mesophilic aerobic count $<10^5$/g *E. coli* <10/g Sulphite reducing anaerobic count <30/g *Staphylococcus* sp. $<10^2$/g salmonellae absent in 25 g
Sheppard (1988) (U.K.)	5 d (including day of cooking and consumption)	Core temperature of 70°C must be achieved. Reheating to start within 30 min of removal from chiller.	No set specifications. Concern should be noted if: Aerobic organisms (except *Bacillus* spp.) present. *Bacillus* spp. $\approx 10^4$/gram *Clostridium* spp. >10/gram

(continued)

Table 15.10. (continued).

References (Country)	Prepasteurization Storage °C	Pasteurization	Cooling	Storage Temp °C	Distribution Temp °C
Schafheitle (1990) (U.K.)	Not stated	A process of 70°C/2 min must be achieved for 5 d shelf life. The following are recommended for extended shelf life Temp (°C) Time (min) 65 354 70 148 75 62 80 26 85 11 90 4.5 95 2	Core temperature of 0 to +3°C to be achieved within 90 min (cooling to begin within 30 min of completion of cooking	0 to +3°C	0 to +3°C

Table 15.10. (continued).

References (Country)	Shelf Life	Reheating	Microbiological Criteria
Schafheitle (1990) (U.K.)	5 d (including day of cooking and consumption). Details are given for adequate processes to achieve a 21 or 42 d shelf life.	A process of 70° C/2 min must be achieved at the centre of the food. Reheating to start within 30 min of removal from chiller. Food to be served within 15 min of heating during which the centre temperature must not fall below 63°C.	The criteria apply to chilled food before it has been reheated (after DOH 1989). Total aerobic count $<10^5$/g • *E. coli* <10/g • *S. aureus* (coagulase + ve) <100/g • *C. perfringens* <100/g • *L. monocytogenes* absent in 25 g • *Salmonella* spp. absent in 25 g

(continued)

Table 15.10. (continued).

References (Country)	Prepasteurization Storage °C	Pasteurization	Cooling	Storage Temp°C	Distribution Temp °C
Agriculture Canada (1990) (Canada)	Prepared product $< +10°C$	Not stated. Pasteurization process should destroy defined level of identified target microorganisms. Reference is made to French regulations.	Core temperature of $\leq +4°C$ to be achieved within 120 min	$-1°C$ to $+4°C$	$-1°C$ to $+4°C$

Table 15.10. (continued).

References (Country)	Shelf Life	Reheating	Microbiological Criteria
Agriculture Canada (1990) (Canada)	Not stated. Reference is made to French legislation as an example of shelf life	Not stated	Immediately after process and cooling. Number of organisms per gram. salmonellae absent Coliform $<10^2$ *E. coli* absent *S. aureus* (coagulase + ve) absent *L. monocytogenes* $<10^2$

Reproduced from Betts (1992) with permission of Campden Food and Drink Association.

with a 42 d shelf life. The manufacturer has to ensure that products remain in compliance with recommended microbiological criteria throughout the shelf life of the product, plus an additional 48 h at 3°C. It is recommended that the actual microbial testing of these products be done at two different temperatures, i.e., 14 d at 4°C followed by 7 d at 8°C for the 21 d shelf life products, and 28 d at 3°C followed by 14 d at 8°C for the 42 d shelf life products.

From a review of all the available literature on sous vide, a number of very important recommendations on process parameters for sous vide products are summarized in the document. They include the following.

(1) Raw materials should generally be stored at 0 to 3°C.
(2) Preparation of raw materials and assembly of sous vide meals should be done at $\leq$ 10°C.
(3) For sous vide products with a shelf life of 1 – 10 d or > 10 d, the pasteurization process should achieve the equivalent of a minimum of 100 min at 70°C or 10 min at 90°C, respectively, at the slowest heating point. Products which receive the latter time/temperature heat treatment, and which are properly stored should be able to achieve a shelf life approaching 42 d.
(4) Post process chilling should start within 30 min after processing and within a further 90 min, the core temperature should reach 0 to 3°C.
(5) The final product should be stored at 0 to 3°C.
(6) For in-house sous vide catering, reheating should be done within 15 min of removal from refrigerated storage and the core temperature of the product should, as a minimum, reach 70°C for 2 min.

Very importantly, it is also strongly recommended that microbiological challenge testing be done for each and every product to ensure safety throughout the storage life of the product.

FUTURE PROSPECTS

It is evident from this chapter that, with the exception of France, there are no countries that have specific regulations dealing with MAP or sous vide products. However, there are a number of countries which have specific documents dealing with recommendations on the safe manufacture and sale of these types of products. As more and more of these foods come onto the marketplace, especially in North America, the possibility that they will be involved in human illness increases. As with other situations, it is only then that regulations are pushed. It is only inevitable that MAP and sous vide foods will be more tightly regulated in the future. However, regulatory agencies should work with industry on any putative regulations to ensure that they are workable and do not stifle the creativity, productivity and viability of the industry. These products have generally been developed because of the consumers' desire for fresh, tasty, convenient and nutritious foods, and they should be strongly endorsed and marketed. With industry working together with government agencies we can expect and hope to see a rapid proliferation of MAP and sous vide products on the consumers' table as we move into the 21st century. Only by applying good science, strict temperature and process controls, along with a well planned out effective and ongoing HACCP program, can companies and consumers alike benefit in the future from this highly desirable new technology.

REFERENCES

Adams, C. E. and Banks, J. G. 1992. "Applying HACCP to sous vide," *Proceedings of ASEPT Conference on Predictive Microbiology and HACCP*, Laval, France, June 1992.

AFDO (Association of Food and Drug Officials). 1990. *Retail Guidelines, Refrigerated Foods in Reduced Oxygen Packages.* Washington, D.C.

Agriculture Canada. 1990. *Canadian Code of Recommended Manufacturing Practices for Pasteurised/Modified Atmosphere Packaged/Refrigerated Food.* Agri-Food Safety Division, Agriculture Canada, Ottawa, Canada.

Anon. 1974. "Réglementation des conditions d'hygiène relatives à la préparation, la conservation, la distribution et la vente des plats cuisinés à l'avance." Arrêté du 26 Juin 1974. République Française Ministère de l'Agriculture.

Anon. 1988. "Prolongation de la durée de vie des plats cuisinés à l'avance, modification du protocole permettant d'obtenir les autorisations." Note de Service DGAL/SVHAIN 881 No. 8106 du 31 Mai 1988. République Française Ministère de l'Agriculture.

Betts, G. D. 1992. *The Microbiological Safety of Sous Vide Processing.* Campden Food and Drink Research Association, Technical Manual No. 39, Chipping Campden, Gloucestershire, U.K.

Codex Alimentarius Commission. 1994. "Proposed draft code of hygienic practice for fish and fishery products in controlled and modified atmosphere packaging," Codex Committee on Fish and Fishery Products. Joint FAO/WHO Food Standards Program. May 2–6, 1994, Bergen, Norway.

Day, B. P. F. 1992. *Guidelines for the Good Manufacturing and Handling of Modified Atmosphere Packed Food Products.* Campden Food and Drink Research Association, Technical Manual No. 34, Chipping Campden, Gloucestershire, UK.

DOH. 1989. *Chilled and Frozen Guidelines on Cook-Chill and Cook-Freeze Systems* HMSO, ISBN 0113211619.

Farquhar, J. and Symons, H. W. 1992. "Chilled food handling and merchandising: A code of recommended practices endorsed by many bodies," *Dairy Food Environ. Sanit.*, 12:210–213.

FIOC (Food Institute of Canada). 1990. *The Canadian Code of Recommended Handling Practices for Chilled Food*. The Food Institute of Canada, Ottawa, Ontario, Canada.

Gaze, J. E. and Brown, G. D. 1990. "Determination of the heat resistance of a strain of non-proteolytic *Clostridium botulinum* type B and a strain of type E, heated in cod and carrot over the temperature range 70 to 92°C." CFDRA Technical Memorandum No. 592.

Health and Welfare Canada. 1992a. "Guidelines for the production, distribution, retailing and use of refrigerated prepackaged foods with extended shelf life," Guideline #7, Food Directorate, Health Protection Branch, Health Canada.

Health and Welfare Canada. 1992b. *Recommended Canadian Code of Hygienic Practice for Modified Atmosphere Packaging of Fresh Fish.* Health Canada, Bureau of Microbial Hazards, Ottawa, Ontario, Canada.

Lambert, A. D., Smith, J. P. and Dodds, K. L. 1991. "Combined effect of modified atmosphere packaging and low-dose irradiation on toxin production by *Clostridium botulinum* in fresh pork," *J. Food Prot.*, 54:94–101.

NACMCF (National Advisory Committee on Microbiological Criteria for Foods). 1990. "Recommendations for refrigerated foods containing cooked, uncured meat or poultry products that are packed for extended, refrigerated shelf life and that are ready-to-eat or prepared with little or no additional heat treatment," Adopted January 31, 1990, Washington, D.C.

NACMCF (National Advisory Committee on Microbiological Criteria for Foods). 1992. "Vacuum or modified atmosphere packaging for refrigerated raw fishery products," Adopted, March 20, 1992. Washington, D.C.

NYDAM (New York State Dept. of Agriculture and Markets). 1992. *Refrigerated Foods in Reduced Oxygen Packages*. Division of Food Safety and Inspection, Albany, NY.

Richmond, M. 1990. *The Microbiological Safety of Food. Part 1.* Report of the Committee on the Microbiological Safety of Food, Chairman Sir Mark Richmond, ISBN 0113212739, HMSO.

Schafheitle, J. M. 1990. *A Basic Guide to Vaccum (Sous-Vide) Cooking*. Bournemouth Polytechnic, Ferm Barrow, Dorset, U.K.

SFIA. 1985. *Guidelines for the Handling of Fish Packed in a Controlled Atmosphere*. Sea Fish Industry Authority, Edinburgh, U.K.

SFIA. 1987. *Guidelines for the Handling of Chilled Fish by Retailers.* Sea Fish Industry Authority, Edinburgh, U.K.

SFIA. 1989. *Guidelines for the Handling of Chilled Finfish by Primary Processors*. Sea Fish Industry Authority, Edinburgh, U.K.

Sheppard, J. 1988. *The Sous-Vide Handbook. A Users Guide to the System.* Convotherm Ltd., London, England.

Smith, J. P. 1990. "A Hazard Analysis Critical Control Point approach to ensure the microbiological safety of sous-vide processed meat pasta product," *Food Microbiol.*, 7(3):177.

SVAC (Sous-Vide Advisory Committee). 1991. "Code of practice for sous-vide catering systems," SVAC, Tetbury, Gloucestershire.

SYNAFAP. 1989. "Guidelines of good hygienic practices for prepared refrigerated meals," 44 rue d'Alésia - 75682 PARIS CEDEX 14.

Index

Milton Keynes UK
Ingram Content Group UK Ltd.
UKHW021903071024
449327UK00021B/1610